The Hadal Zone

Life in the Deepest Oceans

The hadal zone represents one of the last great frontiers in marine science, accounting for 45% of the total ocean depth range. Despite very little research effort since the 1950s, the last 10 years has seen a renaissance in hadal exploration, almost certainly as a result of technological advances that have made this otherwise largely inaccessible frontier a viable subject for research.

Providing an overview of the geology involved in trench formation, the hydrography and food supply, this book details all that is currently known about organisms at hadal depths and linkages to the better known abyssal and bathyal depths. New insights on how, where and what really survives and thrives in the deepest biozone are provided, allowing this region to be considered when dealing with sustainability and conservation issues in the marine environment.

Alan Jamieson is a Senior Lecturer at the University of Aberdeen, based at Oceanlab. His research is focused on the exploration of the hadal zone for biological research. He is the designer and operator of the 'Hadal-Landers' that over the last 7 years have been deployed over 100 times in ultra-deep trenches of the Pacific Ocean.

The Hadal Zone

Life in the Deepest Oceans

ALAN JAMIESON

University of Aberdeen, UK

CAMBRIDGE
UNIVERSITY PRESS

University Printing House, Cambridge CB2 8BS, United Kingdom

One Liberty Plaza, 20th Floor, New York, NY 10006, USA

477 Williamstown Road, Port Melbourne, VIC 3207, Australia

314-321, 3rd Floor, Plot 3, Splendor Forum, Jasola District Centre, New Delhi - 110025, India

103 Penang Road, #05-06/07, Visioncrest Commercial, Singapore 238467

Cambridge University Press is part of the University of Cambridge.

It furthers the University's mission by disseminating knowledge in the pursuit of education, learning and research at the highest international levels of excellence.

www.cambridge.org
Information on this title: www.cambridge.org/9781107016743

First published 2015

A catalogue record for this publication is available from the British Library

Library of Congress Cataloging in Publication data
Jamieson, Alan (Alan J.)
The hadal zone : life in the deepest oceans / Alan Jamieson.
    pages cm
ISBN 978-1-107-01674-3 (Hardback)
1. Deep-sea biology. I. Title.
QH91.8.D44J36 2014
578.77′9–dc23   2014006998

ISBN 978-1-107-01674-3 Hardback

Additional resources for this publication at www.cambridge.org/9781107016743

For Rachel, William and Matthew

Contents

    Colour plates can be found between pages 176 and 177.

Preface

The hadal zone is an enigmatic ecosystem or rather a cluster of deep ocean trench ecosystems. It is not only one of the most extraordinary, extreme marine environments in terms of high hydrostatic pressure, geological instabilities and low food supply, the hadal zone is also a place where life has been found to thrive at such great depths, despite common perceptions to the contrary. The hadal zone represents one of the last great frontiers in marine science, accounting for 45% of the total ocean depth range, yet it receives little or no mention in contemporary deep-sea biology text books. In the 1950s, the hadal fauna were subject to a great deal of attention as a result of the Soviet *Vitjaz* and Danish *Galathea* biological sampling expeditions, the discovery of the deepest trenches and the first manned submersible dives into the trenches. The decade culminated in the first visit by humans to the deepest place on Earth, the Mariana Trench. Despite the myriad of public attention and the advances in our understanding of hadal biology, ecology and geology, interest appeared to dwindle and very little scientific endeavour occurred through the 1970s, '80s and '90s. In fact, during these decades, the hadal zone's main accolade was as a potential dumping ground for pharmaceutical and radioactive waste materials, driven by the anthropocentric opinion of 'out of sight, out of mind'. Thankfully, this exploitation and perturbation of trench habitats did not become common practice and in recent years there has been a renaissance in hadal exploration, almost certainly as a result of technological advances that have made this otherwise largely inaccessible frontier a viable subject for research. Furthermore, as the current resurgence continues to flourish, it is hoped that the hadal zone will, at long last, be placed equally alongside all other marine ecosystems. We face an uncertain climatic future and thus, the ocean must be understood and maintained in its entirety and not categorised by proximity to the nearest human. It is startling how little is known about the hadal environment and this lack of knowledge renders a limited view of the communities that survive at the greatest depths and endure the highest pressures on Earth.

Officially, the hadal zone occurs in areas where one tectonic plate subducts beneath another; where the topography of the vast abyssal plains suddenly plunges to depths of nearly 11 km below sea level. The hadal zone was named after *Hades*, the name of both the ancient Greek kingdom of the underworld and the god of the underworld himself, *Hades* (brother to Zeus and Poseidon). The hadal zone comprises many disjunct trenches, mostly around the Pacific Rim and these trenches are host to most major taxa, some of which flourish even at the greatest depths (e.g. Holothurians, amphipods, bivalves, gastropods). The geological setting of the trenches is notoriously unstable, adding another string to the extreme environment bow. The trenches offer up many biological and ecological conundrums, such as why two seemingly isolated trenches, sometimes tens of thousands of kilometres apart, are both home to the same species that do not inhabit the areas in between? Likewise, how were the trenches ever colonised at all, given that the quantity and distribution of surface-derived food supply varies so drastically from the surrounding plains? Also, what happens to the benthic communities following a catastrophic earthquake? Are there seasons, interannular variability or potential for chronobiology? How connected are the hadal communities with each other and to the wider deep-sea communities? We do not know the answers to most of these questions because hadal science is still in its infancy relative to conventional deep-sea research; however, we now have the technological and scientific support required to expand our knowledge of this unique environment.

There are many aspects of life in the hadal zone that are shared with the wider deep sea, therefore, this book aims to focus specifically on the trenches and the hadal communities therein and should be viewed as a companion resource to other key works. While compiling the information for this book, it became apparent that, despite so many recent developments on the topic, the large amount of sampling undertaken on the Danish and Soviet expeditions of the 1950s, particularly of invertebrates, has not yet been repeated on the same scale. All the data from these expeditions were compiled and documented in the works of the late G.M. Belyaev, from the P.P. Shirshov Institute of Oceanography in Russia, in a book entitled *Deep Ocean Trenches and Their Fauna*. The book was translated into English by the Scripps Institution of Oceanography in the USA and is a fabulous resource and a fantastic compilation of hadal research up until the late 1980s. The intention of this new book is not necessarily to update the work of Belyaev, however, it must be acknowledged that a lot of the information documented in Part III and the updated species list in the Appendix was derived or moderately updated from his work, due to lack of new information on many taxa. The species list in the Appendix and tables throughout are, to the best of my knowledge, correct and I accept any responsibility for errors or reclassifications that may have been missed.

Throughout the book there is also regular mention of 'HADEEP'. HADEEP, the HADal Environments and Education Program are a series of projects that have been running from 2006 to the present. HADEEP was undertaken in collaboration with the University of Aberdeen (UK), the University of Tokyo (Japan) and the National Institute for Water and Atmosphere Research (New Zealand). These projects amassed a large dataset on hadal organisms from many trenches and enabled the compilation of various information databases. Many of these resources are referred to in this book and are referenced to 'HADEEP'. It was during the first HADEEP project that the idea for this book arose.

In terms of organisation, this book is split into four parts, where the first part (History, geology and technology) provides a review of the history of hadal science and exploration (Chapter 1), geography and geology (Chapter 2) and full ocean depth technology (Chapter 3). These three chapters provide an overview of the components that led to the contemporary understanding of hadal science and provide the appropriate background information for reference in the following sections. Part II (Environmental conditions and physiological adaptations) includes chapters on the hadal environment, hydrostatic pressure and food supply to the trenches. In the hadal zone context, hydrostatic pressure and food supply are independent chapters, given their significance in this environment. Part III (The hadal community) comprises four chapters describing our current understanding of the hadal communities from bacteria to fish (Chapters 7, 8, 9 and 10). The final part, Part IV (Patterns and current perspectives), describes more recent developments in ecology and evolution (Chapter 11) and for the first time, attempts to explore some rudimentary ecology. Chapter 12, Current perspectives, details the interactions between humans and the hadal zone, whether good, bad, present or historical. The Appendix is a species list of all known hadal specimens for reference.

The hadal zone is the subject of many things to different people. For example, it has been the subject of curiosity-quenching exploration, scientific endeavour, a potential source for pharmaceutical prospecting and a source of potentially devastating earthquakes and tsunamis. However, to most people, the hadal zone is a dark, mysterious realm that incites inquisitiveness, fascination and the thirst for exploration. Contrary to long-perceived opinion, we now know that the hadal zone is definitely not a deep, dark area of little importance to the world, where nothing but the weirdest and enigmatic creatures simply eke out an existence.

Future efforts are urgently required to comprehensively sample numerous trenches in order to enable global generalisations about life on Earth. We now live in an age where technology is at a level where very few unexplored frontiers remain, but we can now study these remote and extreme environments beyond simple 'flag planting' and try to understand them, maintain them and enjoy them now and in the future. With ever more exploration, visual imagery, experimentation and scientific understanding of these deep environments, it is hoped that the hadal zone will become less 'alien' to the world, while still retaining a sense of majesty and wonder.

Acknowledgements

From the University of Aberdeen, I am indebted to my colleagues Professor Monty Priede, Professor Stuart Piertney and Dr Toyonobu Fujii for their long, frequent and healthy discussions on all aspects of hadal science and their contributions to the book, and to Dr Martin Solan (now at NOCS, UK) and Dr Phil Bagley (now at AkerSolutions, UK) for their part in founding HADEEP. I thank Dr Tomasz Niedzielski (University of Wrocław, Poland) for his GIS contribution. The University of Aberdeen students Nichola Lacey, Thomas Linley, Heather Ritchie, Amy Scott-Murray and Ryan Eustace, and Matteo Ichino (NOCS, UK) are thanked for their enthusiasm, assistance and contributions to this book. I wholeheartedly thank the New Zealand HADEEP contingent: Dr Ashley Rowden, Dr Malcolm Clark, Dr Anne-Nina Lörz and Dr Peter McMillan (National Institute for Water and Atmosphere research; NIWA), Andrew Stewart (Museum of New Zealand, Te Papa Tongerewa) and especially Dr Niamh Kilgallen (now at the Australian Museum). I equally thank the Japanese HADEEP contingent: Dr Kumiko Kita-Tsukamoto, Professor Hidekazu Tokuyama, Professor Mutsumi Nishida, Dr Kota Kitazawa and Dr Robert Jenkins (Atmosphere and Ocean Research Institute, University of Tokyo, Japan). For supply of and permission to use images and other advice and expertise, I thank Professor Julian Partridge and Milly Sharkey (University of Bristol, UK), Dr Torben Wolff and Professor Danny Eibye-Jacobsen (Zoological Museum, Danish Museum of Natural History, Denmark), James Cameron (Lightstorm Entertainment, USA), Charlie Arneson (USA), Dr Eric Breuer and Samantha Brooke (Pacific Island Fisheries Science Center, NOAA, Hawai'i, USA), Peter Sloss (retired; NOAA/NGDC, USA), Professor Doug Bartlett, Kevin Hardy and Professor Lisa Levin (Scripps Institution of Oceanography, USA), Dr Paul Yancey (Whitman College, USA), Dr Andy Bowen (Woods Hole Oceanographic Institute, USA), Dr Andrey Gebruk (P.P. Shirshov Institute of Oceanology, Russia), Professor Ronnie Glud (University of Southern Denmark), Kevin Mackenzie (University of Aberdeen, UK), Dr Robert Turnewitsch (Scottish Association for Marine Science, UK), Professor Andy Gooday (NOCS, UK), Ikuta Azusa (JAMSTEC, Japan), Dr Tomislav Karanovic (Hanyang University, S. Korea) and Fredrik Søreide (Promare, USA). I also thank Dr Gordon Paterson (Natural History Museum, London, UK) and Hannah Tilson (ex-University of Southampton, UK) for use of material from an unpublished Masters thesis. For additional hard work and help in compiling electronic databases, I thank the University of Aberdeen 2013 graduates, Sarah Breimann, Sophie Miles and Bruce Leishman, and the 2013 Brazilian 'Science Without Borders' exchange students, Ingrid Padovese Zwar and Gabriel Stefanelli Silva. A million thanks also go to Dr Rachel Jamieson for her patience, proof-reading and commenting on this entire book through its many drafts. The funding bodies Nippon Foundation (Japan), NERC (UK), Total Foundation (France) and the Marine Alliance for Science and Technology, Scotland (MASTS) are gratefully acknowledged for supporting the hadal research prior to and during the writing of this book.

Part I

History, geology and technology

Introduction

The history of hadal science is full of the legacies of inquiring scientists who dared to explore the unknown and push the boundaries of the seemingly impossible in order to satisfy human curiosity for the natural world. The story is equally fraught with academic minds who, in their time, publicly deemed impossible many of the fundamental facts about the deep sea that are now taken for granted. For example, deep-sea biologists had to contend with the infamous statement of Edward Forbes who claimed that life did not exist at depths greater than 600 m below sea level (Forbes, 1844). Similarly, over 100 years later, Pettersson (1948) also expressed doubt as to the existence of life deeper than 6500 m, ironically on the eve of the discovery of life at 7900 m (Nybelin, 1951). Challenging the likes of Forbes may indeed have inspired the pursuit of life to deeper than 600 m and may ultimately have led to the discovery of life at nearly 11 000 m, full ocean depth. Around this time, hydrologists navigating the seas discovered areas of ocean that were deeper than ever thought possible. These deep areas, now recognised collectively as the 'hadal zone', are extremely deep trenches, located at tectonic plate boundaries. However, even in the mid-1900s, the theory of plate tectonics and continental drift was still disregarded by many academics. In 1939, on the subject of plate tectonics and continental drift, the well-known geologist Andrew Lawson voiced the then current opinion of many when he said; 'I may be gullible! But I am not gullible enough to swallow this poppycock!' (Hsu, 1992).

The discovery and exploration of the hadal zone has been slow relative to that of other deep-sea habitats. The primary reason for this is that the area of seafloor which encompasses the trenches is small relative to the surrounding abyssal plains. Therefore, trenches were less likely to be encountered during standard sounding surveys. Furthermore, early sounding of extreme depths occurred before there were any theories to estimate how deep the ocean really was, and long before the development of plate tectonic theory. Thus, the trenches as we know them today, their existence and formation, were completely unheard of.

Today, we now not only understand tectonics, but we have also felt the presence of the trenches first hand. From a geological context, the trenches have never before been as conspicuous in the public domain. The 2010 Cauquenes earthquake off Chile (magnitude M_w 8.8) and the 2011 Tōhoku-Oki earthquake off Japan (magnitude M_w 9.0) were both the result of the geological activity of hadal trenches (the Peru–Chile and Japan Trenches, respectively). Furthermore, the latter and the 2004 Indian

Ocean earthquake (magnitude $M_w > 9$; triggered by the Java Trench) are remembered for the devastating tsunamis that followed.

From a biological perspective, progress in sampling the hadal zone has been slow and was initially hampered by the technological challenges associated with its sheer distance from the ocean's surface. Equipment had to be lowered through thousands of metres of water, a challenge once again exacerbated by the fact that full ocean depth was still to be determined. Before the onset of ship-mounted acoustic systems (echo-sounders), determining ocean depth would have been an extraordinarily laborious task. The hadal zone also presented a technical challenge in the form of extremely high hydrostatic pressure. Sampling equipment had to be capable of withstanding over 1 ton of pressure per square centimetre in order to resist implosion.

Despite these challenges, we now know the precise locations of the trenches and have made significant progress in understanding the biology and ecology of life in the deepest places on Earth, whilst having developed some sophisticated and innovative technology along the way. The first part of this book provides a brief synopsis of the people, projects and expeditions that paved the way in the exploration of the final frontier in ocean science (Chapter 1), examines the formation and location of the hadal zone (Chapter 2) and reviews the challenges and innovations of technology for full ocean depth (Chapter 3).

1 The history of hadal science and exploration

The history of hadal science and exploration is a peculiar story and, largely due the challenges of sampling at such extreme depths, research effort seems to have occurred in waves. Major events in this history are often as disjunct as the trenches themselves. Around the turn of the twentieth century, early pioneers began sampling at greater and greater depths. Following the burst of curiosity concerning the extent to which animal life could be found and the true depths of the oceans, there was a lull in progress. It was in the 1950s when the first major hadal sampling campaigns began with the extensive series of Soviet RV *Vitjaz* expeditions and the round the world Danish RV *Galathea* expeditions. The *Vitjaz* continued to periodically sample the great depths for some time but, in general, research campaigns were few and infrequent. The first manned dive to the deepest place on Earth took place in 1960 amidst a myriad of public interest. However, it was the first and only time that this submersible ventured to the bottom of a trench.

The 1990s saw new interest in the trenches when the Japan Agency for Marine-Earth Science and Technology (JAMSTEC) developed the first full ocean depth remotely operated vehicle (ROV) named *Kaikō*. *Kaikō* was used numerous times in the trenches and provided scientists with the tools to access full ocean depth. Aside from JAMSTEC, few other academic institutes were involved in hadal research until the mid-2000s.

We are now in an age where more countries are involved in hadal research than ever before, and the number of projects relating specifically to biology at full ocean depth are too many to mention. Scientists from the USA, UK, Japan, New Zealand and Denmark, among others, are prominently involved in active sampling at hadal depths. Coinciding with this work, scientists have supported high profile 'firsts' such as when the *Deepsea Challenger* submersible reached the deepest place on Earth. What actually sparked this recent wave of interest is unclear, but it may be attributed, like most eras in deep-sea research, to the development of new technology. It is also nice to think that this new found interest in the great depths is, in some way, a public response to an ever-changing climate; that people are becoming more aware of the urgency, the means that we possess and responsibility that we now have to investigate the ocean in its entirety; from the air–sea interface to full ocean depth.

1.1   Sounding the trenches

Since the days of Aristotle, the altitude of the land and the depth of the seas have prompted great curiosity. The first deep sounding expeditions began in 1773 with Lord Mulgrave's expedition to the Arctic Ocean, where a depth of 683 fathoms (1249 m) was recorded. In 1817–18, Sir John Ross recorded a depth of 1050 fathoms (1920 m) and collected a sediment sample using a wire deployed 'deep-sea clam' in Baffin Bay, east of Greenland. During the *Erebus* and *Terror* expeditions in 1839–43, Sir James Clark Ross used a 3600 fathoms (6584 m) long wire that was marked every 100 fathoms. The time interval between each mark was noted until the intervals significantly increased, where it was thought that the line had reached the bottom. The same technique was adopted during the British round-the-world expedition on the HMS *Challenger* (1873–76), under the leadership of Charles Wyville-Thomson (Thomson and Murray, 1895). Equipped with 291 km of Italian hemp for sounding wire, the HMS *Challenger* unexpectedly recorded a depth of 4500 fathoms (8230 m) in the northwest Pacific Ocean at latitude 11° 24' N, longitude 143° 16' E, southwest of the Mariana Islands and north of the Caroline Islands in the North Pacific Ocean. This sounding was the first measurement that indicated the existence of extraordinarily deep areas and, in due course, led to the discovery of the Mariana Trench. It was this *Challenger* expedition, the first global marine research campaign, that laid the framework for all future marine research.

At the time of the *Challenger* expeditions, scientists aboard the USS *Tuscarora* employed a similar method for sounding. They used piano wire to record a depth of 4665 fathoms (8531 m) in the Kuril–Kamchatka Trench in the northwest Pacific Ocean, originally coined the *Tuscarora Deep*.

Sir John Murray (1841–1914) documented the first systematic measurements of ocean depth distribution and mean depth of the oceans with which he calculated the first hypsometric curve, thus beginning the process of 'mapping' the oceans in three dimensions (Murray, 1888). Based on the available data of the day, he calculated the volume of the ocean, the volume of the continents above sea level and even the depth of a uniform ocean if the seafloor were level and no continents existed. Following on from the work of Murray (1888), sounding data became more numerous with time. Charts that mapped the oceans, such as those of E. Kossinna in 1921 and T. Stocks in 1938 (cited in Menard and Smith, 1966) were frequently produced and many studies relating to the nature of the seafloor and depth distribution were undertaken (e.g. Murray and Hjort, 1912; Menard, 1958; Menard and Smith, 1966).

Towards the turn of the twentieth century, new sounding devices were developed by the British Royal Navy, notably the *Hydra Rod* (so-called following its design by the blacksmith onboard the HMS *Hydra*) and the *Baillie rod* (named after the navigating lieutenant on HMS *Challenger* who designed it) (Thomson and Murray, 1895). The first depth sounding of greater than 5000 fathoms was recorded during a British expedition on the HMS *Penguin* in 1895 using a Baillie rod lowered with piano wire. They recorded a depth of 5155 fathoms (9144 m) in the Kermadec Trench in the southwest

Pacific Ocean off the north coast of New Zealand. Shortly afterwards, the German vessel *Planet* measured a greater depth in the Philippine Trench, where later, the Dutch would record an even deeper 5539 fathoms (10 319 m) using the first audio-frequency sounding methods onboard the *Willebord Snellius*. Using this primitive but pioneering audio-frequency technique of recording sound echo, the Scripps Institution of Oceanography's USS *Ramapo* measured a depth of 5250 fathoms (9600 m) in the Japan Trench (now known to be the Izu-Bonin Trench, south of the Japan Trench in the northwest Pacific Ocean). Another Scripps vessel, *Horizon*, also recorded 5814 fathoms (10 633 m) in the Tonga Trench (southeast Pacific Ocean) and named the site *Horizon Deep* (Fisher, 1954). Following these new findings, the German vessel *Emden* recorded a measurement of 5686 fathoms (10 400 m), once again in the Philippine Trench. The record for the greatest depth found was broken once more by the USS *Cape Johnson* during World War II, with a reading of 5740 fathoms (10 500 m) in the Philippine Trench off Mindanao, which was for years thought to be the deepest place on Earth (Hess and Buell, 1950). The method of projecting sound and recording the echo to measure the depths, coined 'echo sounding', was developed and quickly superseded wire-deployed methods.

The new echo-sounder method often relied on 'bomb sounding', whereby someone threw a half-pound demolition block of TNT off the ship to create the sound source from which the echo was received onboard the ship via a transducer amplifier (Fisher, 2009). This method, albeit primitive relative to today's technology, was accurate enough to distinguish between the trench floor at the axis and the trench slopes. It was used to sound the maximum depths of the Middle America, Tonga, Peru–Chile and Japan Trenches (Fisher, 2009). It ultimately led to the discovery of the deepest point on Earth: the *Challenger Deep* in the Mariana Trench (nearly 11 000 m; Carruthers and Lawford, 1952; Gaskell *et al.*, 1953).

The ability to accurately sound the depths of the oceans using ship-mounted acoustic systems provided sounding data with relative ease, and with much greater replication and resolution than wire-deployed systems. Such accuracy led to several in-depth reports on the internal topography, morphology and sedimentation of some deep trenches, for example, Fisher (1954), Kiilerich (1955) and Zeigler *et al.* (1957). However, there was still the question of how the trenches were formed. Figure 1.1 shows how these early soundings were interpreted into three-dimensional topography and Figure 1.2 shows the equivalent data using modern sounding methods but based on the same principles.

1.2   Development of plate tectonic theory

The discovery of the deep trenches occurred long before the development of any theories relating to how they came to be. The discovery of continental drift, which, in turn, prompted the discovery of plate tectonics and convergence zones (where trenches occur), was nearly 360 years in the making. Abraham Ortelius (1596) first noted how

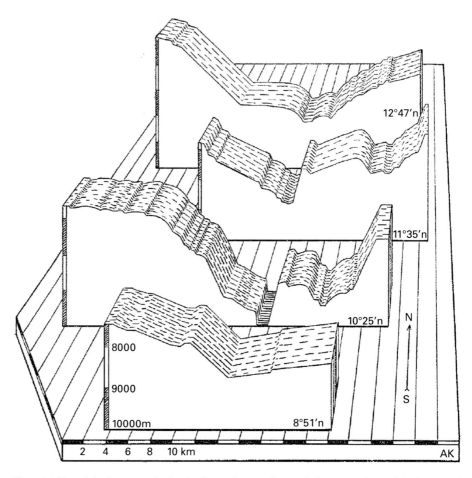

Figure 1.1 Trench bathymetry obtained using early sounding techniques: sections of the bottom of the Philippine Trench. From Kiilerich (1955); reproduced with the permission of *Galathea Reports*).

the continents, in particular South America and Africa, 'seemed to fit together' as if they had once formed a single land mass; an observation reiterated by others in the 1700s and 1800s (Romm, 1994). Around the turn of the twentieth century, Roberto Mantovani suggested the previous existence of a super continent (what is now known as Pangaea). However, credit for the development of continental drift theory, beyond simple observations of a 'jigsaw-puzzle fit', came from the German meteorologist Alfred Wegener (Wegener, 1912; Demhardt, 2005). Wegener hypothesised that the continents had once formed a single land mass prior to splitting apart and drifting to their current locations. The splitting apart of Pangaea was thought to have occurred by volcanic activity and this led Mantovani to suggest that the Earth was expanding (Mantovani, 1909). This was, of course, not the case and over the coming years various theories from Wegener and others were put forth ranging from lunar gravity driven drift, centrifugal pseudo-force and astronomical precession. None of them, however, proposed a sufficiently

Figure 1.2 Examples of modern digital swathe bathymetry showing the three-dimensional bathymetry of the Mariana Trench. Above and below is a frame grab from a computer generated 'fly-through' movie, note the 100-fold vertical exaggeration. Images courtesy of Peter Sloss, NOAA/NGDC (retired).

strong force to explain the drift. In the absence of a driving force for continental drift, the theory was not accepted generally for many years and it sparked lively debates between 'drifters' (supporters of the theory) and 'fixists' (opponents of the theory) (Scheidegger, 1953).

An Australian geologist, Samuel Carey, advocated the continental drift theory of Wegener. Carey provided a mechanism to explain such processes whereby super continents divide and drift and cause the generation of new crustal zones in deep oceanic ridges. His theory, however, still backed the idea of an expanding Earth. Despite the eventual acceptance of plate expansion, the expanding Earth theory was erroneous.

Further support for continental drift came during the late 1950s and early 1960s as bathymetry of the deep ocean provided evidence of seafloor spreading along the mid-oceanic ridges (e.g. Heezen, 1960; Dietz, 1961; Vine and Matthews, 1963). The advances in early seismic imaging techniques along the deep trenches adjacent to many continental margins showed how the oceanic crust could 'disappear' into the mantle, providing evidence, for the first time, that was contrary to the expanding Earth theory. The disappearance of one crust beneath another was termed 'subduction' (Amstutz, 1951).

It was the onset of magnetic instruments (magnetometers) which provided unequivocal evidence of multiple lithospheric plates. Magnetometers, adapted from World War II airborne and submarine detection systems, identified unexpected magnetic anomalies and variations across the ocean floor. They detected volcanic rock (iron-rich basalt) which contained a strongly magnetic mineral (magnetite) that gave the basalt measurable magnetic properties. Furthermore, the Earth's magnetic field was recorded at the time when the newly formed rock was cooled. The magnetic variations turned out to be recognisable patterns and not random. A 'zebra-like' striped pattern emerged when a wide area of seafloor was scanned. These striped patterns were areas of normal polarity alternating with areas of reversed polarity.

The discovery of magnetic striping prompted theories that mid-ocean ridges were structurally weak zones, where the seafloor was being torn apart lengthwise along the ridge crest, pushed apart by magma rising up from the Earth's mantle through these weak zones to create new oceanic crust, a process we now know as 'seafloor spreading' (SFS). The SFS hypothesis represented major progress in the development of the plate tectonic theory.

The official acceptance of plate tectonics (originally termed 'New Global Tectonics') by the scientific community occurred at a symposium at the Royal Society of London in 1965. In addition to the discovery of seafloor spreading at divergent zones and subduction at convergent zones, adding the new concept of transform faults to the general tectonic model provided the final piece of the puzzle completing the explanation for tectonic plate mobility (Wilson, 1965). Two years later, at a meeting of the American Geophysical Union, it was proposed that the Earth's surface was made up of 12 rigid plates that move relative to each other (Morgan, 1968) and this was quickly followed by a complete model based on six major plates and their relative motions (Le Pichon, 1968).

In the 1960s and early 1970s frustration grew as identification of the structures that were diagnostic to the working of subduction processes was greatly inhibited by the technical challenges of extreme water depths and the steep slopes of the trenches (von Huene and Shor, 1969; Scholl et al., 1970). The extreme depths made conventional rock and sediment sampling by dredging difficult, while the steep trench slopes caused a reduction in the resolution of subsurface structures captured by the acoustic imaging techniques. Such techniques were, by then, available to most researchers (von Huene and Sholl, 1991). By the mid-1970s, offshore seismic reflection studies had resolved some of the important issues such as sediment subduction and accretion (Karig and Sharman, 1975; discussed in detail in Chapter 2).

Since the 1980s, advances in offshore geological and geophysical techniques and technologies have permitted the remote exploration of the deep subsurface structures of convergent margins (e.g. the Mariana Trench; Fryer et al., 2002). Swath-bathymetry now readily provides accurate areal images of the morphology of even the most complex topography, such as trenches, and it is able to do so with incredible detail and clarity (see Fig. 1.2).

1.3   Establishing full ocean depth

During the *Challenger II* expeditions in 1951, the vessel returned to the Mariana Trench in the northwest Pacific Ocean, this time equipped with an echo sounder. A depth of 5940 fathoms (10 860 m) was recorded at 11° 19' N, 142° 15' E (Carruthers and Lawford, 1952; Gaskell *et al.*, 1953). In conjunction with the echo soundings, a Baillie rod was deployed, and on the third attempt a sample of 'red clay' was retrieved from the deepest point on Earth, newly named Challenger Deep. Challenger Deep is still regarded as the deepest point on Earth although the exact depth often varies. For example, in 1957, the Soviet vessel *Vitjaz* recorded 11 034 m (and dubbed the area the 'Mariana Hollow'; Hanson *et al.*, 1959), whereas the American MV *Spenser F. Baird* recorded 10 915 m in 1962 and the Scripps INDOPAC expedition recorded 10 599 m in 1977 (Yayanos, 2009). A review by Angel (1982) states the maximum depth as 11 022 m but does not cite the source. JAMSTEC have visited Challenger Deep more times than any other country due to the construction of the full ocean depth rated ROV *Kaikō*, deployed from its mother ship RV *Kairei* (Mikagawa and Aoki, 2001). The Japanese literature states the depth of Challenger Deep as 10 890 m (Taira *et al.*, 2004), 10 897 m (Takami *et al.*, 1997), 10 898 m (Kato *et al.*, 1997, 1998), 10 933 m (Fujimoto *et al.*, 1993) and 10 924 m (Akimoto *et al.*, 2001; Fujioka *et al.*, 2002). On 24 March 1995, the ROV *Kaikō* descended to 10 911 m and placed a plaque bearing the name and date of the dive to officially mark the deepest point on Earth (Fig. 1.3).

Figure 1.3 Video frame grab of the Japanese full ocean depth rated ROV *Kaikō* placing a flag to mark the deepest place on Earth; Challenger Deep, 10 911 m in the Mariana Trench. Image © JAMSTEC, Japan.

Table 1.1 Summary of both sounding and direct measurement of the deepest place on Earth; Challenger Deep in the Mariana Trench. Modified from Nakanishi and Hashimoto (2011).

Sounding

Year	Vessel	Depth (m)	Reference
1875	HMS *Challenger*	8184	Thomas and Murray (1895)
1951	*Challenger VIII*	10 863 ± 35	Carruthers and Lawford (1952)
1957	*Vitjaz*	11 034 ± 50	Hanson *et al.* (1959)
1959	*Stranger*	10 850 ± 20	Fisher and Hess (1963)
1962	*Spencer F. Baird*	10 915 ± 20	Fisher and Hess (1963)
1975	*Thomas Washington*	10 915 ± 10	R.L. Fisher (pers. comm. in Nakanishi and Hashimoto, 2011)
1980	*Thomas Washington*	10 915 ± 10	R.L. Fisher (pers. comm. in Nakanishi and Hashimoto, 2011)
1984	*Takuyo*	10 924 ± 10	Hydrographic Dept. Japan Marine Safety Agency (1984)
1992	*Hakuho-Maru*	10 933	Fujimoto *et al.* (1993)
1992	*Hakuho-Maru*	10 989	Taira *et al.* (2005)
1998	*Kairei*	10 938 ± 10	Fujioka *et al.* (2002)
1998/99	*Kairei*	10 920 ± 5	Nakanishi and Hashimoto (2011)

Selected dives

Year	Vehicle	Depth (m)	Reference
1960	Bathyscaphe *Trieste*	10 913 ± 5	Piccard and Dietz (1961)
1995	ROV Kaikō (test)	10 911	Takagawa *et al.* (1997)
1996	ROV Kaikō (Dive 21)	10 898	Takagawa *et al.* (1997)
1998	ROV Kaikō (Dive 71)	10 907	Hashimoto (1998)
2009	HROV *Nereus*	10 903	Bowen *et al.* (2009b)

In 2009, using a modern, deep-water multi-beam sonar bathymetry system, the American RV *Kilo Moana* sounded 10 971 m at Challenger Deep. The equipment is thought to have accuracy better than 0.2% of the depth, suggesting accuracy to within ±11 m (10 960–10 982 m). The area of Challenger Deep is, of course, unlikely to be flat and this accounts for the variation in exact depth measurements taken (the mean of the above depths is 10 908 m ± 114 S.D.). As technology improves in accuracy and precision, it is reasonable to assume the most up-to-date *in situ* value, in this case the 2009 RV *Kilo Moana* value of 10 971 m. However, a more recent study focused entirely on what the exact depth of the Challenger Deep is and concluded that it consists of three en-echelon depressions along the trench axis, each of which is 6–10 km long (~2 km wide), and each deeper than 10 850 m where the eastern depression is the deepest at 10 920 ± 5 m (Nakanishi and Hashimoto, 2011). Table 1.1 summarises the depth estimates and measurement by both sounding and *in situ* measurements.

Different instruments with varying accuracies and interpretations will undoubtedly produce more depth records in the vicinity of 10 900 m, but the important point is that the trench is 'nearly 11 000 m' deep and furthermore there are four other trenches that are also 'nearly 11 000 m' deep; the Philippine (10 540 m), the Kuril–Kamchatka (10 542 m), the Kermadec (10 177 m) and the Tonga Trenches (10 800 m).

Furthermore, in 2001, a sonar survey performed by the Hawai'i Mapping Research Group (HMRG) identified an area 200 km to the east of Challenger Deep which was 10 732 m, close to or potentially challenging Challenger Deep or Horizon Deep in the Tonga Trench (second deepest place; ~10 882 m). Identification of this area, originally coined the HMRG Deep (now known as the Sirena Deep), suggests that there are multiple sites >10 500 m deep within one trench (Fryer *et al.*, 2002).

In terms of biology, it is the fact that there are multiple areas in the ocean reaching close to 11 000 m deep that is important. The exact depth of the single deepest point is extraneous. Likewise, the deepest point is simply one value of just one parameter of an entire trench ecosystem. Therefore, while the 'top five' deepest trenches represent the deepest 45% of the total ocean depth range, there are still 46 distinct habitats which make up the hadal zone (see Chapter 2).

1.4   First sampling of the trenches

The existence of life in the deep sea was established by Charles Wyville-Thomson, aboard the HMS *Porcupine*, in the late 1860s in the northeast Atlantic Ocean at ~4000 m (Thomson, 1873). The success of the *Porcupine* expedition prompted the round-the-world HMS *Challenger* expedition of 1873 to 1876. The HMS *Challenger* was the first expedition to physically sample the hadal zone by collecting a small sediment grab at 7220 m in the Japan Trench. Analysis of the sediment revealed 14 species of Foraminifera shells, but scientists were unable to conclusively determine whether or not these were hadal species or simply the fragments of species that had fallen from shallower waters. In 1899, a US expedition on the RV *Albatross* trawled at 7632 m in the Tonga Trench, but like the *Challenger* samples, only fragments of a siliceous sponge were recovered (Agassiz and Mayer, 1902).

Success finally came in August 1901, when the expedition on the *Princess-Alice* successfully trawled specimens of Echiuroidea, Asteroidea, Ophiuroidea and demersal fish from 6035 m, in the Zeleniy Mys Trough (North Atlantic Ocean) (Koehler, 1909; Sluiter, 1912). These data suggested that that multicellular organisms were present deeper than 6000 m, contrary to popular perception at that time (Pettersson, 1948).

In 1948, a Swedish-led expedition on the *Albatross* successfully trawled a variety of benthic species from 7625–7900 m, from the Puerto-Rico Trench (Nybelin, 1951). The catch comprised mostly holothurians (with some polychaetes and isopods), and proved unequivocally that life existed well beyond 6000 m (Eliason, 1951; Madsen, 1955).

New evidence of life at these extreme depths spurred a period of intense research. The sudden boom in scientific interest concerning the trenches and the fauna therein was primarily sparked and pioneered by two research campaigns; the Soviet *Vitjaz* expeditions (1949–53, 1954–59) and the Danish *Galathea* expedition (1951–52; Fig. 1.4).

In 1949, the *Vitjaz* trawled 150 benthic invertebrates belonging to 20 species from 10 different classes from 8100 m in the Kuril–Kamchatka Trench (Uschakov, 1952).

Figure 1.4 The Danish vessel *Galathea* in Copenhagen harbour. Image courtesy of Torben Wolff, Zoological Museum, Danish Museum of Natural History, Denmark.

Three years later in the same trench, six large trawl catches were obtained from 6860–9500 m. From 1954 to 1959, the *Vitjaz* expeditions (cruises 19, 20, 22, 24–27 and 29) continued sampling hadal fauna in the Pacific Ocean (mainly Japan, Aleutian, Izu-Bonin, Volcano, Ryukyu, Bougainville, Vityaz, New Hebrides, Tonga, Kermadec and Mariana Trenches; Belyaev, 1989 and references therein). In 1966, on cruise 39 to the Kuril–Kamchatka Trench, a more detailed examination was made of hadal fauna at depths of between 6000 and 9530 m. Three bottom grabs and 17 successful trawls were obtained that yielded exceptionally abundant catches, even from the trench's greatest depth (Zenkevitch, 1967). *Vitjaz* cruise 57 (1975) comprised 30 successful trawls and 12 grab samples, and collected hadal fauna from the Ryukyu, Philippine, Palau, Yap, Mariana, Volcano, Izu-Bonin and Banda Trenches. Later, during the 59th cruise, five trawl catches were obtained from 7500 m in the Japan Trench. After this cruise, the *Vitjaz* remained operational until 1979, but ceased to conduct further work at hadal depths.

From 1949 to 1976, the RV *Vitjaz* sampled hadal depths during 20 expeditions to 16 trenches in the Pacific and Indian Oceans. The scientists aboard managed to amass an exceptionally abundant and diverse collection of hadal fauna, from a grand total of 40 bottom grab samples and 106 successful trawls, 18 of them at depths >9000 m and five at >10 000 m. The vast contribution of the RV *Vitjaz* to hadal research has never been surpassed by any other research vessel to date. *Vitjaz* completed her last cruise in 1979 and returned to the port of Kaliningrad which she had left 34 years before. She can be found their today, residing on the banks of the River Pregolya, fulfilling her new role as a maritime museum.

The Danish *Galathea* expedition successfully sampled depths >6000 m by trawl and sediment grab in six trenches: the New Britain, Java, Banda, Bougainville and Kermadec and Philippine Trenches. The deepest trawl was achieved in the Philippine Trench at 10 120 m and contained holothurians whose capture provided evidence that

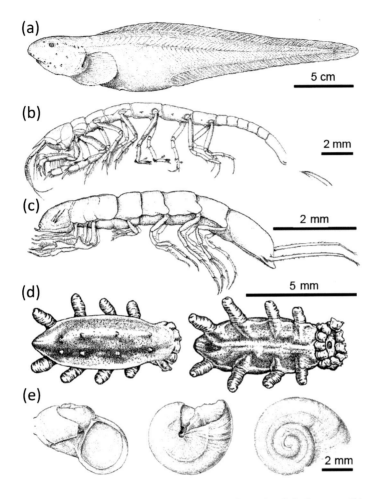

Figure 1.5 Examples of hadal fauna recovered from the *Galathea* expedition. (a) The fish *Notoliparis kermadecensis*, (b) the tanaid *Apseudes galathea* taken from 6770 m in the Kermadec Trench, (c) the isopod *Macrostylis galathea* from 9790 m in the Philippine Trench, (d) the holothurian *Elpidia glacialis* from 8300 m in the Kermadec Trench and (e) the gastropod *Trenchia wollfi* from 6730 m in the Kermadec Trench. Reproduced with the permission of *Galathea Reports*.

life existed everywhere in the oceans, even at depths greater than 10 000 m below sea level (ZoBell, 1952; Wolff, 1960).

The extensive work of the *Galathea* and *Vitjaz* expeditions resulted in the sampling of almost every known major trench, with recovery of multi-celled organisms from every trench at every depth sampled (for examples, see Fig. 1.5). The first comprehensive reports of these expeditions were published in the *Galathea Reports* (available online via the Natural History Museum of Denmark, www.zmuk.dk) and were first summarised by Wolff (1960) and expanded later in Wolff (1970) and Zenkevitch (1954; Zenkevitch *et al*. (1955), respectively. The entire *Galathea* and *Vitjaz* collections were also collated and summarised in Belyaev (1966) and later, incorporating other hadal research up to the mid-1980s (Belyaev, 1989).

In the years after the *Galathea* and *Vitjaz* expeditions, there were several more biological sampling expeditions to the trenches, but each of these focused on a single trench. Although the number of samples taken and expeditions undertaken were relatively low compared to those of shallower zones, there are still too many to mention them all. However, to provide some examples, an American expedition on the RV *James M. Gilliss* successfully obtained two trawls from 7600 and 8800 m and two box cores from 8560 and 8580 m in the Puerto-Rico Trench (George and Higgins, 1979). Further sediment samples were taken from the Puerto-Rico Trench on the USNS *Bartlett* in 1981 (8371–8386 m; Richardson *et al.*, 1995) and the RV *Iselin* in 1984 (7460–8380 m; Tietjen *et al.*, 1989). Occasional trawl sampling by some nations in neighbouring trenches commenced (e.g. Anderson *et al.*, 1985; Horikoshi *et al.*, 1990), but very little trawling occurred at hadal depths after the early 1980s.

In addition to trawling and sediment grabs, another method rose to the fore in the challenging quest to sample the deep ocean; a method that enabled scientists to view specimens and the deep-sea environment *in situ*. Although underwater photography has existed for over 100 years (Boutan, 1900), it was not until the 1940s and 1950s when it became an integral part of seafloor investigations, whether for biological or geological applications (Ewing *et al.*, 1946; Hahn, 1950; Emery, 1952; Pratt, 1962; Emery *et al.*, 1965). The first wave of scientifically useful images from hadal depths were captured in the Atlantic Ocean, in the Puerto-Rico and Romanche Trenches (Pratt, 1962; Heezen *et al.*, 1964; Heezen and Hollister, 1971). Further images were taken in the South Sandwich Trench (Heezen and Johnson, 1965), the New Britain and New Hebrides Trenches (Heezen and Hollister, 1971). All of the images described above were captured between 6000 and 8650 m, using single or dual film cameras lowered to the seafloor on a wire. In 1962, the American PROA expedition on the *Spencer F. Baird* obtained ~4000 images of the hadal seafloor, from depths of 6758 to 8930 m in the Palau, New Britain, South Solomon and New Hebrides Trenches (Lemche *et al.*, 1976).

Over time, the preferred sampling methods for extreme depths shifted from using long wires to the deployment of free-falling baited traps and cameras. For example, the American SOUTHTOW expedition in 1972 to the Peru–Chile Trench deployed a free-fall camera 17 times between 84 and 7023 m that provided the first recorded evidence of live hadal fauna (described in Hessler *et al.*, 1978). The same system, with traps, was deployed from the RV *Thomas Washington* at 9600 and 9800 m on the floor of the Philippine Trench and revealed extraordinarily large densities of scavenging amphipods (Hessler *et al.*, 1978). These results showed that, not only could scavenging Crustacea be easily photographed, but they were readily attracted to and caught in baited traps (Hessler *et al.*, 1972). Similar traps were used to recover necrophagous scavengers from the Philippine Trench (8467–9604 m), the Palau Trench (7997 m) and the Mariana Trench (7218–9144 m) from the same vessel during the EURYDICE (1975), PAP-TUA (1986) and INDOPAC IX (1977) expeditions, respectively (France, 1993). These samples were used to examine gene flow of species between different trenches for the first time. Physiological experiments followed suit in 1980. Hyperbaric traps were employed to recover live amphipods from the trenches under ambient pressures (Yayanos, 1977, 2009). Although the results were rudimentary, they provided new insight into the potential for large vertical migration (Yayanos, 1981).

The year 1995 saw the construction of the first full ocean depth rated ROV *Kaikō* (Kyo *et al.*, 1995; Mikagawa and Aoki, 2001). In addition to the ceremonial marking of the deepest place on Earth, *Kaikō* was also used to collect sediment core samples and deploy baited traps in Challenger Deep. The samples obtained from *Kaikō* (mostly sediment samples) provided specimens for a flurry of research focusing on piezophilic bacteria (Kato *et al.*, 1997, 1998; Fang *et al.*, 2000), Foraminifera (Akimoto *et al.*, 2001; Todo *et al.*, 2005) and microbial flora (Takami *et al.*, 1997), and ultimately led to the discovery of the deepest chemosynthetic-based community, at 7326 m in the Japan Trench (Fujikura *et al.*, 1999; Fujiwara *et al.*, 2001). The ROV *Kaikō* completed 295 dives, with more than 20 dives to full ocean depth, however, it was unfortunately lost in an accident in May 2003 (Momma *et al.*, 2004; Tashiro *et al.*, 2004; Watanbe *et al.*, 2004).

There were a few more opportunistic sediment core and baited trap samples taken in the years following *Kaikō*'s loss. These included sediment samples and amphipods from the Peru–Chile Trench (Danovaro *et al.*, 2002, 2003; Perrone *et al.*, 2002; Thurston *et al.*, 2002), and other amphipod studies from the Kermadec and Tonga Trenches (Blankenship *et al.*, 2006; Blankenship and Levin, 2007), where the latter samples were employed in the first attempt to examine a hadal food-web.

Coinciding with the increasing presence of Japanese technology at hadal depths via the ROV *Kaikō*, others began deploying instruments to measure environmental parameters such as temperature, salinity, pressure, current speed and direction. Using bespoke conductivity, temperature and depth (CTD) probes, profiles of the entire water column were obtained from the surface to >10 800 m in the Mariana Trench (Mantyla and Reid, 1978; Taira *et al.*, 2005) and to 9209 m in the Izu-Bonin (Ogasawara) Trench (Taira, 2006). 'Super-deep' current meters were also deployed to 10 890 m at the same location and remained there for 14 months (Taira *et al.*, 2004). The resulting data revealed that current speeds were low, as had been anticipated, and also showed the presence of tidal cycles at full ocean depth at 14–15 day and 28–32 day spectral periods.

1.5   Exploratory bathyscaphes

In 1956, the first photographs of the hadal zone were taken by Jacques Cousteau at 8000 m, in the Romanche Trough (North Atlantic Ocean; Cousteau, 1958).

The first humans ever to visit Challenger Deep were Swiss scientist Jacques Piccard and US Navy Lt. Don Walsh in the human-occupied vehicle (HOV) or 'Bathyscaph', *Trieste* in 1960 (Piccard and Dietz, 1961). Although the dive represented an enormous achievement in human endeavour, unfortunately it did not offer any scientific insight or herald a new era in trench exploration. The reports of the animals that the men saw on the bottom were somewhat dubious and, despite being rapidly discredited in the scientific literature (Wolff, 1961), the erroneous account of a 'flat fish' which was described as 'white and about one foot long', is sadly still perpetuated to this day (Jamieson and Yancey, 2012).

The *Trieste* dive did, however, confirm that life was present at full ocean depth, but the tall tales of fish at Challenger Deep perhaps dampened interest in the reports from other bathyscaphes at hadal depths. For example, in 1962, the French bathyscaphe *Archimède* made a total of eight dives over 7000 m (including three >9000 m) in the Kuril–Kamchatka Trench and at the junction of the Japan and Izu-Bonin Trenches. In 1967, the *Archimède* achieved a further eight dives off Japan, ranging from 5500 to 9750 m and observed numerous benthic animals and collected several samples using a manipulator arm (Laubier, 1985). The *Archimède* also completed 10 dives in the Puerto-Rico Trench, two of which were dedicated to biological observations at 7300 m (Pérès, 1965). The pilot observed and documented numerous benthic animals including holothurians, isopods, decapods and fish, although no video or still images were taken to substantiate these personal notes and unfortunately this prohibited any accurate identification of species or abundance calculations. The description and behaviour of the organisms reported by Pérès (1965) bore an uncanny resemblance to much more recent video and still images taken from the Japan and Kermadec Trenches at similar depths (Jamieson *et al.*, 2009a, b, 2010; Fujii *et al.*, 2010). This more recent evidence suggests that, sadly, the reports from *Archimède* were far more valuable at the time than was realised, because of the deeper but erroneous claims of the *Trieste*.

1.6   Modern hadal research

The turn of the century saw a renewed interest in hadal research. While opportunistic sampling is still occasionally undertaken (e.g. Blankenship *et al.*, 2006; Itoh *et al.*, 2011), there have been several significant developments in the last decade.

Following the loss of ROV *Kaikō*, JAMSTEC developed the *Kaikō*-7000, a 7000 m rated vehicle (Murashima *et al.*, 2004; Nakajoh *et al.*, 2005) which was shortly superseded by Automatic Bottom Inspection and Sampling Mobile (ABISMO; Yoshida *et al.*, 2009). ABISMO is a much smaller, compact and relatively low-cost vehicle, capable of taking small water and sediment samples only. However, it was successfully deployed to 9707 m in the Izu-Bonin Trench in 2007 and 10 257 m at Challenger Deep in 2008 (Yoshida *et al.*, 2009).

The Woods Hole Oceanographic Institute, USA, has developed a new full ocean depth ROV, capable of conversion to an autonomous underwater vehicle (AUV; Bowen *et al.*, 2008, 2009a; Fletcher *et al.*, 2009). This new 'hybrid' vehicle or HROV, *Nereus*, completed field trials at 10 903 m at Challenger Deep on 31 May 2009 (Bowen *et al.*, 2009b) and now offers new and unprecedented access to trench environments. Capable of performing acoustic and video transects, recovering organisms and sediment samples, *Nereus* provides the potential to map, observe and quantify the distribution of hadal fauna *in situ*.

Coinciding with these technological advances, a collaboration between the Universities of Tokyo (Japan) and Aberdeen (Scotland) began the 5-year HADEEP project in

Figure 1.6 Biological samples recovered from the Japan, Kermadec and Peru–Chile Trenches. (a) The snailfish *Notoliparis kermadecensis*, (b) the lysianassoid amphipod *Eurythenes gryllus*, (c) close up of *E. gryllus*, (d) the pardaliscid amphipod *Princaxelia jamiesoni*, (e) unidentified leptostracan, (f) the gastropod *Tacita zenkevitchi* and (g) the decapod *Hymenopenaeus nereus*. All images courtesy of the HADEEP projects, except (b) Shane Ahyong (Australian Museum, Australia) and (d) courtesy of Tomislav Karanovic (University of Seoul, S. Korea).

2006 (HADal Environments and Educational Program; Jamieson *et al.*, 2009c). HADEEP was the first international campaign to perform repeated, standardised experiments in multiple trenches in order to disentangle the effects of depth (i.e. 'hadal') and individual trench identity (Jamieson *et al.*, 2010). Using autonomous free-fall imaging landers and traps, the project carried out seven trench expeditions covering six trenches (Kermadec, Tonga, Peru–Chile, Mariana, Izu-Bonin and Japan Trenches), and sampled depths ranging between 4000 and 10 000 m. The expeditions obtained hours of *in situ* video footage, thousands of samples (see Fig. 1.6) and *in situ* still images which have provided novel data concerning species distribution, behaviour and physiology. A further collaboration between the University of Aberdeen and the National Institute for Water and Atmosphere research (NIWA) in New Zealand saw the project (dubbed HADEEP2) extended to 2013 and later (HADEEP3 and 4) running until 2015. These projects, still ongoing, have introduced a fleet of baited landers and baited traps to collect samples and record footage from the Kermadec, New Hebrides and Mariana Trenches.

One of the outcomes of the first HADEEP project was 'Trench Connection'; the first international symposium focusing entirely on the hadal zone biology, ecology, geology and technology (Jamieson and Fujii, 2011). It was held at the University of Tokyo's Atmosphere and Ocean Research Institute (AORI) in November 2010. The symposium attracted an international collective of 70 scientists and engineers from six different countries to discuss the latest developments in the exploration and understanding of the deepest environments on Earth. It was the hope of the 'Trench Connection' organising committee that bringing together such a diverse group of scientists would increase the opportunities for hadal research, with the aim to progress hadal science from its infancy into a fully-fledged area of research, integrated within the wider deep sea research community. One such project emerged from the Trench Connection Symposium; The HADES project.

The HADES project is an international collaboration between various institutes in the USA, the UK and New Zealand. This project will bring together a diverse range of disciplines with a suite of technology (traps, cameras and ROV) to perform comprehensive biological surveys in the Kermadec and Mariana Trenches in 2014. This project aims to determine the composition and distribution of hadal species, the role of pressure, food supply, physiology, depth and seafloor topography on deep-ocean communities and the evolution of life in the trenches.

1.7   Terminology

Based on the 1949–53 *Vitjaz* data, an endemism-based biological transition was identified at 6000–7000 m, thus distinguishing the abyssal zone from the 'ultra-abyssal zone' (Zenkevitch et al., 1955). A trench was defined as 'a long but narrow depression of the deep-sea floor having relatively steep sides' (Wiseman and Ovey, 1953) with 'features such as a flat floor' (Menzies and George, 1967), an area of 'high seismic activity' (Ewing and Heezen, 1955) and 'negative gravity anomalies' (Worzel and Ewing, 1954).

Around the same time, the same conclusion regarding biological zonation was drawn by Anton Bruun, leader of the *Galathea* expedition, based on their findings. It was Bruun (1956a) who proposed the term 'hadal' and 'hadal fauna' to describe depths and fauna greater than 6000 m (with 'hadopelagic' to describe mid-water fauna deeper than 6000 m). Russian literature at that time, and sometimes still, refers synonymously to the hadal zone as the 'ultra-abyssal zone' (following Zenkevitch, 1954). Attempts to rename hadal fauna by Menzies and George (1967) as 'trench floor fauna' were rejected by Wolff (1970) on the grounds that it did not reflect the influence of elevated hydrostatic pressure (also stressed by Madsen, 1961; Wolff, 1962; Belyaev, 1966). In addition, the new name implied that fauna originated on the floor of the trench axis and not on the slopes, thus making a distinction between fauna within the trench for which there was no evidence to suggest a difference (Wolff, 1970).

The term 'hadal' is derived from *Hades*, meaning both the Greek kingdom of the underworld and the god of the underworld himself (son of Cronus and Rhea and brother to Zeus and Poseidon). According to myth, the three brothers defeated the Titans and claimed rulership over the underworld, air and sea, respectively. The term can also be loosely translated as 'the unseen', 'abode of the dead' or 'the dominion of Hades'. In modern days, Hades is perhaps more associated with evil but in mythology he was often portrayed as more passive than malicious. It is interesting to note that although he was portrayed as more altruistic than he is now known, he was known to strictly forbid the inhabitants of his dominion to leave, which is a rather apt analogy for hadal fauna endemism; the species inhabiting hadal depths are often confined to one or more trenches and are rarely capable of leaving. Furthermore, Hades was renowned for his overwhelming wrath in response to anyone trying to leave, which is also analogous to the effects of decompression on obligate piezophilic organisms if removed from the hadal zone.

The hadal zone differs somewhat from shallower zones (littoral, <200 m; bathyal, 200–2000 m; and abyssal 2000–6000 m; Gage and Tyler, 1991) because it is not simply a continuation of the preceding deep-sea environments. In fact, the progression of the environment from the continental slopes and rises to the abyssal plains eventually splits into clusters of fragmented and often vastly isolated trenches. Therefore, the term hadal was suggested by Bruun (1956a) over 'ultra-abyssal', so as not to imply a mere extension of the abyssal zone and in order to provide a name as distinctive and consistent with the terms littoral, bathyal and abyssal.

Although the maximum depth of the hadal zone is clearly defined as the maximum known depth, the minimum depth has prompted various discussions. The geographic isolation of the trenches is thought to have promoted a high degree of endemism (Wolff, 1960, 1970; Belyaev, 1989). This has, in turn, been used as an indicator of the lower depth limit of the hadal zone. Wolff (1960) published the first summary of hadal research and, based on the fauna known at that time, suggested 6000 m as the minimum hadal depth. It was also suggested that the depth limit could be 6800–7000 m, but the 6000 m mark which represented 58% (Wolff, 1960) or 56% (Belyaev, 1989) endemism was subsequently recognised as the boundary between the abyssal and hadal zones. Although endemism >6000 m can vary between trenches from 37–81% (Belyaev, 1989), it is generally recognised that 6000–7000 m represents the 'abyssal–hadal transition zone' (Jamieson *et al.*, 2011a). More recently, the 6000 m boundary has, presumably for convenience and to account for the abyssal–hadal transition zone, been revised to 6500 m (UNESCO, 2009). The same recommendations have also readdressed the depth ranges of all vertically stratified zones.

The bioregional classification adopted by the United Nations Educational, Scientific and Cultural Organization (UNESCO, 2009) were based primarily on suggestions regarding regions and provinces by Zezina (1997), Menzies *et al.* (1973), Vinogradova (1979) and

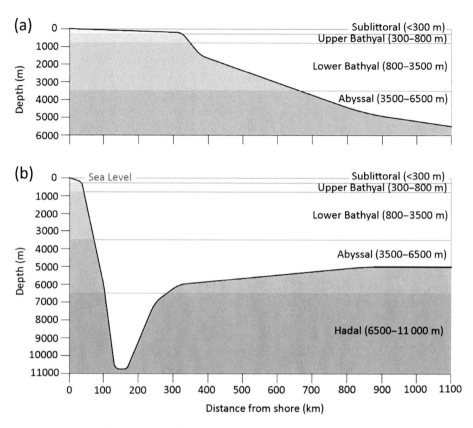

Figure 1.7 Profiles of the seafloor and bathymetric biozones for aseismic zones (a) and seismic zones (b). Vertical exaggeration = 50.

Belyaev (1989), and the newly defined boundaries have been set on the basis of unpublished observations, re-analyses of existing data and more recent data reviewed by an expert group during the United Nations Environment Programme (UNEP) workshop on 'Biogeographic Classification Systems in Open Ocean and Deep Seabed Areas Beyond National Jurisdiction', held in Mexico City, Mexico, 22–24 January 2007. This led to a proposal by Watling *et al.* (2013) that the deep-sea biogeographic classification depth ranges are upper bathyal (300–800 m), lower bathyal (800–3500 m), abyssal (3500–6500 m) and hadal (6500–11 000 m) (Fig. 1.7). The depth zones translate to 2.9, 23.0, 65.4 and 0.21% of the total ocean area as calculated by Watling *et al.* (2013).

The upper bathyal zone comprises the continental margins and generally falls within the Exclusive Economic Zone (EEZ; 200 nautical miles from coast) of many nations. The lower bathyal zone encompasses three physiographic categories; lower continental margins, seamounts and mid-ocean ridges. The abyssal zone accounts for the large majority of the deep-sea floor and the hadal zone, as described previously, is generally restricted to convergent plate boundaries in areas where lithospheric plate subduction occurs. As these plate boundaries tend to form the continental land masses, most of the trenches fall within national jurisdiction (Watling *et al.*, 2013)

A more detailed description of the hadal environment in terms of geographic location and geological formation is given in Chapter 2. However, with regards to terminology it is important to clarify the exact boundaries of the hadal zone for the context of this book. While acknowledging that the oceans do not ever conform to convenient human-contrived nomenclature, but rather form transition zones between one distinct zone and another, hereafter the hadal zone is considered to be the deep subduction trenches and hadal fauna is that inhabiting depths of greater than 6000 m. Although Belyaev (1989), UNESCO (2009) and Watling *et al*. (2013) suggest 6500 m as the lower limit, no significant new or comprehensive dataset has been recently undertaken and, until it does, 6000 m will be used.

2 Geography and geology

The geography and geology of the hadal zone is somewhat unique and is more similar to habitats such as canyons and seamounts than to the other depth-stratified biozones such as the abyssal and bathyal zones. This is because the trenches are enclosed, distinct geological features, often isolated from one another by thousands of kilometres. Furthermore, trenches are more complex than simply 'areas deeper than 6000 m'. They are formed by immense geological forces which, on one hand, provide a unique geological, biological and environmental setting, however, on the other hand, these forces result in the devastating earthquakes and tsunamis for which they are better known.

For the purposes of this book, an analysis of global trench locations and topography was undertaken by Dr Tomasz Niedzielski from the Department of Cartography at the University of Wrocław in Poland, using ArcMap 9.3.1 under ArcInfo licence, provided by ESRI. The abyssal–hadal boundary of 6000 m was automatically selected across contours imposed on imported bathymetry (30 arcsec grid released by GEBCO). The maps were projected to the cylindrical equal area projection with a central meridian 180° and a standard parallel 30° S, with a subkilometre spatial resolution. The analysis of the topography was focused further by slicing the bathymetry into 500 m depth bins. This approach permitted the extraction of habitat size (km^{-2}), projected area (km^{-2}), mean slope (°) and water volume (km^{-3}). These data were used to provide a list of up-to-date locations and depths of the trenches, with additional information on size and volume characteristics. The exact depth of each trench should be taken as indicative since the map resolution may not detect very small depressions within the deepest point.

2.1 Geographic location

With the entire Earth's oceans now mapped, we now know the locations of all the areas of seafloor that comprise the hadal zone. It is accepted that the hadal zone is comprised of deep trenches and troughs; however, the numbers of these habitats cited in scientific literature often varies. For example, Smith and Demopolous (2003) cite a total of 14 trenches, Herring (2002) cites 37 trenches, Jamieson *et al.* (2010) cites 22 trenches (and 15 troughs), Angel (1982) cites 22 trenches (one of which is <6000 m), Vinogradova *et al.* (1993a) list 32 trenches (of which two are <6000 m), whereas Belyeav (1989) cites a total of 55 trenches and troughs. The confusion perhaps arises when considering those areas which are not large trenches formed at tectonic convergence zones: the 'troughs' and other features such as faults or fracture zones that intersect mid-ocean ridges (such as the Romanche trough) and small localised depressions on the abyssal plains. To complicate the matter, the original depth boundary of the hadal zone (6000 m) has been recently reviewed and changed to 6500 m (UNESCO, 2009), automatically omitting some shallower areas.

To address this issue, a digital global bathymetry database (GEBCO) was used to extract all known areas deeper than 6500 m using GIS software (ArcGIS). The results were then categorised into trench or trough using the following criteria:

- Trench: a distinct, single elongated area deeper than 6500 m generally formed by tectonic subduction or fault.
- Trough: large area, or cluster of basins deeper than 6500 m which are not formed at converging plate boundaries but rather basins within an abyssal plain.

Using these criteria, 33 trenches (27 subduction trenches and six trench faults) and 13 troughs were identified, 46 individual hadal habitats in total. An example of the characteristic differences between trench, trough and trench fault topography are shown in Figure 2.1. A list of all the trenches and troughs is shown in Table 2.1 with total projected area, maximum depth and the latitude and longitude of the deepest point. These parameters are also illustrated in Figures 2.2, 2.3, 2.4 and 2.5.

The mean depth of the trenches (including trench faults) and troughs are 8216 m $\pm$ 1331 S.D. and 7229 m $\pm$ 665 S.D., respectively (total mean depth = 7938 m $\pm$ 1257 S.D., Table 2.1, Fig. 2.6). The deepest trench is the Mariana Trench at 10 920 $\pm$ 5 m (Nakanishi and Hashimoto, 2011), but the largest trench in terms of area is the Izu-Bonin Trench which spans 99 801 km^{-2} (maximum depth = 9701 m). The deepest, and also the largest trough is a vast cluster of deep basins in the middle of the North Pacific with a maximum depth of 8565 m and an area of 23 670 km^{-2}. For ease of comparison, the 22 trenches can be categorised into deep, medium and shallow trenches (Fig. 2.6). The medium trenches, of which there are 11, are trenches within 1000 m of the median depth. Therefore, there are five deep trenches, all of which exceed 10 000 m and six shallow trenches.

The total area of the trenches and troughs is approximately 750 000 km^{-2} and 50 500 km^{-2}, respectively, thus, in total, the 'hadal zone' covers 800 500 km^{-2}. The area of the entire ocean is 335 258 000 km^{-2}, therefore, the total area exceeding 6500 m constitutes 0.24% of the entire ocean. Despite comprising a very low percentage of the world's ocean area, the depth range between 6500 and 10 982 m (4482 m) accounts for 41% of the total depth range.

Of the 27 hadal trenches, there is no correlation between depth and descriptors of size (area and length). The longest trenches (in their entirety, regardless of depth) are the Java Trench (~4500 km), the Atacama Trench (3700 km) and the Aleutian Trench (3700 km) which have depths of 7450, 8074 and 7822 m, respectively. In fact, most of the longest trenches are not excessively deep. The shortest five trenches are the Volcano, Yap, Banda, Admiralty and Palau Trenches of which the Palau and Admiralty Trenches are also among the five smallest trenches in terms of area.

The length of the trench is, of course, just an indicator of the size of the geological formation. In terms of percentage of the trench length occurring deeper than 6500 m, for

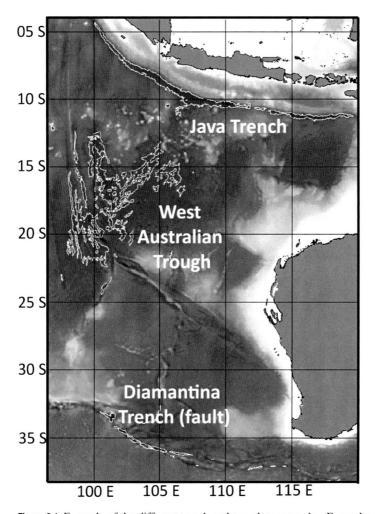

Figure 2.1 Example of the different trench and trough topography. Examples shown are the Java Trench, the West Australian Trough and the Diamantine Trench fault in the Indian Ocean south of the Indonesian island of Java and West of Australia.

58% of the trenches, at least 70% of the trench lies within hadal depths. In contrast, the Middle America Trench and the Diamantina Trench fault have less than 20% of their length exceeding 6500 m and for the New Hebrides Trench only 8% is found below 6500 m. These trenches are predominantly abyssal trenches. Perhaps as expected, the trenches with the largest percentage of seafloor area >6500 m are the deep trenches such as the Mariana, Kuril–Kamchatka, Tonga and Kermadec Trenches (all greater than 90%).

The projected area >6500 m depth gives a far more representative indication of the size of the hadal habitat, irrespective of depth. The largest hadal habitats are found within the Izu-Bonin (99 801 km^{-2}), Kuril–Kamchatka (91 692 km^{-2}), Mariana (79 956 km^{-2}), Tonga (65 817 km^{-2}) and the Aleutian Trenches (63 036 km^{-2}), whereas the smallest are in the Admiralty (4050 km^{-2}), New

Table 2.1 A List of all the trenches, trench faults and troughs deeper than 6000 m with maximum depth (m), the approximate location of the deepest point and ocean location.

	Max depth (m)	Latitude	Longitude	Ocean
Trenches				
Admiralty	6 887	00.5600 S	149.3800 E	Pacific
Aleutian	7 669	50.8791 N	173.4588 W	Pacific
Atacama	7 999	23.3679 S	71.3473 W	Pacific
Banda	7 329	05.3852 S	130.9175 E	Pacific
Bougainville	9 103	06.4762 S	153.9323 E	Pacific
Hjort	6 727	58.4400 S	157.6800 E	Pacific
Izu-Bonin	9 701	29.8038 N	142.6405 E	Pacific
Kuril–Kamchatka	10 542	44.0700 N	150.1800 E	Pacific
Japan	8 412	36.0800 N	142.7500 E	Pacific
Java	7204	11.1710 S	118.4669 E	Indian
Kermadec	10 177	31.9270 S	177.3126 W	Pacific
Mariana	10 920	11.3808 N	142.4249 E	Pacific
Middle America	6 547	13.9097 N	93.4728 W	Pacific
New Britain	8 844	07.0225 S	149.1623 E	Pacific
New Hebrides	7 156	23.0733 S	172.1502 E	Pacific
Palau	8 021	07.8045 N	134.9869 E	Pacific
Philippine	10 540	10.2213 N	126.6864 E	Pacific
Puerto-Rico	8 526	19.7734 N	66.9276 W	Atlantic
Ryukyu	7 531	24.5109 N	127.3602 E	Pacific
San Cristobal	8 641	11.2800 S	162.8200 E	Pacific
Santa Cruz	9 174	12.1800 S	165.7700 E	Pacific
South Orkney	6 820	60.8510 S	41.0442 W	Southern
South Sandwich	8 125	56.2430 S	24.8326 W	Southern
Tonga	10 800	23.2500 S	174.7524 W	Pacific
Vityaz	6 150	10.2142 S	170.1178 E	Pacific
Volcano	8 724	24.3326 N	143.6107 E	Pacific
Yap	8 292	08.4073 N	137.9244 E	Pacific
Trench faults				
Cayman	8 126	19.1700 N	79.8633 W	Atlantic
Emperor	8 103	45.1594 N	174.1444 E	Pacific
Lira	6 881	01.3800 N	150.6500 E	Pacific
Massau	7 208	01.4200 N	148.7400 E	Pacific
Romanche	7 715	00.2226 S	18.5264 W	Atlantic
Vema	6 492	08.9232 S	67.4983 E	Indian
Troughs				
Agulahas	6 787	53.8494 S	26.9643 E	Atlantic
Argentina	6 859	48.8498 S	50.6501 W	Atlantic
Canaries	7 268	24.1248 N	35.6662 W	Pacific
Central	8 211	01.1723 S	168.2845 W	Pacific
Madagascar	7 113	31.3555 S	61.0106 E	Indian
N American	6 922	26.1278 N	55.8783 W	Atlantic
NW Pacific	8 565	39.8184 N	178.8757 W	Pacific
Philippine	7 872	20.8711 N	136.7116 E	Pacific

Table 2.1 *(cont.)*

Troughs				
SE Atlantic	6 559	13.3632 S	1.8716 W	Atlantic
South African	6 509	45.5201 S	14.4328 E	Atlantic
South Australian	6 826	45.0223 S	128.3304 E	Indian
W Australian	7 782	22.2517 S	102.3780 E	Indian
Zeleniy Mys	6 708	14.5810 N	35.2081 W	Atlantic

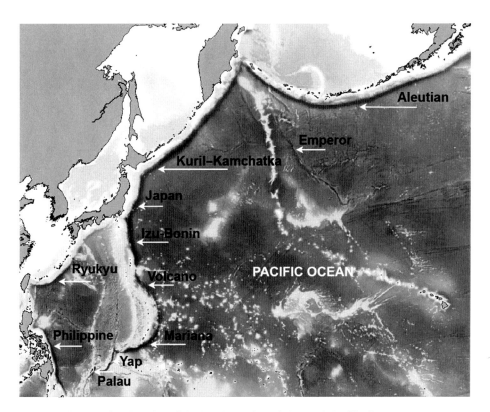

Figure 2.2 Geographical location of the hadal trenches of the North Pacific Ocean.

Hebrides (2439 km^{-2}), the Diamantina (2430 km^{-2}), the Palau (1692 km^{-2}) and the Middle America Trenches (36 km^{-2}).

Of the 33 hadal trenches and trench faults, 26 are located in the Pacific Ocean (79%), three are found in the Atlantic Oceans (9%), two in the Indian Ocean (6%) and two in the Southern Ocean (6%). Of the 13 troughs, six are found in the Atlantic Ocean (46%), four in the Pacific Ocean (31%) and three in the Indian Ocean (23%). These figures are the number of individual trenches and troughs within each ocean. In terms of hadal habitat size distribution, 84% of all hadal habitat is found in the Pacific Ocean (673 855 km^{-2}) compared to 8% in the Atlantic Ocean (64 053 km^{-2}) and 4% in both the Southern and Indian Oceans (31 293 and 31 779 km^{-2}, respectively).

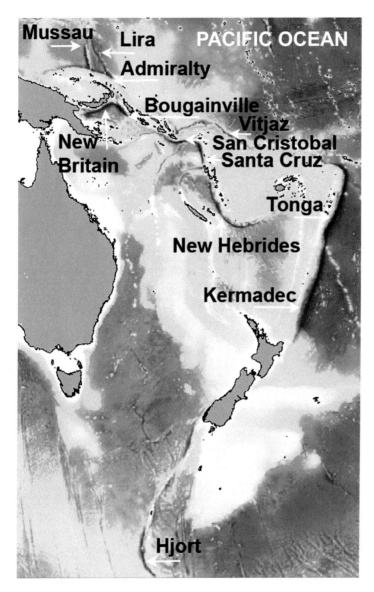

Figure 2.3 Geographical location of the hadal trenches of the southwest Pacific Ocean.

The numerical majority of hadal areas, in particular the deep trenches, are located around the Pacific Rim or, more specifically, around the perimeter of the Pacific tectonic plate where it converges with neighbouring tectonic plates. Due to the immense geological processes responsible for the formation of the deep trenches, and particularly the resulting elevated volcanic activity associated with dynamic convergence, the boundaries of this area are often referred to as the 'Ring of Fire'.

When compiling the definitive list of the names of hadal zones, there are various other criteria and issues to consider. For example, individual trenches and hadal habitats

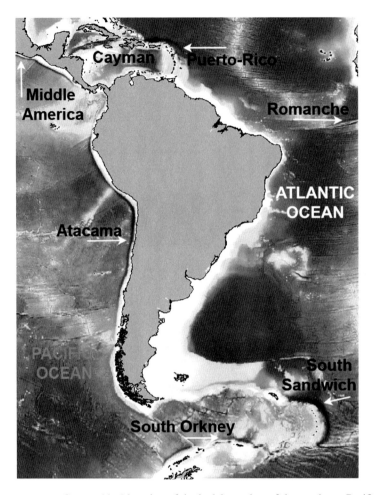

Figure 2.4 Geographical location of the hadal trenches of the southeast Pacific, Atlantic and Southern Oceans.

should be distinct from 'trench systems'. Such a distinction relates to the topographic partitioning <6500 m between two neighbouring trenches. For example, the Tonga and Kermadec Trenches are clearly part of the same geological feature, however, they are partitioned by the Louisville Seamount Ridge, an abyssal partition, and thus represent two separate hadal habitats. A similar partition occurs in the Peru–Chile Trench where the Nazca Ridge intercepts the trench at ~15° S, isolating the northern and southern sectors of the trench by a 5000 m deep partition. The Peru–Chile Trench differs from the Kermadec and Tonga scenario in that the northern sector (sometimes referred to as the Milne-Edwards Trench; e.g. Menzies and George, 1967) is not hadal, whereas the southern sector (known sometimes as the Atacama Trench; e.g. Danovaro et al., 2002) is approximately 8000 m deep. Therefore, in the hadal context it is often referred to as the Atacama Trench. The Mariana Trench is also a trench with a confusing definition. It is often described as the trench that runs in an arc formation from the

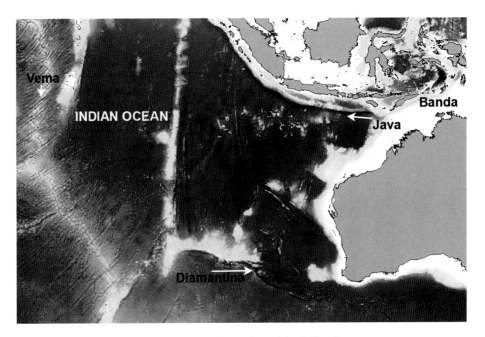

Figure 2.5 Geographical location of the hadal trenches of the Indian Ocean.

northern tip of Yap Trench to the Southern tip of the Izu-Bonin Trench, however, the northern sector of the Mariana Trench is partitioned by depths <6000 m which divide it from the Volcano Trench. Again, from a hadal context the Mariana Trench and Volcano Trench are treated as separate habitats as the partition between the two is as distinct as those between the Palau, Yap, Izu-Bonin and the Mariana Trenches.

Many trenches are referred to by more than one name, depending on the nation reporting on them. For example, in the North Pacific Ocean, the Izu-Bonin Trench is often referred to as the Izu–Ogasawara Trench, the Ryukyu Trench is also known as the Nansei-Shoto Trench and the Yap Trench is also known as the West Caroline Trench. In the South Pacific Ocean, the San Cristobal Trench is also known as the South Soloman Trench (and is also sometimes grouped with the Santa Cruz Trench as one), the Java Trench is also known as the Java or Sunda Trench and the Banda Trench is also known as the Weber Basin. Likewise, the North American Trough is also known as the Nares Deep.

The general tendency of naming deep areas within a single trench as 'deeps' can further confuse the terminology. For example, the Ramapo Deep in the Izu-Bonin Trench (Fisher, 1954), the Meteor Deep in the South Sandwich Trench (Herdman *et al.*, 1956), the Milne-Edwards, Krümmel Deep, Haeckel Deep and Richards Deep in the Peru–Chile Trench (Zeigler *et al.*, 1957) and in the Puerto-Rico Trench, the Gilliss, Brownson and Milwaukee Deeps (George and Higgins, 1979). In other areas such as the Tonga and Kermadec Trenches, the Tonga Trench has a specific name for the deepest point, Horizon Deep (Fisher, 1954), whereas the equivalent deep area in the Kermadec Trench does not. The most famous of these deeps is, of course, the Challenger Deep in the Mariana Trench because it is the deepest area in the world, however,

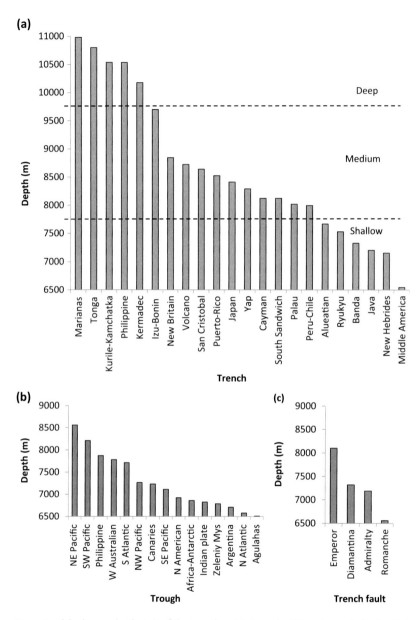

Figure 2.6 Maximum depth (m) of the trenches (a), troughs (b) and trench faults (c) in order of maximum (left) to minimum (right). Deep, medium and shallow trenches are defined as either above or below the medium trenches (= median depth ± 1000 m).

there are also two others nearby in the same trench; the Sirena (HMRG) Deep and Nero Deeps (Fryer *et al.*, 2002).

The concept of coining these areas as deeps was thought to be convenient nomenclature spurred by the desire to declare a new find and does not really offer anything useful scientifically (Wiseman and Ovey, 1954). In fact, in the case where there is more than one deep, such as the Mariana Trench, this tends to result in a disproportionate amount of research effort focused on the deepest area, thus omitting the rest of the often large trench ecosystem.

The issue of coining deeps has been debated for decades. According to an internationally approved definition of deep-sea floor features, a deep is a 'well defined deepest area of a depression of the deep-sea floor which applies when soundings exceed 3000 fathoms [5486 m]' (Wiseman and Ovey, 1953). However, shortly after this publication the British National Committee on Ocean Bottom Features came to the conclusion that although geographical names should be given wherever possible to all major underwater features, deeps that were defined from a morphological point of view were relatively unimportant and therefore newly discovered deeps should remain unnamed (Wiseman and Ovey, 1954). They went on to suggest that the term should fall into abeyance and subsequently the British Committee omitted the deeps from their recognised names of undersea features. Sixty years on and it seems that concept of a deep may indeed be valid as they do represent relatively distinct geomorphological features (Fryer *et al.*, 2002) and unique habitats characterised by a distinct lack of macrofauna, high concentrations of microbial cells and extremely soft sediment accumulations (Danovaro *et al.*, 2003; Glud *et al.*, 2013). Further research is needed to investigate the distinctiveness of the deeps to the surrounding trench habitat and community structure.

There are, of course, a series of trenches close to hadal depths that fall just short of the 6500 m limit. These include the South Shetland, Puyseger and Henry Trenches of the Southern Ocean and the Amirante and Chagos Trenches of the Indian Ocean, among others. Although they do not constitute a hadal habitat, and data is currently lacking, the trench morphology and isolation is such that they may share similar characteristics to the hadal trenches and may provide an ideal setting to test direct effects of trench topography on the distribution of abyssal species.

The biogeographic classification of benthic habitats by UNESCO (2009) categorised the hadal trenches into subregions and provinces, based primarily on Belyeav (1989). Within the provinces trenches are essentially grouped into clusters of adjoined or neighbouring trenches underlying similar hydrographic or productivity regimes. The hadal provinces are listed in Table 2.2.

2.2   Trench formation

The outermost surface of the Earth, the crust and uppermost mantle, act as a single mechanical layer known as the lithosphere. The lithosphere is made up of 14 major tectonic plates and 38 minor plates which constantly move relative to one another above the slightly less dense asthenosphere. The way in which these dynamic plates move relative to one another at the plate boundary are categorised into three types: divergent, transforming and convergent zones (Fig. 2.7). Divergent plates move away from each other and new lithosphere is formed by seafloor spreading (e.g. the Mid-Atlantic Ridge). Transforming plates slide past each other in opposite directions or at different speeds

Table 2.2 Hadal subregions and provinces as classified by UNESCO (2009). The trenches in parentheses are not mentioned directly by UNESCO (2009) and are assumed. The Middle America and Cayman Trenches do not conform to an obvious province.

Subregion	Province	Trenches
Pacific	Aleutian–Japan	Aleutian, Emperor, Kuril–Kamchatka, Japan, Izu-Bonin (Emperor)
	Philippine	Philippine, Ryukyu
	Mariana	Volcano, Mariana, Yap, Palau
	Bougainville–New Hebrides	New Britain, Bougainville, San Cristobal and New Hebrides (Admiralty, Banda, Lira, Massau, Santa Cruz, Vityaz)
	Tonga–Kermadec	Kermadec, Tonga
	Peru–Chile	Peru–Chile/Atacama
North Indian	Yavan	Java, Banda (Vema)
Atlantic	Puerto-Rico	Puerto-Rico
	Romanche	Romanche
Antarctic-Atlantic	Southern Antilles	South Sandwich (South Orkney)

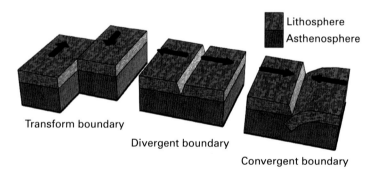

Figure 2.7 The three types of lithospheric plate boundaries: transform boundaries, where plates move parallel in opposite directions; divergent boundaries, where plates spread away from each other; and convergent boundaries, where two plates collide and one is driven (subducted) beneath the other into the asthenosphere.

(e.g. the San Andreas Fault). Convergence zones exist where two plates collide head on and create an underlying subduction zone where sediments, oceanic crust and mantle lithosphere return to and reequilibrate with the Earth's mantle (Amstutz, 1951; White *et al.*, 1970; Stern, 2002; Fig. 2.8).

Divergent boundaries are typified by the presence of heightened volcanic activity, whereas transform boundaries form structural discontinuities which can result in extensive seismic activity. However, convergence and the underlying subduction zones are dynamic plate boundaries associated with both extensive volcanic and seismic activity and are characterised geomorphically by the deep-ocean trenches, which in turn, make up the majority of the hadal zone.

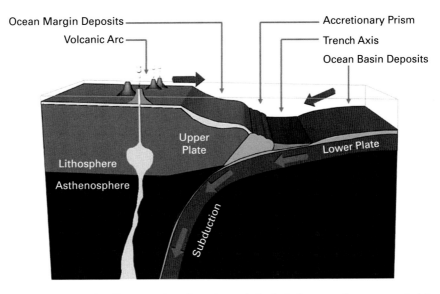

Ocean Margin Deposits

Volcanic Arc

Accretionary Prism

Trench Axis

Ocean Basin Deposits

Upper Plate

Lower Plate

Lithosphere

Asthenosphere

Subduction

Figure 2.8 Principal sedimentary deposits and morphological elements of a convergent plate boundary.

Most of the hadal trenches in their modern form, particularly in the Pacific Ocean, are believed to have formed 65.5 million years ago (mya) during the Cenozoic period when the continents moved into their current position (Menard, 1966; Belyaev, 1989 and Russian citations therein), although it is believed that trenches may have existed for 10^7 years or more. The age of the seafloor can be much older than the age of the trench in its current form. For example, the seafloor age of the Mariana, Tonga and Kermadec, Kamchatka and Atacama Trenches are 150, 120, 70 and <40 million years, respectively (Stern, 2002).

The total global convergence plate margins span >55 000 km (Lallemand, 1999), nearly equal to that of mid-ocean ridges (60 000 km; Keary and Vine, 1990) and are thought to be the most important tectonic features on Earth (Stern, 2002). Generally there are 37 distinct areas that comprise the convergence zones. These areas range in length from 300 km (East Luzon) to 2700 km (Aleutian–Alaska; von Huene and Scholl, 1991; Fig. 2.9a).

The subduction process involves a heavy oceanic plate colliding with, and subsequently being driven beneath, a lighter upper plate (Fig. 2.8). The lower plate is driven down at either high angles (≥30°) or low angles (<30°) (Li *et al.*, 2011). The ocean plates are generally underlying oceans, whereas the light upper plates are topped by a layer of terrestrial (continental) crust and/or an arcuate or linear belt of eruptive centres which form 'volcanic arcs'. Areas where oceanic–continental convergence occurs often result in mountainous areas adjacent to the near-offshore deep trench, for example, the Peru–Chile Trench and the Andes or the Japan Trench and Japan. When the convergence is oceanic to oceanic, volcanic arcs are created. These arcs are formed when the lower plate descends towards the Earth's mantle; the asthenopsheric mantle is sucked towards it, interacting with water and other elements from the sinking plate causing the

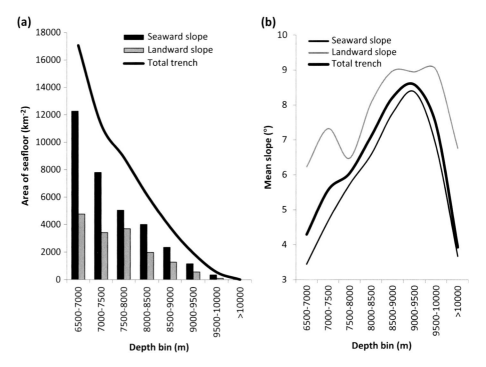

Figure 2.9 Area of seafloor (a) and mean slope (b) for the western slope, eastern slope and entire trench, example shown is data from the Kermadec Trench.

mantle to melt. It is these melts which ascend vertically through the mantle to erupt at arc volcanoes (Stern, 2002). Examples of volcanic arcs can easily be seen to the north of the Aleutian Trench and the Kuril Islands, west of the Kuril–Kamchatka Trench. Although the volcanic arcs represent an important geological aspect of the convergence zones, in the context of hadal biology it is the process of subduction that ultimately forms the hadal habitat. The rate at which the lower plates are subducted varies greatly depending on location and age of the trench (Stern, 2002). Convergence rates through-out the convergence margins range from 10 km per million years (e.g. Manila) to 170 km per million years (the Tonga Trench; von Huene and Scholl, 1991).

Earth appears to be the only terrestrial planet with subduction zones and plate tectonics. Both Mercury and the Earth's moon are tectonically and magmatically dead. Mars appears to have tectonically ceased and is now a single-plate planet (Connerney *et al.*, 1999) and Venus is dominated by thick lithosphere and mantle plumes (Phillips and Hansen, 1998). On Earth, subduction zones produce continental crust which can protrude from the ocean (i.e. continents), and it has been speculated that without subduction zones the Earth's solid surface would be submerged in the oceans and terrestrial life, including humans, would never have evolved (Stern, 2002).

From a biological perspective, the subduction processes affect the trench communities in various ways. The convergence process plunges the seafloor thousands of metres deeper than the abyssal plains which account for most of the Earth's surface.

Furthermore, the act of subduction creates the characteristic trench morphology, typified by a V-shape cross-section which supports sedimentation between the slopes of the converging lithospheric plates (Thornburg and Kulm, 1987). The thickness of the sediment within each convergence zone ranges from 0.4 to 6 km and averages at 1.4 km thick (von Huene and Scholl, 1991; Fig. 2.9). Beyond geographic isolation and creation of the extreme depths, the dynamic plate boundaries undergo continual metamorphosis as tectonic activity shapes the configuration of onshore drainage basins, the pathways of sediment dispersal across the continental margin and depositional processes with the trenches. Over extremely long timescales, sedimentation, sediment subduction and accretion shape the trench floor in a unique way that does not occur in other deep-sea habitats. In addition, frequent seismic activity, varying in severity, can lead to additional and unpredictable events, ranging from turbidity currents to catastrophic sediment slides (Itou *et al.*, 2000; Fujiwara *et al.*, 2011; Oguri *et al.*, 2013).

2.3   Topography

The topography of trenches is unique and the hadal trenches are considerably larger than trenches found in shallower zones (e.g. the Mediterranean Sea; Faccenna *et al.*, 2001; Masson, 2001; Tselepides and Lampadariou, 2004). Trench topography shares similarities with features such as submarine canyons and channels, where sediment accumulation also occurs (Vetter and Dayton, 1998; Tyler *et al.*, 2009; De Leo *et al.*, 2010) but does not exhibit an open end where material can be flushed out into the wider neighbouring abyssal plains (Canals *et al.*, 2006; Arzola *et al.*, 2008), although large depth gradients, steep slopes and sedimentation rates are similar. Trenches also differ from other sloping areas such as continental slopes and rises because in trenches, the area of seafloor diminishes with depth, culminating at a single point (the deepest point), rather than a steady continuation of the slope with depth.

    The trenches are asymmetrical in that the subduction process affects the landward and seaward slopes in different ways. The seaward slopes (on the lower plate) tend to be of a more gradual relief as they constitute what was once the neighbouring abyssal plains gradually travelling down towards the trench axis. However, the landward slopes (the upper plate) tend to be far steeper and complex as they are forced upwards by the lower plate. The steeper landward slopes are generally smaller in terms of area than the seaward slopes.

    Using the Kermadec Trench in the southwest Pacific Ocean as an example, the general shape of the trench is such that the area of seafloor beyond the relatively flat abyssal plains (>6500 m) decreases linearly ($y = -2336.9x + 16733$; $R^2 = 0.9352$) (Fig. 2.9a). For example, the area between 6500 and 7000 m spans 17 057.9 km^{-2} (34.3% of the trench) whereas >10 000 m spans just 15.5 km^{-2}, accounting for just 0.03% of the trench. In fact, the shallowest 50% of the trench in terms of depth (6500–8500 m) accounts for 87.3% of the benthic habitat (43 425 km^{-2}).

    The total area of the trench can be deconstructed into the east (seaward) slope which spans 32 998.5 km^{-2} and the west (landward) slope which is 15 832.9 km^2. These

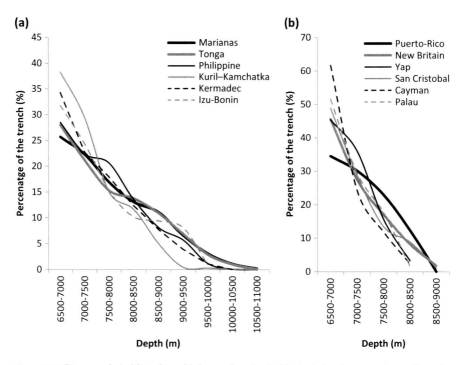

Figure 2.10 Decrease in habitat size with increasing depth, illustrated as in percentage of trench per 500 m depth bin for the six deepest trenches (a) and six randomly selected medium depth trenches (b).

areas account for 66.5% and 33.5% of the trench >6500 m, meaning that the seaward slope is nearly twice as large (due to steeper slopes) as the landward slope in terms of benthic habitat. The mean slope of the entire trench is $6.4° \pm 1.7$ S.D. but can reach up to 48° in places. The mean slope on the west side is $7.7° \pm 1.2$ S.D. and east side is $5.9° \pm 1.8$ S.D. (Fig. 2.9b).

Analysis of the mean slope of the trench clearly shows the boundaries between the relatively flat abyssal plains (mean slope $= 3.2° \pm 0.4$ S.D. between 4000 and 6500 m) and trench slopes (mean slope $= 6.7° \pm 1.7$ S.D.). The trench slopes become steeper with depth as the mean slope ranges from 4.3 to 8.5° from 6500 to 9500 m. Beyond 9500 m the slope begins to decrease indicating a flattening at the deep trench axis (mean slope $= 5.7° \pm 2.5$ S.D.), with a mean slope of 3.9° beyond 10 000 m.

The linear increase in area is common to all trenches, regardless of whether they are deep (10 km+) or mid-depth trenches, as illustrated in Figure 2.10. In the six deepest trenches, the shallowest 1000 m (6500–7500 m) accounts for an average of $55\% \pm 7.3$ S.D. of the entire trench, whilst the deepest 1000 m accounts for $0.97\% \pm 0.4$ S.D. In six randomly selected mid-depth trenches, the upper 1000 m accounts for an average of $77\% \pm 7.5$ S.D. while the deepest 1000 m is $14\% \pm 4.1$ S.D. of the trench. Likewise, the mean slope also tends to follow a similar trend whereby the steepest gradients occur on the mid slopes, again regardless of whether the trenches are deep or mid-depth, see Figure 2.11. On average, in most trenches the maximum mean

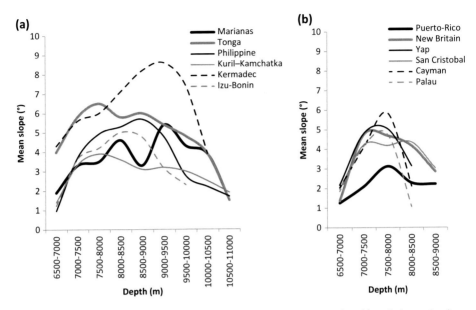

Figure 2.11 Mean slope with increasing depth for the six deepest trenches (a) and six randomly selected medium depth trenches (b).

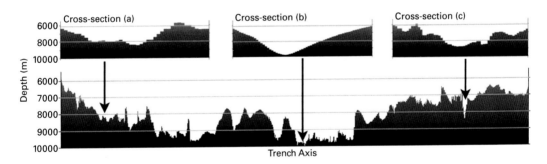

Figure 2.12 Cross-sectional profiles across the trench axis showing plateaus (cross-section (a)), smooth gradients (cross-section (b)) and complex ridges and valleys (cross-section (c)). The example shown is the Philippine Trench in the northwest Pacific Ocean (vertical exaggeration on cross-sections and trench axis are 60 and 120, respectively).

mid-slope gradient is approximately three times that of the slope at 6500 m, whereas the mean slopes at the deepest trench axis are comparable to those at abyssal depths.

These parameters are, of course, averaged and, in reality, the trenches rarely conform to a smooth gradient from the abyssal plains to the deepest point whether it is perpendicular or parallel to the trench axis. Most of the trenches are large geological structures with complex internal topography. Extracting bathymetric profiles along the trench axis or cross-sections perpendicular to the trench axis can reveal just how heterogenic the topography is (see Fig. 2.12). Within the trenches there are areas particularly on the landward side which vary greatly in terms of gradient and include large mounds, escarpments, plateaus, dips and depressions. These features also occur, albeit to a lesser extent, on the seaward slopes. Such complexity in topography will undoubtedly cause a similar heterogeneity in substrata, where softer, finer sediments may accumulate in localised pockets within depressions, whereas the steeper rockier outcrops are likely to be devoid of sediment. The cross-section profiles of the Philippine Trench shown in Figure 2.12 show three different characteristics; cross-section (a) features a large plateau, cross-section (b) conforms to a more classic representation of a trench and cross-section (c) shows a highly variable topographical structure.

2.4   Sedimentation and seismic activity

Deep-sea sediment is composed mainly of clayey, terrigenous material derived from continental erosion, carbonate and siliceous material supplied by pelagic and benthic organisms, and metal oxides and ash of volcanogenic origin (Howell and Murray, 1986; Hay *et al.*, 1988).

The rates of sedimentation vary within different sections of the slopes and trench floor and between trenches, depending on the nature of the relief and geographic location. Deposits located far from continents or adjacent to poorly drained land masses are typically only 200–600 m thick. The supply rate and the size of particles decreases with distance from land (Thistle, 2003), therefore within trenches adjacent to large continental land masses, sedimentary deposits are dominated by coarser grained sand and silt of terrestrial origin. The thickness of these layers is typically greater than 500 m but can reach as deep as 5–6 km (Fig. 2.13). In addition, the persistent rain of particulate organic matter (POM) from the surface layers and turbidites are deposited along the trench axis and may thicken the overall trench floor layer of sediment to several kilometres or more (von Huene and Scholl, 1991).

The complex internal topography within the trenches results in highly variable sediment depositional morphology, such as trench fans, axial channels, sheeted basins, ponded basins and axial sediment lobes (e.g. the Peru–Chile Trench; Thornburg and Kulm, 1987). In addition, these spatially variable depositional bodies change over time due to continual subduction and accretion of sediment. Sedimentation does not always occur on the steep projections of the slopes but rather on the numerous areas with rocky outcrops (Belyaev, 1989). Rock fragments are frequently deposited on the trench floors, having fallen from the upper slopes during sediment and rock slides caused by the high seismic activity in the trenches. Trenches are considered to be regions where the sedimentation rates are considerably high (Belyaev, 1989). For example, the sedimentation rate on the floor of the Kuril–Kamchatka Trench varies from 5–10 to 50–1000 mm per 100 years which is considerably greater than the levels reported at abyssal depths.

It is the characteristic V-shape of the trenches, formed by the subduction process which creates a funnelling or 'sediment trap' effect resulting in an obvious accumulation of sediment at the trench axis that may not have otherwise occurred on the flatter abyssal plains (Nozaki and Ohta, 1993; Danovaro *et al.*, 2003; Glud *et al.*, 2013). Given

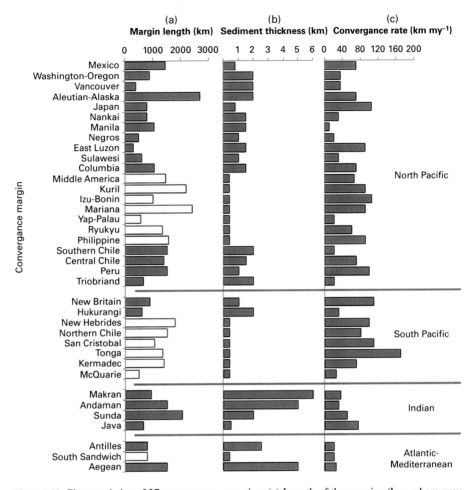

Figure 2.13 Characteristics of 37 convergence margins. (a) Length of the margins (km; where grey bars = Type 1 margins and white bars = Type 2 margins); (b) sediment thickness (km); and (c) convergence rates (km my^{-1}). Data taken from von Huene and Scholl (1991).

the enclosed setting of individual trenches, material that is deposited into a trench remains within it, slowing migrating downwards, driven by gravity towards the deepest trench axis (Oguri *et al.*, 2013).

Sediment subduction occurs where sediment remains attached to the subducting oceanic plate and is driven under the resistive prism-shaped buttress of consolidated sediment and rock on the frontal edge of the upper plate. Most of the material subducted is, therefore, derived either directly or indirectly from continental denudation. This process is categorised into two types: type 1 margins, where distinct accretionary prisms occur, and type 2 margins, where only small accretion occurs.

At type 1 margins, sediment subduction occurs at the seaward face of an active buttress of consolidated accretionary material that has accumulated in front of a core buttress of framework rocks. In the case of small-to-medium sized prisms, ~20% of the incoming trench floor is skimmed off and the remaining 80% is subducted. In the case of large prisms, ~70% of the incoming trench floor section is subducted. The estimated rate of solid-volume sediment subduction at convergent ocean margins is $1.5 \text{ km}^{-3} \text{ y}^{-1}$ (von Huene and Scholl, 1991). At type 2 margins, nearly all incoming sediment is subducted beneath the landward trench slope, generally resulting in a thinner layer of sediment than at type 1 margins. Type 1 areas constitute ~56.5% of the total convergent margin (type 2 = 43.5%).

Nozaki and Ohta (1993) retrieved deep gravity and piston cores from 9750 m in the Izu–Ogasawara Trench. They found that the lower sediments were dominated by layers of material derived from terrestrial courses and the upper layers were of contemporary marine origin. They also found that the sediment on the upper layer of the Izu–Ogasawara Trench was similar to those recently deposited on the northwest Pacific region and had a sedimentation rate of an order magnitude greater than at 8800 m in the neighbouring Japan Trench. The source of the sediment was found to be the Japanese Islands or the nearby Asian continent, transported via turbidity currents which may have developed recently through submarine canyons on the western flank of the trench. This in turn may have been effected by changes in turbidity source strength or deep-water circulation patterns between the two trenches and is likely to have had occurred for long periods of time to account for the observed differences.

The oceanic crust of the Pacific plate near the Izu–Ogasawara Trench is ~120 million years old (Heezen and McGregor, 1973). Therefore, Nozaki and Ohta (1993) estimated that if the average sedimentation rate is 0.5 cm per 10 000 years for the North Pacific Ocean, then the sediments above the crust near the trench are 600 m thick. Assuming that all sediments carried by the tectonic movement of ~10 cm per year are redistributed to the flat trench floor of ~5 km in width, then the sediment accumulation rate is 1.2 cm y^{-1}, which is several times faster than is actually observed. This strongly suggests that the sediments on the oceanic converging plate are largely subducting, adhered to the plate, and only a small fraction can be redistributed in the subduction zone. Therefore, the V-shaped trench morphologies are maintained through continual removal of sediments by subduction.

The subducting action at convergence zones is not only responsible for the removal of material but also results in juvenile crustal growth and terrane accumulation. The subduction process either builds up new terrestrial crust through arc volcanism or builds new areas of crust through the piling up of accretionary masses of sediment deposits and fragments of thicker crustal bodies scraped off the subducting lower plate (von Huene and Scholl, 1991). The principal contributor of input material to accretionary mass is the lower plate, which is constructed of igneous rocks from the oceanic crust and its overlying cover of ocean basin sediment (Hay *et al.*, 1988). These convergence processes continually create new regions of continental crust because some of the igneous, sedimentary and thicker crustal masses attached to the subducting oceanic plate are mechanically scraped off and accreted to the seaward edge of the upper (terrestrial) plate (Howell, 1989). Arc magmatism also supplies igneous rocks from the underlying mantle providing new material to the terrestrial layer, one of the most important processes in sustaining the Earth's stock of terrestrial matter (Reymer and Schubert, 1984).

Interestingly, it has been estimated that during the latter part of the Earth's history, the volume of subducted material reaching mantle depths equals estimates of new igneous masses to the layer of terrestrial rocks (von Huene and Scholl, 1991).

The lateral transport of sediment particles originating from the continental shelf and slope into the trench (Monaco et al., 1990; Biscaye and Anderson, 1994) are very important in biogeochemical cycling (Honda et al., 1997; Ramaswany et al., 1997; Otosaka and Noriki, 2000). A vertical array of sediment traps moored in 9200 m in the Japan Trench revealed a higher vertical particulate flux at the deeper traps suggesting there is a significant addition to the deeper depths from horizontal sources (Lerche and Nozaki, 1998).

In addition to the 'routine' input of material, sedimentary perturbations occur by episodic lateral transport of particulate material from coastal regions to the deep sea, initiated by earthquakes (Heezen and Ewing, 1952; Garfield et al., 1994; Thunell et al., 1999). This process is also observed directly and indirectly in trenches (Itou et al., 2000 and Fujioka et al., 1993, respectively). Sediment sides and slumps can occur on slopes as gentle as 2° resulting in the relocation of large volumes of sediment, hundreds of metres thick and thousands of metres long. Flows of slower moving debris can occur on slopes as gentle as 0.5° (Gage and Tyler, 1991). The gradients of the trench slopes are much greater than this, thus the magnitude of seismically induced sediment perturbation is exacerbated in trenches.

In general, earthquakes are limited to the uppermost 20 km, however, earthquakes in subduction zones occur at much greater depths. Planar arrays of deep earthquakes along the path of the subducting plate occur as deep as 660 km, at the edge of the mesosphere. Seismologists use the seismic moment magnitude scale (MMS) to measure earthquake magnitude in terms of the energy released (measure in M_W). The magnitude is derived from the seismic moment of the earthquake, which is equal to the Earth's rigidity multiplied by the average amount of, and size of slip on the fault during an earthquake. As an example, the Mariana and Izu-Bonin Trenches $= 7.2\,M_W$, whereas the Tonga and Kermadec Trenches $= 8.3\,M_W$. The highest values on the MMS are for the Japan and Kamchatka Trenches at $8.3–9.0\,M_W$ and the Atacama Trench off Southern Chile at $9.5\,M_W$ (Stern, 2002). These high levels of seismic activity associated with deep trenches have been recently demonstrated in all these areas in the 2010 Cauquenes earthquake in the Atacama Trench off Chile ($M_W = 8.8$), the 2011 Christchurch earthquake in New Zealand ($M_W = 6.3$) and the 2011 Tōhoku-Oki earthquake in the Japan Trench ($M_W = 9.0$).

Due to the unpredictable nature of earthquakes, there are a few direct observations of a seismically induced turbidity flow, although there are historical reports suggesting such events in areas of similar topography (e.g. Prior et al., 1987; Porebski et al., 1991; Thunell et al., 1999). However, Itou et al. (2000) happened to have long-term sediment traps moored at 6150, 2950 and 350 m above the bottom (depths $= 1000$, 4200 and 6800 m, bottom depth $= 7150$ m) in the Japan Trench over the course of the 28 December 1994 Sanriku-Oki earthquake (magnitude $= 7.7$) and subsequent aftershocks. They reported a distinct increase in non-biogenic material at 4200 and 6800 m immediately after the earthquake struck (Fig. 2.14). Although the composition (Mn/Al) of the

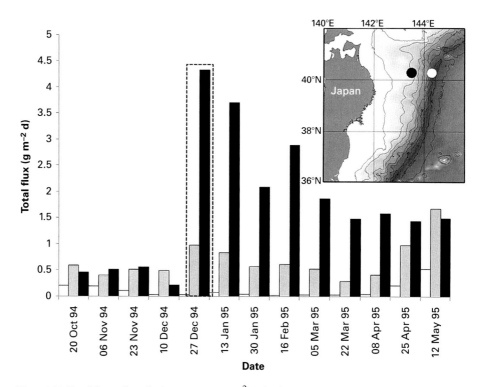

Figure 2.14 Total flux of particulate matter (g m^{-2} d) in the Japan Trench before and after the 27 December 1994 Sanriku-Oki earthquake (black dot on map) as collected by sediment trap (white dot on map) at depths of 1000 m (6150 mab; white bars), 4200 m (2950 mab; grey bars) and 6800 m (350 mab; black bars). Data taken from Itou *et al.* (2000).

particulate material differed between these sediment traps, implying difference source areas, they concluded that the material originated from surface sediments, transported from the eastern slope of the Japan Trench.

The first major scientific account of the 11 March 2011 Tōhoku-Oki earthquake (magnitude = 9.0) was published by Fujiwara *et al.* (2011), who described the location and mechanism of the earthquake and resultant tsunami, and documented the resulting physical changes in the trench morphology. They believe that the earthquake was caused by a fault rupture extending to the shallower end of the subduction zone of the Japan Trench. By comparing a new post-earthquake acoustic multi-beam survey to one carried out in 1999, they found that the rupture had extended to the trench axis. Furthermore, they found that the seafloor elevation (depth) throughout the landward side was shallower by 7 to 10 m, and ± 50 m at the axes due to a submarine landslide. They also estimated a horizontal displacement of 56 m towards the east-southeast. It is clear from these results that with such huge forces and instability underlying the trenches, even the shape of the hadal habitats are continually changing. Four months after the Tōhoku-Oki earthquake, a 30–50 m thick nepheloid layer was still present and the top 31 cm of sediment in the trench axis revealed three recent deposition events characterised by elevated ^{137}Cs levels (originating from the Fukushima Dai-ichi nuclear disaster) and alternating sediment densities (Oguri *et al.*, 2013).

    The location of Japan is such that it experiences a very high number of earthquakes and as a nation, *in situ* monitoring of these events is performed routinely and perhaps more widely than by any other nation. As such, it is likely that the interaction between seismic activity and the trenches will be fully understood through research undertaken by the Japanese in their adjacent waters.

2.5   The hado-pelagic zone

The enclosed nature of trench topography also encompasses the water mass within it; the hado-pelagic zone. As the area of the benthic habitat decreases with depth, so do the three-dimensional pelagic habitats. In the five deepest trenches, the volume of water decreases relatively from ~30% of the pelagic waters being between 6000 and 6500 m to just less than 5% beyond 9000 m. In fact, most of the deepest areas (>10 000 m) account for <1% of the pelagic habitat (Fig. 2.15). Whilst the hydrography and

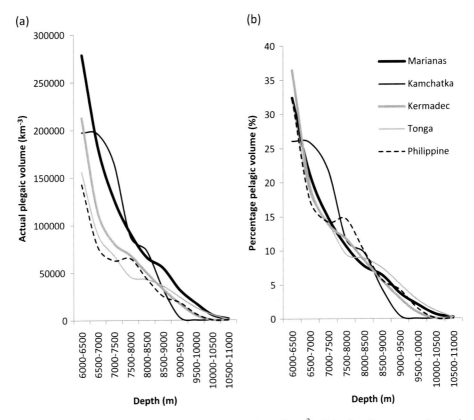

Figure 2.15 (a) The decline in actual hado-pelagic volume (km^{-3}) of the five deepest trenches and (b) the relative decrease in percentage (%).

environmental characteristics of the hado-pelagic zone are relatively well sampled (Johnson, 1998; Kawabe *et al.*, 2003; Taira *et al.*, 2004, 2005), from a biological perspective, the hado-pelagic fauna are virtually unknown. This is mostly a result of the technical challenges associated with mid-water trawling at such extreme depths. There are only a handful of vertical plankton hauls from the hado-pelagic zone and these date back to the 1950s and 1960s (Vinogradov, 1962). More recently there have been some studies of the microbial communities from 6000 m in the Puerto-Rico Trench (Eloe *et al.*, 2010, 2011). Endemism at hado-pelagic depths was estimated at 41% by Belyaev (1989), a value somewhat lower than that of benthic fauna (56%). However, this was based on very few samples that did not cover the entire depth range of the hado-pelagic.

3 Full ocean depth technology

Scientific endeavour into the world's ecosystems, whether on land or in the sea, tends to occur in three stages: exploration, observation and experimentation (Tyler, 2003). The discovery and subsequent exploration of marine ecosystems, whether serendipitously or otherwise has been ongoing since the birth of marine science and provides the rudimentary groundwork for all future research. As detailed in Chapter 1, this discovery and exploration has culminated in the mapping of the world's oceans in their entirety. In shallow and coastal seas, the observational and experimental stages of research have been ongoing for more than a century and are a well-established practice due to the relative ease of access compared to deep environments. In shallow-water ecosystems, the fauna and environmental correlates are readily accessible for study, both *in situ* and in the laboratory, a luxury rarely obtainable in deep-sea research. The inherent nature of the deep sea does not lend itself readily to examination using the experimental tools and procedures of typical laboratory settings and research is often further hampered by the great distances from shore and depths from the surface. As a result, deep-sea sampling has lagged behind that of coastal and inshore research. The understanding of the deep-sea environment has greatly increased in the last century, particularly in the last 50 years, due to the ever-increasing sampling effort driven by the ever more apparent importance of the oceans on our planet.

It has recently been said that 'if there are "ages" in science, the deep-sea biological community is in the later stages of the age of observation' (Tyler, 2003). While this is certainly true for the general deep-sea environment, hadal research is more likely placed in the early stages of observation. The technical challenges of exploration, observation and experimentation that have hindered research at bathyal and abyssal depths are exacerbated at hadal depths. It is the technological capabilities that underpin progress in understanding these deep environments, as even the most basic scientific question is unlikely to be answered without a heavy reliance on technology.

With literally hundreds of different applications and avenues of research, the diversity and capability of technology, methods and techniques for studying the deep sea are immense. Methods for the study of life in the hadal zone are mostly derived from systems and techniques used in the wider deep sea that have been modified for extension into the deep trenches. Whilst some of the technology itself, methods in delivery and analytical techniques have evolved rapidly, others have remained relatively unchanged for decades. Many of the basic operational principles are shared between the shallow and deep environments, with only minor or extrapolated modifications for use in the hadal zone. There are a plethora of instruments, methods and techniques currently available to deep-sea researchers, so much so, it is not possible describe them all here, although there are comprehensive reviews of such methods (e.g. Eleftheriou and McIntyre, 2005; Humphris, 2010). Hereafter is a description of the instruments and methods that have been or are currently commonplace in sampling specifically hadal depths.

3.1   The challenge of wires

A common denominator in the sampling of bathyal, abyssal and hadal depths is the use of research vessels by which to deliver technology to the deep sea, irrespective of depth. This prompts the first major challenge in sampling the hadal zone: depth, or perhaps more accurately 'distance from the surface'. Regardless of the environmental conditions found at depth, this challenge is prompted by the sheer distance to which some sampling devices must be lowered in order to reach the seafloor at full ocean depth.

Towing or lowering instrument packages or sampling devices on thousands of metres of wire, often without any visual reference, requires specialised research vessels that carry a wire sufficiently long to reach the bottom. In addition, the vessel must have the capability to withstand the weight of such a length of wire, plus the load exerted during hauling, the drag on the wire and the weight of the equipment on the end of it. There are currently few research vessels with the capability of sampling using wires beyond 6500 m.

Time can also be a limiting factor when using wire-deployed instruments at hadal depths. Sampling with sufficient replication to quantitatively sample an area of seafloor is inhibited by the time it takes to lower a package to the deepest trench floor and back. For example, lowering an instrument at 50 m min^{-1}, which is relatively fast, will take over 7 h to reach full ocean depth and return, plus the time it takes to sample on the bottom. A more controlled descent of 30 m min^{-1} will take over 12 h of paying out and hauling in wire for just one sample. These times are based on simply lowering a wire straight down and back up. If a package is being towed, then these times, and the length of required wire, are increased furthermore. If a wire of equal thickness along its length is dragged through the water without touching the bottom, it will form a straight line where the angle is determined by the speed of the ship, the weight of the wire and the drag on the wire (which to some degree offsets the weight of the wire). As the ship's speed increases, the wire will approach a more horizontal position, i.e. it rises ever further away from the seafloor. If pulling a trawl across the seafloor, the wire directly in front of the trawl must be horizontal, therefore, the entire length of wire cannot be straight. The length of wire must be theoretically calculated to account for the extra length needed to form the arc. For abyssal trawling the length of wire required is typically between two and three times the depth of descent, depending on the type of trawl.

In the example of beam trawling, the length of wire that needs to be paid out is often roughly twice the depth of water (at typical towing speeds of 2 knots), in order to enable

Table 3.1 Specifications of the 12 000 m trawling wire from the *Galathea* expedition. Modified from Kullenberg (1956).

Wire diameter (mm)	Length (m)	Cumulative length (m)	Weight in water (kg m^{-1})	Cumulative weight (kg)	Breaking strain (kg)
9.3	3600	3 600	0.26	936.0	7 140
11.6	1750	5 350	0.41	1653.5	11 100
13.2	770	6 120	0.53	2061.6	12 600
14.7	1330	7 450	0.66	2939.4	15 600
17.1	1730	9 180	0.89	4479.1	21 000
19.6	1080	10 260	1.18	5753.5	25 000
20.2	980	11 240	1.25	6978.5	29 200
21.8	760	12 000	1.45	8080.5	33 900

the trawl to be pulled along the seafloor behind the ship. Such an exercise would require 22 000 m of wire to be paid out at full ocean depth. The length of wire would therefore take nearly 15 h to pay out and haul in, plus another 2–4 h on the seafloor to collect the samples, totalling at least 17 h for just one trawl. Therefore, the number of samples obtained per day is far lower than that which can be achieved in shallow waters. With such long deployment times the reliability of the equipment is ever more pertinent. The problems of lower replication with time are further hampered by low densities, small body sizes, high species richness and the many rare species found in the deep sea. For scientific credibility, the presence of these factors in the deep sea means that more numerous samples must be obtained relative to shallower environments, in order to confidently describe the inhabiting fauna.

These estimates of wire lengths, towing speeds and times are, however, extrapolations from abyssal trawling. Given that there have been no trawling campaigns undertaken at hadal depths since the 1950s, there is currently only one detailed description, based on experience, of how to trawl at these depths. This is provided by Kullenberg (1956) based on his experiences during the *Galathea* expeditions.

Kullenberg (1956) detailed his experience in calculating the required length of wire to be paid out for different diameters of wire. To tow a trawl at a depth of 5000 m with 9, 12 and 16 mm wire requires 9600, 7900 and 6700 m of wire, respectively (pay out to depth ratios of 1.9:1, 1.6:1 and 1.3:1). This suggests that heavier wire (16 mm wire) is more suitable, as less wire is required due to the extra weight pulling the wire down, relative to thinner wires. However, as in most deep-sea trawling, the wires are never of a uniform thickness but rather are tapered, starting with a thin section which increases in stages. This prevents the wire from parting under the combined weight of itself and the tension during operations. The *Galathea* carried one such wire which was 12 000 m long. The tapered section and other characteristics are detailed in Table 3.1. The trawling wires used on the Soviet research vessels were steel cables ranging from 15.5–16 mm diameter to 6.8–7.2 mm diameter on the end.

Kullenberg's mathematical calculation for the length of a cable (shown in Fig. 3.1) was successful, as shown by the achievements on the *Galathea* expeditions. His

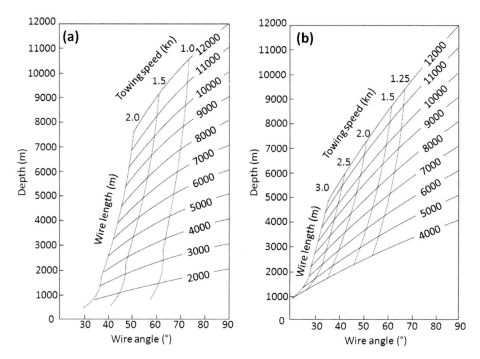

Figure 3.1 The theoretical calculation charts for estimating the required length of wire to trawl at the extreme depths as used on the *Galathea* expeditions, where (a) is for otter trawls and (b) is for sledge trawls. Modified from Kullenberg (1956).

calculations are based on a matrix of the inclination angle of the wire leaving the ship, the depth and the ship's speed. Calculating this correctly is of the utmost importance as insufficient length of wire could mean that the trawl does not reach the bottom. Conversely, the use of excessive wire lengths can result in entanglement and the formation of knots and kinks that could potentially break the wire. His calculations show that, for example, if 12 000 m of wire were paid out with an otter trawl, it would reach ~11 000 m when travelling at approximately 1 knot (wire angle = ~70°, which is near vertical). If the ship's speed increases to 2 knots, the wire would only reach short of 8000 m and the angle would increase to ~50°.

Even when theoretically ignoring the challenge of long wires breaking, the time it would take to pay out the necessary wire compared to more conventional deep trawling depths becomes extremely long and would thus consume more ship-time per trawl (Fig. 3.2).

Despite the trawling successes of the *Galathea* expeditions the Soviet opinion, based on the experience of the *Vitjaz* and *Akademik Kurchatov* expeditions, was that often, the wire paid out does not correspond to the theoretical calculation due to changes in current speed and direction at different depths. The Soviets used a simpler method based on Pythagoras' theorem, where the depth was the vertical leg of a right-angle triangle and the cable length was the hypotenuse, which again relied on the monitoring

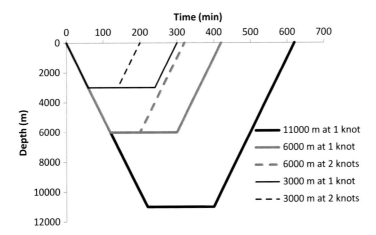

Figure 3.2 The difference in time taken to trawl at 11 000 m, 6000 m and 3000 m assuming pay out speed of 50 m min^{-1}, a towing speed of one knot (solid lines) with the trawl on the seafloor for 3 h. The equivalent trawl distance at 2 knots, which is not possible for 11 000 m operations, is indicated by dashed lines.

of the wire inclination angle at the stern of the ship. With ship-side wire inclination angles up to 30–40°, they found from experience that increasing the wire length by 20–30% of the calculated length was the optimal. Presumably the additional 20–30% accounted for the bend in the wire.

To overcome the challenges of extremely long wires, long deployment times and availability of capable vessels, there are a number of instruments and sampling devices that can be operated using the free-fall method. This method relies on two major components: a basic delivery system and a scientific payload. The delivery system comprises expendable ballast weights designed to sink the package, timing devices or acoustically releasable devices to jettison the ballast weights when desired, and subsurface buoyancy to float the instrument back to the surface (see Tengberg *et al.*, 1995; Bagley *et al.*, 2005 for detailed examples). Therefore, a relatively small instrument package or sampling device (e.g. cameras or traps) are deployed from a ship and sink to the seafloor where they carry out pre-programmed tasks and either return to the surface at a specified time or on command from the ship. Since the free-fall technology used is separate from the ship, this method enables relatively small vessels to sample at full ocean depth, or alternatively, one vessel can deploy multiple systems to operate simultaneously on the seafloor. This method also enables samples to be taken over timescales that are far longer than the duration of the voyage itself (e.g. 12 months). Also, instruments that are unattached to the ship are also independent of any movement from the ship and can therefore perform precise measurements on the seafloor.

This method has become commonplace in hadal sampling and has been used extensively to deliver baited traps to hadal depths (Hessler *et al.*, 1978; Blankenship *et al.*, 2006; Yayanos, 2009; Kobayashi *et al.*, 2012), baited video and still cameras (Jamieson *et al.*, 2009a, b, c, d, 2011a, b; Fujii *et al.*, 2010).

3.2   The challenge of high pressure

The second major challenge for deep-sea research involves coping with the immense hydrostatic pressure found at depth. Sampling methods such as trawling and sediment sampling tend not to suffer any adverse effects from ambient hydrostatic pressure found at depth, since these systems do not include any air cavities that could potentially implode under pressure. Exposure to high hydrostatic pressure must be considered during the technical design of deep-sea instruments containing electrical, electronic or optical components. With pressure up to ~1.1 ton cm^{-2} at the deepest ocean depth (11 000 m), a considerable crushing force will be exerted across any pressure differentials, for example, 'dry' instruments containing air cavities such as cameras, lights, batteries or data loggers. These components must be assembled in sufficiently strong housings to resist these enormous pressures. This can be achieved by inserting the components into metal, water-tight cylinders with sufficient structural integrity to withstand the pressure at its intended operational depth. Alternatively, instruments that do not include any air cavities (such as a lead–acid battery) can be pressure compensated whereby the device is mounted inside a housing and flooded with an inert fluid, usually certain types of oil, that maintain electrical contact. A water-tight flexible membrane can compensate for any minor changes in volume with increasing pressure.

The design of full ocean, depth rated, pressure housing is relatively elemental and can be extrapolated from shallower rated designs. Increasing the operational depth range of an instrument requires an increase in the wall thickness of the housing and thus its weight also increases. Consequently, instruments deployed in the deep sea using the free-fall method require greater buoyancy than their shallower counterparts and this subsequently increases the cost. The issue of housing weight can be overcome by the selection of appropriate materials such as titanium which, due to an excellent strength to weight ratio, will produce a smaller and lighter housing relative to stainless steel, albeit considerably more expensive.

Resisting hydrostatic pressure using metal housings is relatively rudimentary, however, optical devices such as cameras or lights require a transparent window in the housing and this poses further challenges. The design of viewports requires transparent components capable of withstanding high pressure. In shallower zones, this is typically achieved through the use of acrylic (or Polymethylmethacrylate; PMMA) plastic, borosilicate glass or sapphire, each with its own disadvantages and advantages. There are three main types of viewport: plain disc, bevelled disc and hemisphere. Acrylic is cheap in terms of material supply and machining, and it is relatively insensitive to imperfection in manufacture. However, it does suffer from considerable plastic flow prior to fracture, i.e. baroplastic characteristics (Gonzalaez-Leon *et al.*, 2003). Extensive testing of acrylic windows by Gilchrist and MacDonald (1980) showed that permanent plastic deformation occurred when pressurised beyond 83 MPa (~8300 m) and catastrophic failure occurs at 140 MPa (~14 000 m), suggesting acrylic is not an ideal solution for hadal applications (Fig. 3.3a). Borosilicate glass differs in that there is little warning prior to a failure due its brittle characteristics. In addition, glass, like acrylic, requires extremely thick viewports to withstand the pressure, often resulting in very

(a)   (b)

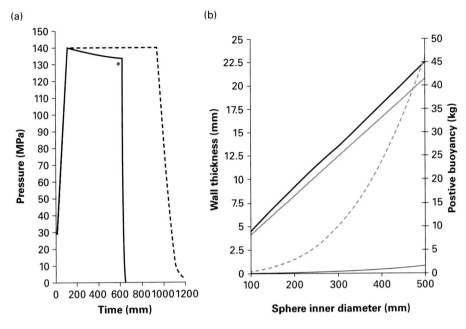

Figure 3.3 Engineering challenges at extreme pressures. (a) Pressure testing of an acrylic viewport (solid line) and a sapphire viewport (dashed line) whereby the equivalent full ocean depth pressure causes the acrylic to deform (indicated by drop in pressure) prior to failure (indicated by *), an effect which does not occur in sapphire. (b) Required wall thickness of a sphere rated to 11 000 m operations of varying diameter made in titanium (black lines) and borosilicate (grey lines), the amount of positive buoyancy (dashed lines) increases in borosilicate but is always <3 kg in titanium regardless of size.

large cumbersome components. Sapphire, however, albeit very expensive, difficult to machine and restricted to available shapes (limited to plain disc design), has excellent mechanical and optical properties, and its incorporation into housings means that the physical size of the viewport is reduced. In large submersible applications, where housing size is not an issue, acrylic and borosilicate glass can be used, while in smaller camera applications, sapphire offers a more practical solution (as used in the Hadal-Landers; Jamieson *et al.*, 2009c, d).

The buoyancy provisions for deep-sea, free-fall instruments are similar to those used in abyssal applications. Typically, there are two types of buoyancy: glass (borosilicate) spheres (Pausch *et al.*, 2009) and syntactic foam (Gupta *et al.*, 2001). Syntactic foam is made from microscopic glass spheres encapsulated in an epoxy resin matrix. The resulting product is robust, insensitive to sudden failure and can be moulded into complex shapes. It is generally used in submersible applications (e.g. HROV *Nereus*, Bowen *et al.*, 2009a; ROV *Kaikō*, Mikagawa and Aoki, 2001; ABISMO, Yoshida *et al.*, 2009). The net buoyancy per volume, however, is much lower than that of glass spheres. Glass spheres comprise two evacuated hemispheres secured inside plastic shells that are typically 43 cm in diameter. They are commonly used in smaller lander operations (e.g. Hadal-Landers; Jamieson *et al.*, 2009c, d) and are relatively cheap to produce compared to syntactic foam. The spherical shape is ideal in resisting pressure due to uniform loading. Glass spheres are, however, prone to implosion if not carefully handled and monitored throughout multiple pressure cycles. Other solutions, such as titanium spheres should, in theory, eliminate the implosion risk; however, the required volume of material needed to withstand pressure at hadal depth negates any positive buoyancy (Fig. 3.3b). In even smaller vehicles, such as single traps (e.g. PRATS; Yayanos, 2009), reservoirs of paraffin oil-based liquids such as Isopar-M™ are used for buoyancy (Yayanos, 1976). Once again, the problem with using floatation modules containing materials such as Isopar-M™ comes down to the weight of the relatively large volume of material required to meet the buoyancy demands. The weight factor thus limits the use of this type of buoyancy to smaller vehicles.

3.3   Trawling and coring

Coring and, particularly, trawling require prior knowledge of the seafloor, more so in trench environments than on the abyssal plains. Trenches are characterised by steep slopes and often complex terrain with spatially variable substrata, all of which can affect the efficiency of the equipment and, ultimately, may damage or endanger it. Echo-sounders are used to clarify the topographic nature of the trench prior to trawling, in order to select a suitably flat trawl path. They are used continuously during the trawl to monitor the course in the event of changes in depth or relief. Likewise, the echo-sounder is used to locate a suitably flat coring station prior to coring operations.

There are two basic types of benthic trawl that have been used regularly in hadal applications: beam trawls and otter trawls. The beam trawls are used primarily for sampling benthic megafauna, whereas otter trawls are more suited to the collection of bentho-pelagic fauna.

The most common beam trawl used in sampling at hadal depths is the Agassiz trawl (also called a Blake or Sigsbee trawl, or a Sigsby–Gorbunov beam trawl by the Soviets). These trawls consist of two D-shaped runners joined by 2.5–3 m long struts that create a fixed trawl mouth area of between 1.5 and 2.1 m^2. In good conditions (low relief and soft sediment), both the *Galathea* and *Vitjaz* expeditions sometimes used a 6 m wide trawl, which was essentially two beam trawls coupled together side by side, to double the trawl mouth area. These trawls have a typical mesh size of 20 mm and a cod end (the sample bag) lined with 10 mm netting.

The main advantage of beam trawling is that the fixed mouth design is relatively easy to deploy. Unlike some other trawls it is not susceptible to net collapse or entanglement and can therefore be towed very slowly on rough or unfamiliar ground, such as in the trenches. The fixed mouths allow the trawl to be lowered almost vertically from the ship and dragged very slowly across the seafloor. The ability to tow the trawl very slowly becomes ever more pertinent with increasing depth, as illustrated by Kullenberg's calculation of wire out (Fig. 3.1). Various safety features are often incorporated into the trawl for use in the event of an entanglement or snagging. Features such as weak links are designed to sheer in certain circumstances, transferring the strain to the cod end.

Beam trawls are designed to sample benthic fauna but are less effective at catching large, mobile animals such as fish. Fish are typically caught using otter trawls, like those employed on the *Galathea* expeditions. Otter trawls have larger nets than beam trawls and a non-fixed mouth that is pulled and held open by two steel boards attached to the main wire. The advantage of otter trawls in deep-sea applications is that they sample a greater volume of water per unit towing time (Stein, 1985). However, otter trawls are susceptible to net collapse at low towing speeds (<2 knots) and, although they do not feature in trench applications as significantly as beam trawls, they have been used successfully. For example, George and Higgins (1979) used a 40 foot otter trawl with a 500 μm mesh cod end at depths of 8580 m in the Puerto-Rico Trench. They towed the trawl at 1.5 knots for 4 h covering 7–11 km of seafloor.

Otter trawls usually include a ballast weight of greater than 100 kg in order to maintain a positive bottom contact. The moment when a trawl or coring device hits the seafloor is difficult to ascertain. Sudden drops in tension on the winch are not readily detectable as the weight of the package is minute relative to the weight of the actual cable. Therefore, for years, scientists relied on the 'wire out calculation' to determine bottom contact time. A more precise method was developed using an acoustic signalling device (or 'pinger') that transmits a signal to the ship on contact with the seafloor. This method was first used at hadal depths onboard the American RV *Thomas Washington* in the Philippine Trench at 9600 m. Most modern methods of deep-sea trawling use pingers or slightly more sophisticated 'trawl monitors' to continuously monitor depth, contact, angle and spread. The idea of monitoring bottom contact was developed further by the inclusion of a freely rotating wheel known as a trawl-graph, which recorded the length of the path traversed by the trawl over the seafloor. This method provided sufficient data from which to establish approximate calculations of the trawled area and subsequently of the biomass and abundance per unit area, albeit only under favourable conditions (Zenkevitch *et al.*, 1955). To cope with more complex macro- and micro-relief of the seafloor, and any major changes in depth, two or three trawl-graphs were used simultaneously across the mouth of the trawl. Unfortunately this method produced such varied results that the trawl-graphs were abandoned in future trench trawling endeavours.

The simple pingers of the 1960s and 1970s were not as effective as modern trawl monitor systems when towing equipment. Today, such monitors contain multiple tilt sensors to relay angle, pitch, twisting and depth data directly to the ship, in a series in time-delayed pulses. Modern bottom trawl tilt switches can show sudden changes in angle when the package is on the bottom, and indicate accidental lifting from the seafloor. However, since no major trawling campaigns have been conducted at hadal depths in the last few decades, modern trawl monitors have yet to be employed at these depths.

More recent (post *Galathea* and *Vitjaz*) reports of trawling at hadal depths do exist, for example, Horikoshi *et al.* (1990), but details of the actual operation are lacking.

Aside from bottom trawling, hado-pelagic 'vertical' trawls (or Bogorov–Rass closing nets) were used in the Soviet expeditions. Vertical trawls are lowered to the desired start depth, opened and hauled vertically to the desired end depth and then closed. This technique was used to obtain quantitative distribution patterns of plankton from the surface to 8000 m over the Kuril–Kamchatka, Mariana, Bougainville and Kermadec Trenches (Vinogradov, 1962).

In addition to the different types of trawl, many other devices are used for sampling the deep sea, including equipment specifically designed to collect the fauna and material in sediments. Sediment grabs and corers are mechanical devices that are lowered to the seafloor on a wire and are triggered, on contact with the bottom, to collect a sediment sample. The earliest methods for sounding the deep trenches included small, single corers, such as the Hydra-rod or the Baillie rod (Thomson and Murray, 1895). Although these grabs were initially used to confirm bottom contact, subsequent development led to the corer as a method of collecting sediments and organisms from the deep-sea floor (Thorson, 1957).

The most common types of grab in use today, in relatively shallow environments, are the Peterson, Van Veen and Day grabs. At hadal depths, the *Galathea* expedition used the 0.2 m^2 Petersen bottom grab (Bruun, 1956a) and the Soviet expeditions used the 0.25 m^2 bottom grab 'Okean-50' (Belyaev, 1989) and managed to obtain samples from 9340 m in the Philippine Trench and 9540 m in the Mariana Trench. Although each grab differs slightly in design (surface area and sample volume), all rely on a trigger mechanism that shuts a pair of jaws on contact with the seafloor, a principle based on the original Peterson grab.

Although highly efficient in coastal applications, these sediment grabs are subject to the same challenges as the other wire-deployed systems described earlier. Deploying these grabs in the deep sea is extremely time consuming relative to the sample size obtained. Over such long distances from the surface, the grabs can suffer severe wash-out during the long ascent from the seafloor. Furthermore, the efficiency and effectiveness at quantifying the infaunal communities from such samples has been questioned for decades. Bow waves can easily sweep away light-bodied organisms and surface sediments from the samples (Wigley, 1967) and the depth of seafloor penetration of the grab is dependent on substrate type and, as a result, organisms that burrow below the grab's penetration depth may not be sampled (Smith and Howard, 1972). Given these issues, sediment grabs were superseded, first by box corers and then by tube corers.

Today, the use of grabs is relatively uncommon in deep-sea research. Nevertheless, they are frequently employed as precision tools for use by ROVs for the virtually undisturbed sampling of benthic organisms. These slightly modified grabs are used to target organisms precisely, particularly large and often fragile epifauna, or specific areas of seafloor (e.g. the Ekman type grab; Rowe and Clifford, 1973). ROV delivery and operation eliminates the bow wave problem, sediment disturbance and problems associated with full ocean depth-capable wires.

The grab, or multiple grabs, is carried to the seafloor on the ROV tool tray. The ROV uncouples the grab and gently places it over the desired organism or area of seafloor. The grab is then gently inserted into the seafloor, enclosing the animal, the underlying sediment and overlying water. The grab is closed and retracted and secured back on the tool tray ready for return to the surface.

Sediments grabs were superseded in the 1970s by spade corers, now commonly known as 'box corers' and these became the standardised method of quantitative sediment sampling in the deep sea (Hessler and Jumars, 1974). Box corers did exhibit the same level of disturbance to which grabs were prone, but retrieved much larger, deeper and less disturbed sediment samples.

Box corers comprise a central column that is weighted with lead and is slotted and gimballed within an outer frame. On the end of the central column is a detachable, square, bottomless, metal box. Prior to operation, the spade is held horizontally by a spring-loaded bolt. When the outer frame reaches the seafloor, the weighted inner column drives the box into the sediment. The spring-loaded pin is withdrawn and a short length of cable is released causing the spade to swing 90° through the seafloor and under the box, thus enclosing the relatively undisturbed sediment within the box. Once the corer is hauled back to the ship, the sediment-filled box is detached, sampled and processed.

Box corers were used extensively in the 1980s in the Puerto-Rico Trench, from depths of 7460 to 8380 m (Tietjen, 1989) and 8371 to 8386 m (Richardson *et al.*, 1995) in order to investigate the benthic assemblages of nematodes and macro- and meiofauna, respectively. Other box core samples have been obtained from 7298 m in the Aleutian Trench (Jumars and Hessler, 1976) and 9600 m in the Philippine Trench (Tendal and Hessler, 1977; Hessler *et al.*, 1978).

Since its invention, there have been many iterations of the box core based on the same principle (e.g. Jumars, 1975; Gerdes, 1990). In general, box corers were found to be superior to grabs in terms of the state and volume of sample collected. However, box corers were still susceptible to the sweeping away of fine surface sediments and delicate organisms by the waves which also contaminated the overlying water (Bett *et al.*, 1994; Shirayama and Fukushima, 1995). They were eventually superseded by the multiple tube corer.

In the late 1990s, the spade corer principle was revisited in the form of a smaller programmable 'auto-corer', which was deployed on the end of a mooring line. The auto-corer was set on a time-delay and programmed only to trigger 7 hours after reaching the seafloor at a reduced speed to minimise sediment disturbance (Danovaro *et al.*, 2002, 2003). Although this method was successful in retrieving 6 cm diameter cores from 7800 m in the Atacama Trench, it was a bespoke design and is not commonplace in hadal sampling.

The surface sediment disruption caused by box corers led to the development of the multiple corer, known as 'multicorers' (Barnett *et al.*, 1984) or in later versions, the 'megacorers' (Gage and Bett, 2005). The multicorer comprises an array of up to 12 core tubes of typically 25.1 cm^{-2}. The core tubes are attached to a central shaft within an outer frame and are driven into the seafloor using the same principle as the box corer, however, the tubes are hydraulically dampened to further minimise disturbance. The top and bottom of the core is sealed upon extraction from the seafloor.

Multicorers (and box cores) are monitored through the water column by pingers coupled to the wire (or on the corer itself). They descend with a payout speed of ~50–60 m min^{-1} until close to the seafloor where they are slowed down to

$\sim$10–15 m min^{-1} prior to penetration. The core is left for a brief period to allow for the slow penetration before a slow pull out and hauling commences. Once onboard the ship, the cores are removed individually and either preserved as whole or extruded and sliced into desired depth horizons. There are currently several different types of multiple tube corer available, based on the same basic principle but with variation on number and size of core tubes.

The challenges of using extremely long wires to reach hadal depths has prevented any significant use of wire-deployed coring systems in trenches, mainly due to the lack of suitable ships. One of the few examples of mulitcorer use in the hadal zone is documented by Itoh *et al.* (2011), whereby two sets of cores were successfully recovered from $\sim$7000 m in the Kuril–Kamchatka Trench. JAMSTEC have also developed an 11 000 m rated free-falling sediment corer, called ASHURA (Murashima *et al.*, 2009). This vehicle is equipped with three hydraulically dampened cores and high definition (HD) camera to image the sediment surface (Glud *et al.*, 2013).

Nevertheless, many more cores have been retrieved from hadal depths through the use of ROV 'push cores' (e.g. Kato *et al.*, 1997; Takami *et al.*, 1997). Push cores are smaller than those on multicore systems and the tubes are typically 58 mm diameter $\times$ 300 mm long with a non-return or manual valve at the top and a 'T-handle' for ROV handling. Most deep-water ROVs can take tens of push cores on one dive, secured to the tool tray. The ROV can gently insert each push core tube into the sediment creating very little disturbance, while the water vents out through the valve on the top. The manual valve is then closed by the ROV or, using a flutter valve, the core is simply pulled straight out allowing suction to close the valve. The core is then removed from the seafloor and placed onto a bung inside a designated quiver, ready for retrieval at the surface.

The ROV method of push coring allows precise, targeted coring on a small scale (e.g. cold seep bacterial mats; Van Dover and Fry, 1994) and can provide the means to perform transects of tens to hundreds of metres. This push coring technique has been used in Challenger Deep to sample Foraminifera (Akimoto *et al.*, 2001; Todo *et al.*, 2005), microbial flora (Takami *et al.*, 1997) and bacteria (Kato *et al.*, 1998; Fang *et al.*, 2000).

3.4   Cameras and traps

In the absence of opportunities to trawl, baited traps are simple and effective methods of recovering animals from hadal depths. Baited traps are biased towards mobile bait-attending fauna and, in one respect, this can be seen as a disadvantage, however, baited traps are capable of recovering a greater sample of scavenging crustaceans in terms of numbers and diversity than either trawling or coring. Their greatest advantage, however, is that when these traps are deployed using the free-fall method, they can be delivered from smaller research vessels and are relatively cheap to construct relative to other methods. Furthermore, baited traps can be integrated into the scientific payload of other vehicles such as baited camera landers, or they can comprise one of a selection of tools used by an ROV. As baited traps are relatively small packages, they can also be constructed in multiple compact units and deployed quickly across an area, thus increasing sample size and replication.

The samples obtained from traps can be used for the same diverse range of downstream scientific applications as any other physical sampling methods, for example, basic taxonomy, population genetics, physiological measurements, stomach content analysis (diet), chemical composition and length–weight (biomass) relationships. Trap samples are also employed to contribute to the accuracy of the baited camera method because, on occasion, a species captured on film may be too small to identify confidently from photographs only. For these reasons, the baited trap is currently the most favoured method for sampling at hadal depths and, consequently, there is currently a disproportionately large sampling bias towards scavenging or bait-attending fauna, particularly amphipod crustaceans.

Free-fall baited traps have been used extensively for many years to recover mobile scavengers at bathyal and abyssal depths (Paul, 1973; Shulenberger and Hessler, 1974; Isaacs and Schwartzlose, 1975; Thurston, 1979; Stockton, 1982), and there has been, since these early years, an increase in their use at hadal depths (Hessler *et al.*, 1978; France, 1993; Thurston *et al.*, 2002, Blankenship *et al.*, 2006; Jamieson *et al.*, 2011a; Eustace *et al.*, 2013).

Baited trap designs can vary in size, volume, number per package or in operational features (e.g. closable), but most are based upon the simple principles of funnel traps (Fig. 3.4a). The most common baited trap is the small invertebrate funnel trap, designed to trap small scavenging crustaceans. These cylindrical traps comprise mesh funnels on one or both ends, with bait inside the cylinder. The bait releases an odour plume that is detected by scavenging animals that follow the plume upstream into the trap. The animal is funnelled into the trap through the large opening, it then passes through the smaller trap entrance and into the cylinder, whereupon it feeds on the bait. The funnel exits are not easily located from the inside as they contact the internal structure of the cylinder and therefore, although some may escape or may be lost during the ascent to the surface, most occupants are successfully recovered. This method is highly effective in capturing a sufficient number of samples in hadal applications (Fig. 3.4b). To reduce any potential loss during ascent and recovery of the system, some traps have used converted closing water samplers (Niskin bottles) to form a closable cylinder (e.g. Blankenship *et al.*, 2006).

These simple funnel traps can be deployed by attachment to larger deep-submergence system (e.g. a lander; Fujii *et al.*, 2010; Jamieson *et al.*, 2011a), or deposited directly on the seafloor individually using an ROV or can be dedicated free-falling systems such as those used by Blankenship *et al.* (2006) or the *Latis* system used in HADEEP (Jamieson *et al.*, 2013; Fig. 3.4c and d, respectively).

The baited trap has been further developed into a more sophisticated 'hyperbaric trap'. Hyperbaric traps entice animals into them in the same way as conventional baited traps, however, the body of the trap is constructed like an open pressure housing. Once the trap leaves the seafloor, the internal chamber is closed either by a piston (e.g. Yayanos, 1977; Fig. 3.5) or by a cantilever mechanism (MacDonald and Gilchrist,

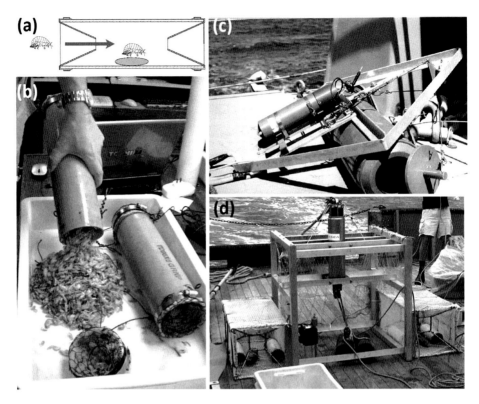

Figure 3.4 Small invertebrate funnel traps. (a) Basic principles of funnel trap whereby an amphipod, for example, is lured through the funnels into the trap with bait and is unable to relocate the entrances. (b) Example of trap efficiency whereby thousands of hadal amphipods (*Hirondellea gigas*) are emptied from a small trap after just 12 h at 9316 m deep (Izu-Bonin Trench; as used in Eustace *et al.*, 2013). (c) Example of full ocean depth rated trap used by Blankenship *et al.* (2006). (d) The *Latis*, large fish trap and funnel trap array as used by Jamieson *et al.* (2013). Images (b) and (d) are courtesy of HADEEP, University of Aberdeen. Image (c) courtesy of L. Levin, Scripps Institution of Oceanography, USA.

1980) and the animals are sealed inside and subsequently recovered at their ambient pressure. This technique was used to recover the first (and only) live organism from hadal depths; a lysianassoid amphipod from 10 900 m in the Mariana Trench (Yayanos, 1981, 2009). Maintaining the ambient pressure is a difficult task due to volume changes during the ascent through the warmer surface layers. Additional thermo insulation around the traps aims to counter this effect. Decompressing organisms from such depths has not yet led to the survival of these deep-sea species at atmospheric pressure, but rather, the process has been used to investigate pressure tolerance and the potential for vertical migration (e.g. MacDonald, 1978; Yayanos, 1981).

The small invertebrate funnel trap was further developed in the construction of the free-falling *Latis* trap (Jamieson *et al.*, 2013). The *Latis* had an array of four conventional invertebrate traps but its main load comprised two larger baited fish traps and pressure sensors to confirm depth of capture. The two fish traps (40 × 40 × 100 cm

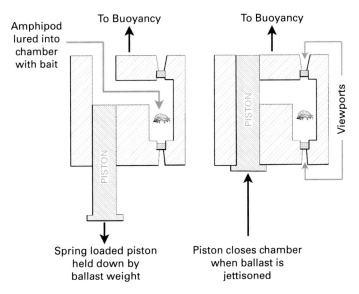

Figure 3.5 Simplified schematic of pressure retaining amphipod trap used by Yayanos (1977). The amphipods are lured into a chamber with bait and upon jettisoning ballast weights a piston seals the chamber maintaining ambient pressure on recovery. The design also has two viewports through which to observe the live specimens. For full detailed description see Yayanos (1977).

cuboid) had a square funnel opening of 14 × 14 cm, recessed 25 cm into the trap. The traps were lined with 1 cm mesh and baited with ~1 kg of jack mackerel inside each trap. The traps were specifically designed to target hadal snailfish by creating a relatively wide and easily accessible opening that was in full contact with the seafloor. Also, in each trap were two baited invertebrate funnel traps (12 cm diameter × 30 cm length) to catch smaller organisms which may have washed out of the larger trap on recovery.

The *Latis* successfully collected thousands of scavenging amphipod samples over 12 deployments between 6097 and 9908 m in the Kermadec Trench, in 2011 and 2012. In addition, the large fish traps successfully recovered nine specimens of the hadal snailfish *Notoliparis kermadecensis* from ~7000 m (the first samples collected in 59 years). The larger traps were also responsible for recovering nine specimens of the 'supergiant' amphipod *Alicella gigantea* from 6295–7000 m (the first samples in the southern hemisphere and at hadal depths), amongst various other invertebrate specimens (Jamieson *et al.*, 2013).

The earliest imaging systems used to capture the seafloor and epifauna at hadal depths comprised either black and white or colour film cameras that were lowered by wire to within a few metres of the bottom (e.g. Pratt, 1962; Heezen *et al.*, 1964; Heezen and Johnson, 1965; Heezen and Hollister, 1971). These systems were generally mechanically triggered to take a photo on contact with the seafloor. The camera systems later evolved into stereoscopic cameras with dual capabilities; one black and white and one colour film camera, which could be towed across the seafloor using a pinger to control the altitude. This system was used on the PROA expedition on the *Spencer F. Baird*, in multiple trenches of the western Pacific Ocean from 6758 to 8930 m (Lemche *et al.*, 1976). The cameras were towed 1–2 m above the seafloor taking images at 10–15 s intervals. Due to variation in altitude, the field of view ranged from 0.5 to 10 m². The system was very successful in progressing from the single images obtained by previous systems to being able to map areas of several hundred to over 2000 m² of seafloor.

The challenges and access to sufficiently long wires more or less halted the use of towed camera systems at hadal depths by the 1970s, although variations on such systems are still regularly used at bathyal and abyssal depths today (e.g. Rice *et al.*, 1979; Barker *et al.*, 1999; Ruhl, 2007; Jones *et al.*, 2009). Camera systems were, thereafter, deployed using the free-fall methods and known as 'free-vehicles' or 'landers'. Imaging landers were pioneered in the 1960s (Isaacs and Schick, 1960; Isaacs and Schwartzlose, 1975) and were used primarily not to image the seafloor and associated epifauna over relatively large distances, but rather to attract mobile fauna to the camera using bait and observe them over time, usually ~12 h. A good example is Hessler *et al.* (1978) and Jamieson *et al.* (2011a) who combined time-lapse cameras with baited traps to both collect scavenging fauna and observe the time course of bait interception and consumption. The cameras are typically mounted vertically, looking down at the seafloor where the bait is placed centrally in the field of view. Images are taken at regular intervals (e.g. 1 min). The resulting images can provide a wealth of information to complement trawl and towed camera studies. For example, observations of live animals over time can provide data on locomotion speeds of slow moving megafauna, species interactions and the presence of taxa that are not readily sampled by other methods due to avoidance (e.g. decapods; Jamieson *et al.*, 2009b) or not recovered in large numbers which would thus otherwise result in an underestimation of the importance of such faunal groups (e.g. amphipods; Hessler *et al.*, 1978).

The use of imaging landers is further enhanced by the use of video systems that operate with the same time-lapse facility as still cameras (e.g. 1 min on, 4 min off; Jamieson *et al.*, 2009a, b). Video footage further complements other sampling methods by recording moving images of live fauna that can be used for behavioural analysis and some physiological studies; for example, decapod predatory behaviour (Jamieson *et al.*, 2009b), fish tail-beat frequency (Jamieson *et al.*, 2009a; Fujii *et al.*, 2010), general locomotion and burst escape responses, as exemplified by the Hadal-Landers (Jamieson *et al.*, 2011b; 2012a, b; Aguzzi *et al.*, 2012; Table 3.2; Fig. 3.6).

Two Hadal-Landers were designed and constructed in 2007 as part of the first HADEEP project (Jamieson *et al.*, 2009c) and combined they completed 34 deployments spanning five trenches and the central Pacific abyssal plains. Hadal-Lander A had a colour video camera and Hadal-Lander B had a 5 megapixel still camera. Both landers were also equipped with conductivity, temperature and depth (CTD) sensors. Both systems had cameras placed vertically at a height of 1 m off the seafloor, with fields of view of 0.35 and 0.29 m^{-2}, respectively. Bait, typically mackerel or tuna, was placed in the centre of the field of view on a scale bar. The landers descended to the seafloor at speeds of ~40 m min^{-1}, aided by three, negatively buoyant, steel ballast weights. At the end of the deployment time (12–24 h), these ballast weights were jettisoned by acoustic

Table 3.2 Specification of the Hadal-Lander developed with the HADEEP projects.

Lander	Hadal-Lander A	Hadal-Lander B	Hadal-Lander C
Type	Baited Video, CTD	Baited Stills, CTD	Baited Video
Year of construction	2007	2007	2012
Delivery system			
Acoustic releases	Oceano 2500-Ti UD (×2)	Oceano 2500-Ti UD (×2)	Oceano 2500-Ti UD (×2)
Buoyancy	17′ glass spheres (×13)	17′ glass spheres (×9)	17′ glass spheres (×X)
Total positive buoyancy	247 kg	171 kg	190 kg
Ballast weight (wet)	135 kg (45 kg × 3)	135 kg (45 kg × 3)	120 kg (120 kg × 1)
Vehicle weight (in water)	180 kg	110 kg	120 kg
Total weight (descent)	68 kg $-ve$	74 kg $-ve$	60 kg $-ve$
Total weight (ascent)	67 kg $+ve$	61 kg $+ve$	70 kg $-ve$
Descent velocity	46 m min^{-1}	34 m min^{-1}	36 m min^{-1}
Ascent velocity	54 m min^{-1}	34 m min^{-1}	35 m min^{-1}
Scientific payload			
Camera	Hadal-Cam 12000	Kongsberg OE14–208	Hadal-Cam 12000 mk2
Camera resolution/ format	704 × 576 pixels (MPEG2)	5 megapixel (JPEG)	704 × 576 pixels (MPEG4)
Camera sample interval	1 min every 5 min	1 min	1 min every 5 min
Camera sample number	120	2000	>350
Battery	12V Lead–Acid	24V Lead–Acid	12V Lead–Acid
Camera field-of-view	68 cm × 51 cm (0.35 m^{-2})	63 cm × 47 cm (0.29 m^{-2})	Oblique; 1 m across centre
Camera orientation	Vertical (1 m)	Vertical (1 m)	Oblique (1.5 m)
CTD	SBE19*plus* V2	SBE19*plus* V2	SBE 39 PT only
CTD resolution (S,T,P)	0.4 ppm, 1 × 10^{-4}°C, 0.002%	0.4 ppm, 1 × 10^{-4}°C, 0.002%	n/a
CTD sample interval	10 s	10 s	n/a
Water sampler	2-l Niskin	2-l Niskin	2-l Niskin
Current meter	n/a	Aanderaa Zpulse TDD 266	n/a
Funnel Traps	30 cm Ø × 40 cm (×3) 10 cm Ø × 30 cm (×1)	None	18.5 cm Ø × 50 cm (×2)
Bait	~1 kg mackerel/tuna	~1 kg mackerel/tuna	~1 kg mackerel/tuna

command from the surface and the landers ascended by virtue of an array of 17 inch glass buoyancy spheres on a 100 m long mooring line. Both systems were pre-programmed and autonomous and the data was downloaded upon recovery.

Unfortunately, after 11 successful deployments, Hadal-Lander A was lost at sea in 2009 following an error in deployment at 7000 m in the Kermadec Trench. This was the lander that filmed the deepest fish ever found, at 7703 m in the Japan Trench (Fujii *et al.*, 2010). Hadal-Lander B was lost three years later at 9500 m in the Kermadec Trench due to suspected implosion of the glass sphere buoyancy modules. This lander, however, achieved 23 successful dives and revealed to the world new finds such as the deepest fish in the southern hemisphere (Jamieson *et al.*, 2011a) and *in situ* images of the supergiant amphipod *Alicella gigantea* (Jamieson *et al.*, 2013). Despite the

Figure 3.6 The Hadal-Landers used in the HADEEP Project. (a) Hadal-Lander A being deployed to 10 000 m in the Tonga Trench and (b) is the new Hadal-Lander C. Images courtesy of J.C. Partridge (University of Bristol, UK) and HADEEP, respectively.

unfortunate end of both landers, the data collected resulted in 14 scientific publications within 5 years. A new Hadal-Lander has been designed, constructed and already used at depths between 1000 and 6500 m in the Kermadec Trench with further, deeper studies scheduled for 2014 in the HADES project.

The lander principle was adopted for full ocean depth operations by the Scripps Institution of Oceanography (USA). Following on from the institute's pioneering 'free-vehicle' developments in the 1960s and '70s (Isaacs and Schick, 1960; Phleger and Soutar, 1971), a series of low-cost compact lander design were developed based on the Deep Ocean Visualisation Experimenter (DOVE) instrument platform (Hardy *et al.*, 2002; Fig. 3.7). This series of landers include:

- DOV *Mary Carol* in 2001; upgraded to 10 000 m operations in 2002 which completed dives to the Puerto-Rico (8400 m) and Aleutian Trenches (7200 m).
- DOV *Bobby Ray* in 2006; successfully retrieved water samples from the Puerto-Rico Trench (8400 m; Eloe *et al.*, 2010).
- DOV *Patty* in 2011; deployed in Sirena Deep, Mariana Trench (10 800 m) but subsequently lost in Typhoon Muifa.
- DOV *Karen* in 2011; contained a single core tube, which retrieved sediment from 5400 m in the Philippine Sea.

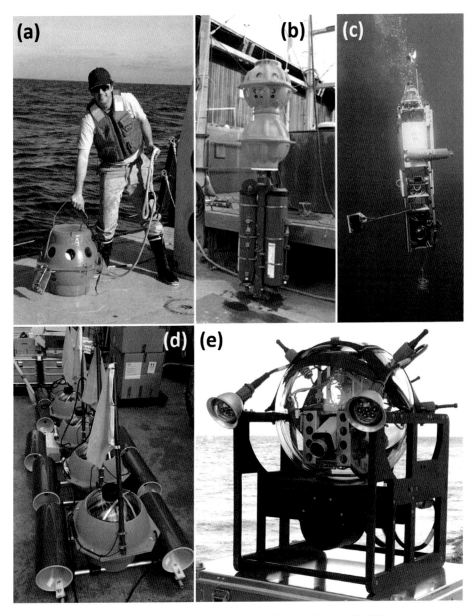

Figure 3.7 Low-cost, compact landers for full ocean depth operations. (a) DOVE designer
Kevin Hardy demonstrating size and weight benefits of the compact design; (b) DOV *Bobby Ray*
(water sampler); (c) DOV *Mike* (*Alpha Lander*; camera and traps); (d) *Obulus* I–V (baited traps);
and (d) Promare's *11k* (camera). Images courtesy of K. Hardy, Scripps Institution of
Oceanography, USA (a–c), HADEEP (d) and F. Søreide, Promare, USA (e).

- DOV *Michelle* in 2011; comprised dual sediment core tubes and rated for
  11 000 m operations but was lost during a dive to 5400 m in the Philippine Sea.
- DOV *Mike* or *Alpha Lander* in 2012; used in the *Deepsea Challenge* expedition
  and was successfully deployed in the New Britain Trench and both Challenger
  and Sirena Deeps in the Mariana Trench.

A similar design was deployed in 2008 in the Puerto-Rico Trench, equipped with an acoustic current meter (Schmidt and Siegel, 2011). However, despite recording a full 8350 m vertical profile of three-axis velocity, temperature and pressure, the instrument flooded after 75 min on the seafloor.

A similar design principle was used in HADEEP in 2011 to supplement small invertebrate sample numbers from the larger imaging landers and traps. These systems, again comprising one 17 inch glass sphere, comprised two baited funnel traps per system and used a timed burn-wire release mechanism to jettison gravel-filled sand bags at a predetermined time. Five systems were built, called *Obulus I–V*, and were deployed at 6968, 6999, 7014 and 8148 m in the Kermadec Trench. Although these four deployments were highly successful in recovering hadal amphipods, one system was lost and two suffered water ingress and thus the *Obulus* array has been temporarily abandoned due to unreliability at depth.

Another low-cost imaging lander has also been developed recently by Promare (USA; Fig. 3.7; Søreide, 2012). Called '11k', the basic vehicle consists of a glass sphere from Nautilus Marine Service GmbH (Bremen, Germany) that contains proprietary software and hardware, as well as an assortment of off-the-shelf components, such as an HD video camera and lithium-ion battery pack from OceanServer Technology Inc. (Fall River, Massachusetts). In addition, the system includes a high-resolution, full ocean depth pressure sensor custom made by Presens AS (Oslo, Norway), and full ocean depth LED lights and a drop-weight system developed by Promare. 11k weighs just 60 kg and measures $50 \times 50 \times 75$ cm. The vehicle has already been deployed to ~8000 m in the Puerto-Rico Trench (Søreide and Jamieson, 2013), and filmed and recovered footage and samples of hadal amphipods (notably *Scopelocheirus schellenbergi*; Lacey *et al.*, 2013). 11k is also scheduled to be used in an AUV or ROV mode in the future (Søreide, 2012).

Other less well-documented imaging systems have been reported, for example, Kobayashi *et al.* (2012), Glud *et al.* (2013) and Oguri *et al.* (2013). These studies used an 11 000 m rated camera system, named 'ASHURA' (Murashima *et al.*, 2009), which was deployed to 10 897 m in the Mariana Trench and recovered specimens of the amphipod *Hirondellea gigas* and sediment samples (Kobayashi *et al.*, 2012). In the Glud *et al.* (2013) and Oguri *et al.* (2013) studies, they used the system equipped with HD video camera, halogen lights, a CTD and three sediment corers, but detailed specifications have not been published.

With modern electronics and data storage capacity ever increasing, the number of still images and length of video footage that can be easily obtained is accelerating, resulting in a greater number of images being obtained from hadal depths. Furthermore, the use of high-resolution digital still photograph and HD video form key components of modern exploratory vehicles (discussed below).

3.5   Biogeochemistry instruments

Lander- or ROV-deployed instrumentation for biogeochemical research is widely used in the deep sea (Tengberg *et al.*, 1995, 2005). However, there is currently only one system rated for hadal operations (Fig. 3.8). Glud *et al.* (2013) adapted a transecting

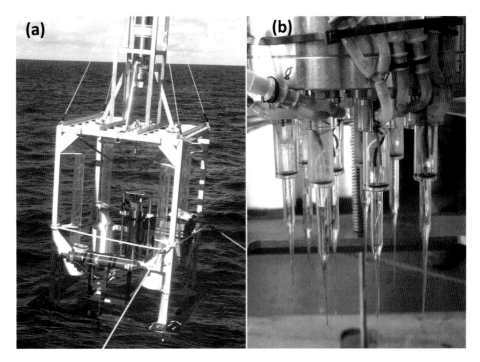

Figure 3.8 The micro-electrode lander used by Glud *et al.* (2013) to measure sediment O_2 profiles at nearly 11 000 m in the Mariana Trench; (a) is the entire lander and (b) is close up of the micro-electrode array. Images courtesy of R.N. Glud, University of Southern Denmark.

deep-sea micro-profiler lander (Glud *et al.*, 2009), capable of measuring *in situ* O_2 profiles at full ocean depth (Fig. 3.7). This profiling lander measured the distribution of O_2 across the sediment–water interface at a resolution of 0.5–1 mm at nearly 11 000 m in Challenger Deep twice. The lander had eight Clark-type oxygen micro-electrodes and a resistivity probe to identify the sediment–water interface.

3.6   Remotely operated vehicles (ROVs)

Even today, very few ROVs are capable of exploring and sampling at full ocean depth. The first ROV of this kind to be developed was the ROV *Kaikō*, constructed in 1993 by JAMSTEC (Kyo *et al.*, 1995). Following over 20 dives to full ocean depth, the ROV *Kaikō* was lost in 2003. For the 15 years that followed, there were no operational, full ocean depth ROVs in circulation until the construction of the HROV *Nereus* by the Woods Hole Oceanographic Institute (WHOI, USA), with collaboration of the Johns Hopkins University and the US Navy Space and Naval Warfare Systems Center San Diego (Fig. 3.9a; Bowen *et al.*, 2008). Only two other systems have ever been used at full ocean depth; the shallower rated *Kaikō 7000 II* (Fig. 3.9b; Murashima *et al.*, 2004) and the compact full ocean depth rated crawler ABISMO (Fig. 3.9c; Yoshida *et al.*, 2009).

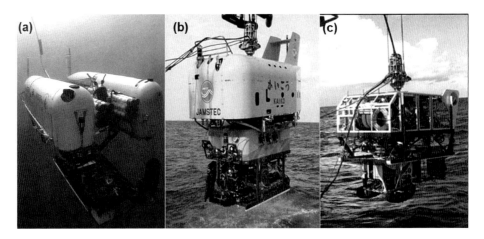

Figure 3.9 Full ocean depth remotely operated vehicles. (a) WHOI's HROV *Nereus*, (b) JAMSTEC's ROV *Kaikō 7000 II* with launcher and (c) JAMSTEC's ABISMO ROV/Crawler system with launcher. Images © WHOI, USA (a) and JAMSTEC, Japan (b, c).

The full ocean depth rated ROV *Kaikō* was constructed in 1995 for scientific research (Kyo *et al.*, 1995; Mikagawa and Aoki, 2001). *Kaikō* was a two-body system where the actual ROV was coupled to a launcher system during descent. The negatively buoyant launcher aided the whole system to descend rapidly as it was lowered on the main umbilical (4.5 cm diameter). The launcher itself had limited operational capabilities on the seafloor, but could be used independently as a towed system equipped for acoustic surveillance (Barry and Hashimoto, 2009). *Kaikō* was connected to the launcher via a 250 m long tether (3 cm diameter), permitting a relatively short but unconstrained range from the launcher. Once uncoupled, the ROV could manoeuvre to explore and sample the seafloor under complete control from the mother ship; RV *Kairei* (Mikagawa and Aoki, 2001). The vehicle's scientific payload comprised several charge-coupled device (CCD) and wide-angle video and stills cameras, multiple high-intensity lights and other sensors, such as a scanning sonar, altimeter, compass and pressure sensors. It had two highly dexterous manipulator arms (six axes and seven axes of motion) for collecting biological or geological samples, or other equipment from the front-mounted sample basket. Despite a successful 8-year reign as the only full ocean depth exploratory vehicle, amassing 295 dives (20 to full ocean depth), the vehicle (not the launcher) was accidentally lost in May 2003 off Shikoku Island in an emergency ascent during Typhoon Chan-Hom (Momma *et al.*, 2004; Tashiro *et al.*, 2004; Watanbe *et al.*, 2004). Using the existing launcher, a 7000 m rated *Kaikō* (*Kaikō 7000 II* modified from the existing *UROV7K*) was later introduced in 2004 (Murashima *et al.*, 2004; Barry and Hashimoto, 2009). The *Kaikō 7000 II* completed several dives to ~7000 m but served as only a temporary replacement and was superseded by the ABISMO (Yoshida *et al.*, 2009). The ABISMO comprises a similar vehicle and launcher system to *Kaikō*, but the vehicle is much smaller with far fewer sampling capabilities; it was specifically designed to retrieve small sediment samples and for seafloor inspection with a pan and tilt video camera. Water samples can be taken using a Niskin bottle array on the launcher. ABISMO has the capability of switching between conventional ROV mode and crawler mode. As a crawler, it drives across the seafloor on treads, as opposed to free-floating using vertical thrusters. In 2007 it reached 9760 m in the Izu–Ogasawara Trench and, in 2008, it collected small sediment and water samples from 10 257 m in the Mariana Trench (Itoh *et al.*, 2008; Yoshida *et al.*, 2009).

The HROV *Nereus* is a novel, deep submergence vehicle designed to perform scientific surveys and sampling to full ocean depth (Bowen *et al.*, 2008, 2009a, b; Fig. 3.8). *Nereus* operates in two different modes. (1) As an untethered autonomous underwater vehicle (AUV) capable of exploring and mapping the seafloor with sonars and cameras for broad area surveying. (2) It can be readily converted to operate as a tethered ROV for close-up imaging and sampling. The hybridisation of AUV and ROV form the basis of the term HROV. The ROV configuration features a novel lightweight fibre-optic tether connected to the surface vessel for high bandwidth real-time video and data telemetry. This enables high-quality remote-controlled teleoperation by a pilot on the mother ship. The development of the lightweight fibre-optic tether was driven by the limitations of steel reinforced cables, which are only self-supporting to ~7000 m, and alternatives such as Kevlar, which present problems in the form of poor hydrodynamics and large cable handling systems. The ultra thin umbilical tether (0.8 mm diameter) was only achieved by designing the vehicle to be self-powered, meaning that power is supplied via a battery pack on the vehicle itself. This excludes the need to transmit power through the umbilical. The only disadvantage to being self-powered is the trade-off between operational depth and flexibility and the issues of limited power and re-charge time between dives. The HROV is equipped with a suite of sensors and sampling devices for both biological and geological sampling, mainly still and video imaging systems (with variable output LED illumination), push core sediment samplers, water samplers with *in situ* temperature sensors, a sample box in which organisms or rock samples can be stored after collection by the manipulator arm, a magnetometer, a salinity–temperature–pressure sensor and high resolution acoustic bathymetry. It also has the capacity for a further 25 kg of equipment.

Following on from its first successful field trial in shallow water in November 2007 (Bowen *et al.*, 2008), the HROV *Nereus* successfully descended to over 10 000 m (max. depth 10 903 m) in the Mariana Trench in May 2009 (Bowen *et al.*, 2009b; Fletcher *et al.*, 2009). The HROV is still fully operational and is scheduled to make up to 30 dives between 4000 and 11 000 m in the Mariana and Kermadec Trenches in 2014 as part of the HADES project.

3.7   Manned submersibles

To date, only two manned submersibles (previously known as 'bathyscaphes' and often referred to as HOVs) have reached full ocean depth, the *Trieste* and the *Deepsea Challenger*, which reached the Challenger Deep in the Mariana Trench in 1960 and 2012, respectively. Another submersible, operational in the 1960s called *Archimède*, was used in three different trenches. There is also the Chinese *Jiaolong* three-man submersible rated to 7000 m (Liu *et al.*, 2010) and the Japanese *Shinkai 6500*; a three-man submersible capable of diving to 6500 m (Nanba *et al.*, 1990).

The history of manned submersibles at hadal depths is interesting from the point of view that there have been so few, and as a technology, they have largely been stuck in the 'record-setting' phase. For example, the *Trieste* is known for completing the 'first manned dive to full ocean depth', followed 52 years later by the *Deepsea Challenger*, which was hailed as 'the first *solo* dive to full ocean depth'. Likewise, the Japanese *Shinkai 6500* website declares that it can go 'deeper than any other manned submersible for academic research around the world today'. In fact, this accolade has recently been beaten by the Chinese *Jiaolong*, which it is said 'has the greatest depth range of any manned research vehicle in the world'. Technical achievements aside, these submersibles have yet to produce any comprehensive scientific findings from the maximum depths that they advertise having achieved. It is hoped that once manned submersibles such as *Deepsea Challenger* and *Jiaolong* are beyond the demonstration phase, they will yield new and exciting data from the deepest places on Earth.

The *Trieste* was designed by the Swiss scientist Auguste Piccard, based on previous designs by Piccard (*FNRS-2*) and was constructed in Italy. It was named *Trieste* after the city and seaport in northeast Italy on the Adriatic Sea where the major components were manufactured. It was originally operated by the French Navy and later by the US Navy in 1958.

The submersible consisted of a 15 m long buoyancy chamber filled with 85 000 l of gasoline, water ballast tanks, 9 tons of expendable magnetic iron pellet ballast weight and a separate 2.16 m diameter pressure sphere on the underside, from which a crew of two would operate it. The pressure sphere was later upgraded to 12.7 cm thick in order to withstand the hydrostatic pressure at hadal depths. The new sphere weighed 13 tons in air and 8 tons in water. The crew of the *Trieste* could look outside the pressure sphere via a tapered acrylic viewport. The illumination was provided by quartz arc-light bulbs, which without any modification were able to withstand the pressures at hadal depths. This design permitted a free descent, rather than a lowering via cable like previous 'bathyspheres'.

The expendable iron ballast weights were used to control the descent speed and once fully jettisoned by electromagnets they initiated the ascent to the surface. The electromagnetic expulsion of the ballast weights also facilitated an emergency ascent in the event of power failure.

The submersible was first operated in August 1953 in the Tyrrhenian Sea near Capri. It was later used in a series of deep-water tests in the Pacific Ocean, which culminated in the 10 916 m dive to Challenger Deep in the Mariana Trench on 23 January 1960, as part of Project Nekton. The submersible was deployed from the USS *Wandank* (ATA-204). The two-man crew comprised Auguste Piccard's son, Jacques Piccard, and US Navy Lieutenant Don Walsh. This marked the first visit to the deepest point on Earth.

The descent to Challenger Deep took 4 h 48 min at a descent rate of 0.9 m s^{-1} and was not without incident. At a depth of approximately 9000 m one of the acrylic viewports cracked, shaking the entire submersible. The two men spent barely 20 min at the ocean floor, eating chocolate bars for sustenance. Visibility at the seafloor was almost entirely obscured by resuspended sediment. They did, however, record seeing animal life on the bottom, including an erroneous account of a flatfish (Jamieson and Yancey, 2012). The ascent to the surface took 3 h 15 min.

Unfortunately, the submersible was never again used at hadal depths, nor was any useful scientific data gained from the dive. In 1963, it was sent to the Atlantic Ocean to search for the missing submarine USS *Thresher*, which it found, wrecked off the coast of New England, at 2560 m. After intensive modifications, followed by decommissioning in 1966, the *Trieste* was exhibited in the National Museum of the US Navy at the Washington Navy Yard in 1980.

The manned submersible *Archimède* was operated by the French Navy and designed by Pierre Willm and Georges Houot. It was christened on 27 July 1961 in Toulon, France, at the French Navy base. Like the *Trieste*, it used 160 000 litres of gasoline buoyancy and weighed 61 tons. It completed its first test dives to 1500 m in 1961, although these were unmanned. Shortly after, it successfully achieved a speed of 3 knots at 2400 m in the Mediterranean Sea, followed by a deeper dive the following year to 4799 m off Japan. It became the first submersible to reach the deepest part of the Atlantic Ocean: 8390 m in the Puerto-Rico Trench. The first credible scientific accounts from manned submersibles at hadal depths were reported by J.M. Pérès in the mid-1960s, from 7300 m in the Puerto-Rico Trench from *Archimède* (Pérès, 1965). Although he did not have any operational imaging equipment onboard, he gave detailed eye-witness accounts of various hadal fauna such as isopods, decapods and fish.

In 1962, the *Archimède* reached 9560 m in the Kuril–Kamchatka Trench in the northeast Pacific Ocean, followed by a descent to 9300 m in the Japan Deep in the Izu-Bonin Trench. The submersible was operational until the 1970s.

The more recent *Deepsea Challenger* (DCV 1; Fig. 3.10) was a 7.3 m long, manned submersible built by the Canadian film director James Cameron to revisit Challenger Deep, 52 years after the last manned submersible to do so, *Trieste*. It was constructed in Australia, in partnership with the National Geographic Society and other commercial support. The development of the submersible heralded new materials including specialised structurally sound syntactic foam buoyancy, designed to produce positive buoyancy at full ocean depth. The structural integrity of the new foam was such that it enabled thruster motors to be mounted within it, but without the aid of metal structures. Like the *Trieste*, the submersible had a pressure sphere, 1.1 m diameter and 64 mm thick, where the single pilot would be housed. The sphere was coupled to the base of the submersible, which weighed over 10 tons. Once subsurface, the *Deepsea Challenger* operated in a vertical attitude, and carried 500 kg of expendable ballast weight to aid the descent and, once jettisoned, ascent to the surface was initiated.

Figure 3.10 The *Deepsea Challenger* submersible piloted by James Cameron on its 2012 dive to the Mariana Trench. Image courtesy of Charlie Arneson.

After various shallow-water dives, the submersible, piloted by Cameron was successfully deployed to 7260 m and 8221 m in the New Britain Trench (southwest Pacific Ocean). Here, he was reported to have seen anemones and jellyfish that were consistent with other findings from that trench (Lemche *et al.*, 1976).

On 26 March 2012, the *Deepsea Challenger* successfully reached 10 898.4 m in the Challenger Deep, after a 2 h 37 min descent from the surface vessel. It remained there for approximately 3 h before successfully returning to the surface.

There are another two manned submersibles capable of reaching depths greater than 6000 m, but only marginally. In 1991, JAMSTEC in Japan began using the *Shinkai 6500* submersible (Takagawa, 1995). Although only capable of descending into the very upper hadal depths, it is the deepest diving submersible for academic research in operation. In 2012, it had already completed 1300 dives and has recently received a major upgrade.

Since 2010, the Chinese have been using a new three-man submersible called *Jiaolong*, rated to 7000 m (Liu *et al.*, 2010). In July 2010, it completed various shallower dives to 3759 m in the South China Sea and 4027 m in the northeast Pacific Ocean the following year. In 2012, the *Jiaolong* descended to 6965 m, followed shortly after by a descent to 7062 m in the Mariana Trench, in the western Pacific Ocean.

The only other manned submersible with potentially hadal capability is the new DSV *Alvin*, not yet in operation and currently under construction by WHOI in the USA. Like the *Shinkai 6500*, the new *Alvin* may ultimately be rated to 6500 m (Monastersky, 2012) and may therefore oversee future operations in the upper hadal zone.

Part II

Environmental conditions and physiological adaptations

Introduction

The hadal trenches are geographically disjunct deep-sea ecosystems. Many have trench-specific geographical settings and thus often unique environmental conditions. Considering all trenches to be a single habitat, i.e. simply 'hadal', it is likely to confound interpretation of environmental drivers. Many environmental aspects are applicable to all hadal environments (e.g. increase in hydrostatic pressure with depth); however, there are those which only apply to a specific trench. Certain environmental conditions exhibit inter-trench variation, a result of the interactions between, for example, the local hydrography (temperature, salinity and oxygen), trench topography, seismic activity, substrata and hydrostatic pressure. Direct measurements of some of these parameters across multiple trenches are nonexistent, sparse, spurious or noncomparable. With the exception of topography and seismic activity, the hadal trenches experience similar conditions to the surrounding abyssal zone, for example, absence of light, low temperatures, salinity and oxygen.

Coping with, or rather adapting to, high hydrostatic pressure is perhaps one of the most important prerequisites to survival in the hadal zone. High-pressure (and low-temperature) adaptations are common to all deep-sea organisms, but in the case of pressure, none more so than to those inhabiting the trenches. Furthermore, the process of ossification (the creation of hard shells) cannot occur beyond the carbonate compensation depth (CCD; >4000–5000 m) and this, in addition to other stresses, has prompted further adaptation enabling organisms to compensate for the high pressures with softer, more organic, physiological structures (e.g. Todo *et al.*, 2005).

Coupled with the need to adapt to these environmental conditions, food supply is also a biological challenge and most hadal animals are directly or indirectly reliant on food derived from the surface waters. Extreme depth can also be expressed as extreme distance from the surface, which means that by the time organic matter has reached hadal depths, it is likely to be greatly reduced in both quantity and quality. Despite these seemingly impossible odds, the hadal zone is inhabited by representatives of most major taxa (Wolff, 1960) and within those taxa there are representatives of all feeding guilds; filter-feeders (e.g. Oji *et al.*, 2009), deposit feeders (e.g. Hansen, 1957), scavengers (e.g. Blankenship *et al.*, 2006), plant and wood consumers (e.g. Kobayashi *et al.*, 2012), chemosynthesisers (e.g. Fujikura *et al.*, 1999) and predators (e.g. Jamieson *et al.*,

2009b). This, in itself, is testament to how well the deep-sea community has evolved and adapted in response to the challenges presented.

In this part, the general environmental conditions with examples of adaptations are discussed (Chapter 4), with a special emphasis on high hydrostatic pressure and its effects (Chapter 5) and an overview of the food supply to the trenches (Chapter 6).

4 The hadal environment

Menzies (1965) made the first attempt at reviewing the environmental conditions in the deep sea. Whilst considered a milestone review of its day, it was largely based on limited point-source observations. The review was later updated and revised in light of subsequent global studies of a more holistic nature (e.g. Tyler, 1995; Thistle, 2003). The Tyler (1995) review highlights the heterogeneity of conditions on both small and large scales and temporal variations, in contrast to the previous perception of a physically stable and unchanging environment (Sanders, 1968). To add further to the reviews of Menzies (1965) and Tyler (1995), this chapter will detail what is currently known about the environmental conditions required for the existence of life in the hadal trenches.

4.1 Deep-water masses and bottom currents

There are two forces driving deep-water currents in the deep sea; thermohaline and tidal (Tyler, 1995). The water masses flowing through the deep oceans are illustrated in the 'great ocean conveyer belt' diagram, published in *Natural History* in 1987 (see Broecker, 1991; Rahmstorf, 2006). Although this diagram has become famous, it represents a simplistic illustration of ocean circulation for the layperson. However, as a basis for understating the global circulation of water masses throughout the world's trenches, it does serve as an ideal starting point. While there are now multiple complex studies into deep-water masses and their interaction with other stratified water masses and underlying topography, on a global context the diagram illustrates the main features: cold surface waters in the Antarctic descend and become deep bottom waters that flow up the western Pacific Ocean, through most of the world's trenches. As this water flows north it warms, causing it to rise towards the surface again in the North Pacific Ocean before flowing west towards the Indian Ocean. This warmed surface water flows up the Atlantic Ocean, cools once more and descends to the bottom and flows southward back down the Atlantic as bottom water. Thus, the 'thermohaline conveyer belt' leads to a net transfer of heat from the south to the north (Berger and Wefer, 1996).

In reality, however, deep-water mass circulation is far more complex (e.g. Kawabe *et al.*, 2003; Fig. 4.1). The deep bottom water masses, driven by the thermohaline, spread slowly towards lower latitudes, following the deepest topographical features. The Earth's rotational forces cause the water to flow against the western boundary of

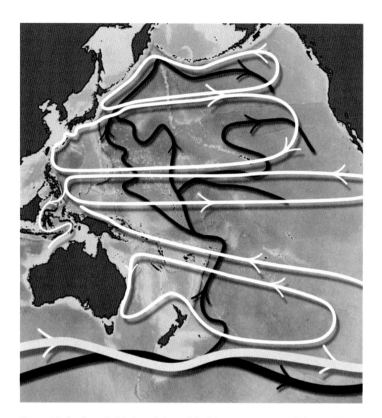

Figure 4.1 Surface (white) and deep (black) water masses of the Pacific Ocean. The contrast between the surface and deep-water masses are the influence of the topography on the deep-water mass as it flows over trenches, ridges and around basins. Figure based on Kawabe and Fujio (2010), illustrated by Amy Scott-Murray, University of Aberdeen, UK.

basins or trenches. Therefore, in the Atlantic Ocean, Antarctic bottom water travels northwards, across the abyssal plains situated between the eastern slope of South America and the Mid-Atlantic Ridge (Broecker *et al.*, 1980). Once at the equator it diverges with one branch flowing into the deep northwest Atlantic, and the other flowing through the Mid-Atlantic Ridge (through the Romanche Trench fault); both northwards and in part southwards into the abyssal plains of the east Atlantic (Angel, 1982). As the waters are driven through the South and North Atlantic they become more saline through evaporation (Schmitz, 1995). On reaching high latitudes in the north, this water cools and sinks by vertical convection to form the northeast Atlantic Deep Water (NEADW). Once in the northwest Atlantic basin it mixes with low-salinity water from the Labrador Sea to form the North Atlantic Deep Water (NADW). The NADW then flows into the South Atlantic and eventually back to the Pacific Ocean via the Circumpolar Current (Worthington, 1976).

The flow of water around the Pacific Ocean is also predominantly driven by thermohaline circulation (Broecker, 1991; Rahmstorf, 2006), as no deep water can be formed in the North Pacific Ocean (Stommel, 1958; Warren and Owens, 1985).

At depths exceeding 3000 m in the Pacific Ocean, the cold (0.5–1.5°C) and relatively saline, Antarctic bottom water is found. These waters are mostly formed during sea ice formation during Austral winter (Tomczak and Godfrey, 1994). This deep-water mass circumnavigates the Southern Ocean and flows up the western boundary of the South Pacific Ocean in deep western-boundary currents, whilst also spreading slowly eastward into the deep Pacific Ocean. There are two major water masses present in the deep Pacific Ocean; the Lower Circumpolar Deep Water (LCDW) and the North Pacific Deep Water (NPDW) (Siedler et al., 2004). The LCDW (Warren and Owens, 1985, 1988; Owens and Warren, 2001; Kawabe, 1993) enters the Pacific Ocean from the south and flows northwards and clockwise through the southwest Pacific trenches (i.e. the Kermadec and Tonga Trenches; Warren, 1981; Johnson, 1998). Thus, these trenches, which are found at the most southerly latitudes, are the coldest in the Pacific Ocean due to this incursion of cold, deep Antarctic water. Through the Samoan Passage the water flows northwest and enters the east Mariana Basin from the east side, before branching into both a northward and westward flow. The northward branch travels north through the Izu-Bonin, Japan and Kuril–Kamchatka Trenches before heading southwards around the Emperor Seamounts and then eastwards around the Aleutian Trench. The deep waters of the North Pacific Ocean are the furthest removed from their Antarctic origin and, therefore, it is at least 1000 years since these waters acquired oxygen from the surface. The flow of water then returns westwards, back around the Aleutian Trench and Kuril–Kamchatka Trench and then southwards through the Japan and Izu-Bonin Trenches.

Data from current meters situated near the seafloor within trenches are generally sparse, however, there are several reports of large-scale velocities within the water masses in the vicinity of and overlying the major trenches of the western Pacific Ocean; the Kermadec Trench (Whitworth et al., 1999), Mariana Trench (Taira et al., 2004), Izu-Bonin Trench (Fujio et al., 2000), Japan Trench (Hallock and Teague, 1996) and Aleutian Trench (Warren and Owens, 1985, 1988), as is summarised in Johnson (1998).

A current meter array, comprising 20 moorings over 22 months, was deployed transecting the Kermadec Trench at 32.5° S within the deep western-boundary current (Whitworth et al., 1999). The current meters were placed at 2500, 4000 and 6000 m. A deep cyclonic circulation was found over the Kermadec Trench, superimposed on the general northward flow of the deep western-boundary current. At the deepest sites, >4000 m, the flow is strong (up to 9 cm s^{-1}) and generally northward over the west side of the trench but switches to a weaker (~2 cm s^{-1}) southward flow over the eastern side.

Three current meter moorings have been deployed across the southwest end of the Mariana Trench at ~142.6° E for up to 424 days, at depths ranging from 6095 to 9860 m (Taira et al., 2004). The velocity data again revealed that the flow was cyclonic in sense around the trench axis (which, in the Mariana, runs approximately east–west). On the north side of the trench (at 6960 m), the waters flowed westward at ~1.3 cm s^{-1}, whereas at the centre of the trench (10 286 m), the velocity was just 0.1 cm s^{-1}. Over the southern flank (6520 m), the water flowed eastward at a rate of ~0.5 cm s^{-1}.

Similar findings are documented for the Izu-Bonin (Izu–Ogasawara) Trench (Fujio et al., 2000). This study comprised 30 current meters transecting the trench at ~34° N for an average of 401 days, at depths ranging from 3830 to 8961 m. The mean direction of flow was closely correlated to the bathymetric contours, and cyclonic circulation around the trench axis was observed. A southward flow of 3.6, 4.6 and 2.4 cm s^{-1}, over depths of 4500, 6000 and 9000 m, respectively, was recorded on the western flank of the trench. At the centre, a southerly flow of 0.8 cm s^{-1} was found, whereas on the eastern flank a northward flow of 3.0 and 12.8 cm s^{-1} was recorded over depths of 9000 and 6000 m, respectively.

Hallock and Teague (1996) reported results from a deep current meter mooring array from the Japan Trench at around ~36° N, and once again, they observed a cyclonic sense of deep circulation. These data showed evidence of a relatively low flow rate of 1.3–1.6 cm s^{-1}, nominally south flowing, on the west side of the trench over depths of 3300 and 4600 m. Similarly, the mean flow of water over the Japan Trench generally follows the bathymetric contours, as in the Kermadec, Mariana and Izu-Bonin examples. On the eastern flank, a northward flow of 5.2 cm s^{-1} over depths of 6400 m has been observed, while other data from RAFOS floats recorded a stronger northward flow on the west and a weaker southward flow on the eastern side of the trench (Johnson, 1998).

In the Aleutian Trench, vertical profiles of mean current velocity have shown that it is moderate (~3 cm s^{-1}) and flowing westward, once more in line with bathymetric contours and forced against the Aleutian Island Arc (Warren and Owens, 1985, 1988). Over the trench and on its southern flank, a slightly weaker velocity of ~2 cm s^{-1} was recorded, with an eastward flow.

These velocity measurements from long-term studies over the trenches of the western Pacific Ocean are consistent with cyclonic gyres of at least a few cm s^{-1} magnitude.

The near-bottom currents in the deep sea flow more slowly than those in shallower environments (Thistle, 2003). At bathyal and abyssal depths, near seafloor current velocities are typically ~10 cm s^{-1} and 4 cm s^{-1}, respectively, with little variation on short timescales (Eckman and Thistle, 1991). Current speeds across the vast abyssal plains of the Pacific Ocean, which often surround trenches, are indeed low, since the shear stresses are inadequate to displace most sediment types (Smith and Demopoulous, 2003). There are, however, certain scenarios which may induce high-energy flows, such as eddies (Hollister and McCave, 1984), submarine canyon flushing (Vetter and Dayton, 1998) or through channels and over peaks (Genin *et al.*, 1986). However, most of these high current flow phenomena are restricted to depths much shallower than the trenches.

Direct measurements of water flow within the hadal trenches are limited. The majority of recorded measurements have been collected over long periods of time within the overlying water column and were taken in order to examine cyclonic circulation in trench localities, e.g. the Kermadec Trench (Whitworth *et al.*, 1999), Mariana Trench (Taira *et al.*, 2004), Izu-Bonin Trench (Fujio *et al.*, 2000), the Japan Trench (Hallock and Teague, 1996) and Aleutian Trench (Warren and Owens, 1985, 1988). However, measurements taken on a shorter, higher resolution are lacking, and it is measurements of this kind that could provide an indication of the kind of flow that is routinely experienced by organisms living on the trench floor.

Table 4.1 Summary of near-bottom current speed data from the Kermadec Trench (~32° S).

Depth (m)	Date	Altitude (mab)	Bottom time (hh:mm)	Current speed (cm s^{-1} ± S.D.)
6116	Feb 2012	2	15:52	4.1 ± 3.0
6475	Feb 2012	2	14:42	3.6 ± 2.9
6980	Nov 2011	1	25:00	1.6 ± 0.9
7501	Nov 2011	1	09:09	1.7 ± 3.0
8631	Feb 2012	2	14:16	3.9 ± 2.8
9281	Nov 2011	1	09:59	0.4 ± 0.3

The pore waters within seafloor sediments do not move and therefore current speeds must decrease to $0 \, \text{cm s}^{-1}$ at the sediment–water interface (Vogel, 1981). This suggests that the current speeds directly above the seabed are much less than those recorded from high above the seafloor, or even within a metre above bottom. This decrease in speed towards the sediment–water interface is sufficient to prevent the erosion of surface sediment or perturbation of epibenthic organisms (Thistle, 2003).

There is sufficient evidence to suggest that near-bed flow at hadal depths is indeed slow. At the deepest place on Earth, Challenger Deep, Taira *et al.* (2004) documented current speeds from sensors tethered about a hundred metres above the seafloor. Despite long periods during which the current speeds were too low to record (37.5% of the time), speeds of up to $8.1 \, \text{cm s}^{-1}$ were otherwise measured. Further measurements were made at 7009 m and 6615 m, where the mean speeds were found to be $0.7 \, \text{cm s}^{-1}$ and $0.5 \, \text{cm s}^{-1}$, respectively.

Closer to the seafloor, Schmidt and Siegel (2011) recorded near-bottom (within 2 m) current speeds of between 1 and $5 \, \text{cm s}^{-1}$, at 8350 m in the Puerto-Rico Trench. In the Philippine Trench, Hessler *et al.* (1978) reported rather variable maximum current speeds of $11.8 \, \text{cm s}^{-1}$ at 9605 m and an exceptionally high $31.7 \, \text{cm s}^{-1}$ at 9806 m. More recently, a new acoustic current meter was deployed six times over two expeditions, within 4 months of one another, at different depths in the Kermadec Trench, at ~32° S (unpublished data, HADEEP). Again, mean current speeds between 6000 and 9000 m were in the region of $1–5 \, \text{cm s}^{-1}$. These data are summarised in Table 4.1 and an example of the data obtained is shown in Figure 4.2.

The water flow driven by the thermohaline is also subject to tidal fluctuation. Tidal periodicity has been well documented at bathyal and abyssal depths (Gould and McKee, 1973; Magaard and McKee, 1973; Elliott and Thorpe, 1983). Lunar and semi-lunar cycles were detected in current flow over long periods in the Mariana Trench (Taira *et al.*, 2004) and in the pressure and current flow measurements obtained from the Kermadec Trench (unpublished data, HADEEP). Many marine organisms possess clock mechanisms that can track the undulations of the tidal cycle (Guennegan and Rannou, 1979; Naylor, 1985; Palmer and Williams, 1986; Wagner *et al.*, 2007). Furthermore, many marine species are also known to display cyclic behaviour that is synchronised to tidal cycles (Blaxter, 1978; 1980; MacDonald and Fraser, 1999; Pavlov *et al.*, 2000). While circadian clocks are principally governed by the light/dark cycle, tidal clocks

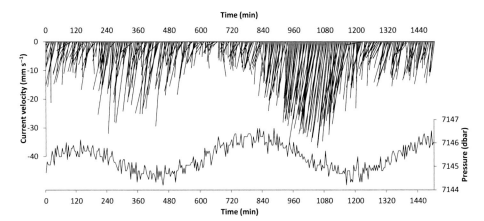

Figure 4.2 Example of near-bottom current speed and direction data (whisker plot, length denotes strength, angle denotes direction) and pressure data (modified from Jamieson *et al.*, 2013).

(Zeitgebers) are entrained by a suite of cycles associated with the tides, including current flow and hydrostatic pressure (Reid and Naylor, 1990). Tidal synchronisation suggests that marine organisms are capable of detecting often minute changes in hydrostatic pressure (Fraser, 2006). It has been demonstrated that small amplitude, slow cycles of hydrostatic pressure are readily detected by decapod crustaceans (Fraser and MacDonald, 1994; MacDonald and Fraser, 1999; Fraser, 2001; Fraser *et al.*, 2001) and fish (Fraser and Shelmerdine, 2002; Fraser *et al.*, 2003).

The presence of tidal (~12.4 h; Fig. 4.2), semi-lunar (~15 days) and lunar cycles (~28 days; Taira *et al.*, 2004), and presumably also daily (~24 h) and lunar-day (~24.8 h) cycles, suggests that chronobiology at hadal depths is likely, although not yet proven.

4.2   Temperature, salinity and oxygen

Temperature per se does not ultimately drive species zonation but it is definitely one of the fundamental abiotic factors involved (Danovaro *et al.*, 2004; Carney, 2005). Small temperature changes can influence the vertical or horizontal (latitudinal) distribution of species (Peck *et al.*, 2004; Brown and Thatje, 2011). Unlike pressure, temperature is not linear with depth and can vary at intra- and inter-trench levels. In the ocean, subsurface temperature generally decreases with depth but varies depending on latitude and region (Mantyla and Reid, 1983). At the surface, a mixed layer is commonly found, caused by friction from surface winds and additional turbulence. This is followed by a thermocline, where the temperature decreases rapidly before reaching the deep ocean (>1000 m). At bathyal (1000–3000 m) and abyssal depths (3000–6000 m), the rate of temperature change slows considerably and as a result 75% of deep water has a temperature of <4°C (Knauss, 1997). The bottom water temperatures in the hadal trenches are considered exceptionally stable in comparison with surface waters (Belyaev, 1989).

The temperature variation with depth is trench-specific, ranging from the warmest trench, the Cayman Trench, 4.46 to 4.49°C (6200–6900 m), to the coldest trench, South Sandwich Trench, −0.27 to −0.09°C (6047–7390 m). However, it is more typical for bottom temperatures to vary between 1 and 4°C, not including the South Sandwich Trench which represents the only sub-zero trench in the world and the Banda and Cayman Trenches which are generally warmer that the other trenches.

Hydrostatic pressure effects always act to increase the *in situ* temperature at a rate of ~0.16°C per 1000 dbar (~1000 m) increase in pressure and this is known as the adiabatic temperature gradient. Adiabatic heating is the result of a compressibility effect, whereby water molecules under mounting pressure increase their temperature without gaining heat from the ambient environment. This increase in temperature is generally masked by the decrease in temperature down through the thermocline. In terms of water masses and stratification, in general, the rate of temperature decrease with depth is reduced in the deep ocean. When the decrease in temperature with pressure in deep waters becomes equal to the adiabatic temperature increase, the minimum *in situ* temperature occurs. The minimum *in situ* temperature does not, therefore, occur at full ocean depth, but rather, it occurs in the abyssal zone (Fig. 4.3a).

In the southwest Pacific Ocean, the *in situ* temperature minimum is ~1.05°C at approximately 4881 dbar and 1.07°C at 4746 dbar overlying the Kermadec and Tonga Trenches, respectively. In the central Pacific Ocean overlying the Mariana Trench, the temperature minimum is 1.47°C at 4820 dbar. In the northwest Pacific Ocean the minimum temperature is 1.49°C at 4170 dbar and 1.44°C at 4017 dbar overlying the Japan and Izu-Bonin Trenches.

Below the minimum *in situ* temperature, the adiabatic increase in *in situ* temperature is larger than the decrease in temperature associated with the water mass stratification in the deep water, and this causes the *in situ* temperature to rise. Therefore, the bottom waters at the very bottom of the trench are actually slightly warmer than at the abyssal–hadal boundary, although the actual temperature range within individual trenches is low (between 1 and 2°C). Of the five trenches which exceed 10 000 m in depth, the total temperature range between 6000 and 10 000 m is approximately 0.85°C, with a general rise of 0.16°C per 1000 m depth (Table 4.2). This heating effect towards the deepest point of the trench results in the *in situ* temperature at full ocean depth equalling that of the bathyal zone and the coldest environments are found in the abyssal zone (Jamieson *et al.*, 2010; Fig. 4.3b). The temperature variation from the bathyal to hadal is low compared to the surface waters and seasonal variation thereof (Sanders, 1968). The rate of the increase in bottom temperature with depth is constant (0.16°C per 1000 dbar), but the actual *in situ* temperature differs depending on the locality of the trench and hydrological regime. For example, the *in situ* bottom temperature in the Peru–Chile Trench is slightly warmer than that of the Japan and Izu-Bonin Trenches, which in turn, are slightly warmer than the Kermadec and Tonga Trenches, although adiabatic heating trends are equal. Likewise, this trend is also visible in the potential temperatures from the three locations (Fig. 4.3c).

Table 4.2 The minimum and maximum bottom temperature of the five deepest trenches including temperature range and temperature rise per 1000 m.

Trench	Upper trench temperature (°C) at depth (m)	Lower trench temperature (°C) at depth (m)	Total temperature range (°C)	Temperature rise per 1000 m
Mariana	1.57 at 6000[a]	2.4 at 10910[a]	0.83	0.169
Tonga	1.18 at 6252[b]	1.91 at 10787[b]	0.73	0.161
Philippine	1.85 at 6000[a]	2.56 at 9864[a]	0.71	0.184
Kuril–Kamchatka	1.65 at 6000[a]	2.15 at 9000[a]	0.5	0.167
Kermadec	1.17 at 6000[c]	1.8 at 9856[b]	0.63	0.163

Data taken from: [a] Belyaev (1989), [b] Blankenship and Levin (2006) and [c] Jamieson *et al.* (2011a).

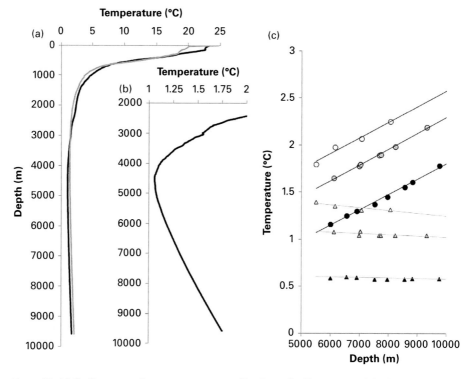

Figure 4.3 (a) Surface to seafloor temperature profiles from the Tonga Trench in the southwest Pacific Ocean (black line) and the Izu-Bonin Trench in the northwest Pacific Ocean (grey line). (b) Close-up of the lower Tonga Trench profile between 1 and 2°C clearly showing the temperature minima at abyssal depths and the adiabatic heating with depth thereafter. (c) The *in situ* bottom temperature with depth (circles) and potential temperature (triangles) for the Kermadec/Tonga Trenches in the southwest Pacific Ocean (black), the Japan/Izu-Bonin Trenches in the northwest Pacific Ocean (grey) and the Peru–Chile Trench in the southeast Pacific Ocean (unshaded). Note the baseline temperatures vary but the adiabatic heating rate with depth is the same. Data obtained by HADEEP.

The adiabatic heating effect can be theoretically removed by conversion of *in situ* to 'potential temperature' (denoted as θ) which determines the true characteristics of the water, thus permitting the tracking and comparison of large water masses by their temperature signal. In very deep regions, such as hadal trenches that may be filled with water from a single source with constant temperature and salinity at the top of the trench, the *in situ* temperature increases with pressure at the adiabatic gradient, but the potential temperature remains constant, reflecting the true temperature of the source water. Hydrostatic pressure increases with depth, therefore, in the hadal trenches (6000 to ~11 000 m), the difference between *in situ* temperature and potential temperature can be considerable. Polynomial algorithms for potential temperature were determined by Bryden (1973) and later developed by Fofonoff (1977) using a fourth order Runge–Katta integration algorithm to allow the smallest possible integration error when calculating the potential temperature. This integration error is less than 0.1×10^{-3}°C for $\Delta p = 10\,000$ dbar (Fofonoff and Millard, 1983).

Generally, the potential temperature decreases until it reaches an isothermal value (where $\Delta \theta = 0$) within the water column, resulting in thick layers of constant potential temperature. In the Kermadec and Tonga Trenches these layers can reach 1178 and 1811 dbar thick, respectively. Further north in the Japan and Izu-Bonin Trenches, layers can be up to 436 dbar and 514 dbar thick, respectively. The thickness of the layers increases with depth; ~80 dbar per 1000 dbar in the northwest Pacific and 500 dbar per 1000 dbar in the southeast Pacific trenches. These thick layers of potential temperature have hitherto gone unrealised as they occur in the trenches beyond the depth of conventional abyssal sampling and little to no layer of potential temperature can be observed <6500 m. The pressure where potential temperature becomes isothermal also increases with depth. By applying linear regressions, the pressure where the potential temperature layers begin increases by ~900 dbar per 1000 dbar depth in the northern Pacific trenches and ~600 dbar per 1000 dbar depth in the southern Pacific trenches.

In summary, the trenches support two temperature gradients that must be considered: the intra-trench adiabatic warming (hydrostatic pressure effect) and the overarching warming with the south to north flow of the thermohaline circulation.

At hadal depths, the *in situ* temperature increases as a result of hydrostatic pressure and the compressibility of the water. The potential temperature decreases, eventually culminating in thick layers of constant potential temperature and layers of homogeneous water masses flowing through the trenches. From these data and from previous studies (Kawabe *et al.*, 2003; Fujio and Yanagimoto, 2005), it appears that the potential temperature reaches an isothermal level at depths where pressures exceed 6500 dbar.

Salinity is an important physiological parameter to consider in shallower environments, particularly in coastal and inshore settings (Thistle, 2003). However, in the deep sea, salinity is relatively constant at ~35‰ and is, in fact, one of the most constant features of the deep-sea floor (Tyler, 1995). The exceptions are the Red Sea, Mediterranean Sea and some areas of the Gulf of Mexico. The salinity within the trenches remains similar to typical abyssal plain values (34.7‰) and is unaffected by pressure. In the Pacific Ocean, salinity decreases from the surface from ~35.6‰ to ~34.4‰ at ~1000 m

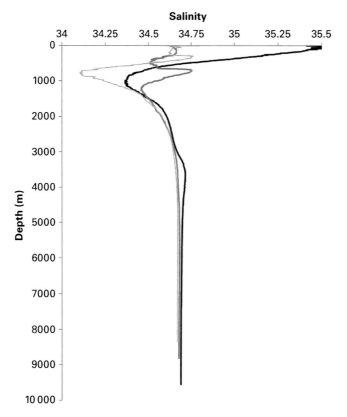

Figure 4.4 Surface to seafloor salinity profiles for the Kermadec Trench (southwest Pacific Ocean; black line), Peru–Chile Trench (southeast Pacific; dark grey line) and the Izu-Bonin Trench (northwest Pacific Ocean; light grey line). Note salinity becomes constant at ~3000 m. Data obtained by HADEEP.

and stabilises at a constant 34.7‰ beyond 3000–4000 m (Knauss, 1997; Fig. 4.4). The majority of hadal trenches are located in the Pacific Ocean and have salinities of typically 34.7‰ with only occasional fluctuations. The Banda Trench hosts the lowest salinity at 34.58–34.67 ‰ (Belyaev, 1989), while the highest salinity readings are found in the tropical Atlantic Ocean: Romanche Trench (34.67–34.96‰), Puerto-Rico Trench (34.80–34.89 ‰) and the Cayman Trench (34.99–35.00‰). Salinity profiles extending into hadal depths have been obtained from the Japan Trench (Fujio and Yanagimoto, 2005), Tonga Trench (Taft *et al.*, 1991), Mariana Trench (Taira *et al.*, 2005), Kermadec Trench (Jamieson *et al.*, 2011a), Izu-Bonin Trench (Taira, 2006) and the Peru–Chile Trench (unpublished data, HADEEP), and a constant salinity of 34.7‰ was found at all depths. Generally, salinity does not vary more than 0.2‰ and even within the aforementioned examples of high and low salinity, the variation does not exceed 0.42‰. Salinity is not, therefore, thought to have any ecological consequences, nor affect the dispersion of even the most stenohaline species (Belyaev, 1989; Tyler, 1995).

Most circulating bottom waters in the deep sea are known to be oxygenated (Tyler, 1995). The oxygen concentration is a result of the time since the parcel of water was at the surface and the reduction of oxygen thereafter by microbiota and benthic fauna. Therefore, deep waters with the lowest subsurface age (North Atlantic and Antarctica) have the highest concentrations, typically 6–7 ml l^{-1}. In areas with longer subsurface ages, such as the North Pacific Ocean, oxygen concentrations are lower, typically <3.6 ml l^{-1} (Mantyla and Reid, 1983).

Waters with the highest oxygen content at hadal depths are found in the South Sandwich Trench, Cayman and Puerto-Rico Trenches, where oxygen concentration measurements range from 4.9 to 6.9 ml l^{-1} (65–70% saturation; Belyaev, 1989). The South Pacific and Indian Ocean trenches (Kermadec, Tonga and Java) have slightly lower oxygen concentrations, at 4.0–4.7 ml l^{-1} (55–63% saturation). The lowest reported oxygen concentration of 2.03–2.38 ml l^{-1} (27–32% saturation) was in the Banda Trench. All the other trenches are characterised by intermediate values of oxygen, albeit there are very few measurements recorded in recent decades, therefore, most oxygen values are derived from samples from the 1950s and 1960s.

These data, summarised by Belyaev (1989), show fluctuations in different areas within a trench (and between trenches), seasonal variation and bathymetric variation. For example, in the Philippine Trench, the oxygen concentration was found to vary from 2.26 to 3.60 ml l^{-1} (30 to 47% saturation), based on 45 measurements (Belyaev, 1989). Other measurements of 2.36–4.32 ml l^{-1}, 2.99–3.92 ml l^{-1} and 3.07–4.42 ml l^{-1} were reported from the Kuril–Kamchatka, Aleutian and Mariana Trenches, respectively. In the Palau Trench, the RV *Vitjaz* 1957 expedition recorded an oxygen concentration of 3.66–3.71 ml l^{-1}, however, an RV *Spencer F. Baird* expedition measured a much lower oxygen concentration of 0.92–1.35 ml l^{-1}, just 5 years later. Belyeav (1989) states this example as the only incidence of such a severe drop in the oxygen measurements between expeditions. Nevertheless, even such a low oxygen concentration did not appear to affect the trench communities as numerous photographs of the bottom revealed an abundant and diverse benthic fauna (Lemche *et al.*, 1976). The same was found in the Banda Trench, despite exhibiting the lowest known oxygen levels of the trenches.

The important point regarding oxygen is that none of the trenches or areas within trenches has been found to be oxygen limited. Similarly, studies that model deep-water mass flow around the Pacific Ocean, in particular, have shown that bottom waters are not stagnant and, therefore, the water clearly circulates, allowing the trenches to ventilate (Johnson, 1998). Cold, surface-derived, bottom water is well oxygenated and the ventilation rates are sufficient to ensure that hadal fauna do not experience any significant stress from low oxygen levels (Angel, 1982). Trench water ventilation has also been suggested by Nozaki *et al.* (1998) who found that trench water masses were mixing relatively freely by isopycnal mixing with the bottom water overlying the northwest Pacific abyssal plain.

There are, however, areas in the deep sea where oxygen concentrations are low enough to be problematic for fauna (<0.2 ml l^{-1}). Examples of such areas include oxygen minimum zones, however, these are typically restricted to depths of <1500 m (Sanders,

1969; Wishner *et al.*, 1990) or localised, acute zones in the immediate exit of hydrothermal vents (Tyler, 1995), which have, to date, not been reported at hadal depths.

4.3   Adaptations to low temperatures

While hydrostatic pressure is known to determine the vertical distribution of species, temperature has long been regarded as the pivotal factor that drives their latitudinal distribution (reviewed in Pörtner, 2002). However, temperature is a major co-varying environmental factor, and is also considered to contribute significantly to vertical zonation patterns. Although the temperature experienced at hadal depths does not differ much from the shallower abyssal or lower bathyal zones, adaptation to these cold waters (typically <4°C; Thistle, 2003) is still an essential prerequisite for hadal fauna. The relationship between pressure and temperature on biological processes is, at best, difficult to unequivocally disentangle, however, there have been a variety of studies which partly provide new information concerning the incorporation of temperature effects on observed biological trends.

A decrease in temperature causes a decrease in chemical reaction rates and rates change by a factor of two to three for each 10°C temperature change (Carney, 2005). Successful adaptation to low-temperature and high-pressure habitats involves increased enzyme concentration, adoption of enzymes with greater efficacy and inclusion of modulator compounds that facilitate enzyme reactions (Somero, 1992; Samerotte *et al.*, 2007). Low temperatures are also known to exacerbate the effects of pressure on shallow-water invertebrate fauna (e.g. Young *et al.*, 1997; Villalobos *et al.*, 2006; Oliphant *et al.*, 2011) but, given the low variation in temperature across the hadal depths, its effects are still unknown but likely to be minimal.

Brown and Thatje (2011) investigated the physiological tolerances of fed and starved specimens of the deep-sea lysianassid amphipod *Stephonyx biscayensis* from depths of *c.* 1500 m, at varying temperature and pressure regimes, by measuring the rate of oxygen consumption. Acclimation to atmospheric pressure or starvation was found to have no significant interaction with temperature and/or pressure. Interestingly, the effect of hydrostatic pressure on respiration rate was found to be dependent on temperature: pressures of ~200 atm were tolerated at 1 and 3°C, 250 atm was tolerated at 5.5°C and at 10°C, 300 atm was tolerated. These variations in tolerance were found to be consistent with the natural distribution range for this species. Therefore, the pressure tolerance is not an effect of simply hydrostatic pressure itself, but is heavily influenced by the ambient temperature; higher temperatures can enable tolerance of higher pressure (Brown and Thatje, 2011).

To further understand the mechanisms that drive the distribution and abundance of trench fauna, the energetic demands of the individuals and populations must be considered. The flow of energy and materials in an ecosystem can be modelled based on metabolic rates, which are largely dependent on both ecological roles and the ambient environmental variables (Childress and Thuesen, 1992; Smith, 1992; Christiansen *et al.*, 2001; Smith *et al.*, 2001). It is often assumed that energetic demands of deep-sea animals can be extrapolated from the data modelled from shallow living analogues (e.g. Mahaut *et al.*, 1995), or derived from a few measurements of representative taxa (e.g. Smith, 1992; Smith *et al.*, 2001). Some marine animals such as fishes, crustaceans and cephalopods exhibit metabolic rates that are an order of magnitude lower in deep-sea species compared to their shallow-water counterparts (Drazen and Seibel, 2007; Seibel and Drazen, 2007). However, after the effects of temperature and body size are taken into consideration, there are some instances of no apparent difference in metabolism (e.g. some fishes, amphipods and crabs; Seibel and Drazen, 2007; Drazen *et al.*, 2011). It is thought unlikely that hydrostatic pressure can account for these data since enzymes that are adapted to high pressures can be inefficient (Somero and Siebenaller, 1979), leading to a lower metabolic rate.

The constant levels of enzymatic activity in the brains and hearts of fishes (Childress and Somero, 1979; Sullivan and Somero, 1980; Siebenaller *et al.*, 1982) confirms that capacity adaptations allow organisms to maintain a certain level of performance regardless of depth (Hochachka and Somero, 2002).

Some fishes and crustaceans have shown no changes in metabolic rate under varying pressure (Meek and Childress, 1973; Childress, 1977; Belman and Gordon, 1979), suggesting that the lower, routine metabolic rates of some deep-sea animals have evolved to match a low food supply (Childress, 1971; Smith and Hessler, 1974; Collins *et al.*, 1999; Treude *et al.*, 2002). Some deep-sea animals respond to low food resources by depressing routine metabolism (Sullivan and Smith, 1982; Christiansen and Diel-Christiansen, 1993; Yang and Somero, 1993) and the low metabolic rate in some abyssal amphipods has, in fact, been hypothesised to be a low food adaptation (Treude *et al.*, 2002).

Interestingly, *in situ* observations of fish (*Notoliparis kermadecensis*, Liparidae) swimming behaviour from *c.* 7000 m in the Kermadec Trench, where food supply is potentially greater than on the abyssal plain, suggest relatively high activity when compared to shallower-water fish of the same family (Jamieson *et al.*, 2009a). Likewise, the locomotive speeds of a hadal holothurian, *Elpidia atakama* (Elpidiidae) was measured at over 8000 m in the Peru–Chile Trench and found to be equal to or slightly faster than those of abyssal holothurians from ~4000 m (Jamieson *et al.*, 2011b). Other measurements of activity from hadal depths include those of the isopod *Rectisura* cf. *herculea* (Munnopsidae) from 6945 and 7703 m in the Japan Trench (Jamieson *et al.*, 2012b) and of two species of amphipod (*Princaxelia abyssalis* and *P. jamiesoni*; Pardaliscidae) from 7966 and 8798 m in the Tonga and Kermadec Trenches, and 7703 and 9316 m in the Japan and Izu-Bonin Trenches, respectively (Jamieson *et al.*, 2012a). Although comparative data from shallower depths is distinctly lacking, there were no observable reductions in activity, despite the extremely high pressure.

An alternative explanation is the Visual Interactions Hypothesis (VIH; Childress, 1995; Seibel and Drazen, 2007). In the surface waters, animals are able to detect both predators and prey at a great distance due to the brightly lit environment. Higher metabolic rates and greater locomotory capabilities for long chases or escapes are essential for catching prey and avoiding predators in such a well-lit habitat. In the deep sea where solar light no longer penetrates (Warrant and Lockct, 2004), predators and prey do not interact over such long distances or as frequently, relaxing the requirement of high locomotory capacity, which, in turn, reduces metabolism. Seibel and Drazen (2007) support the VIH, in that sighted taxa exhibit metabolic reductions with depth, but nonvisual groups such as holothurians do not. The VIH would predict that nonvisual hadal animals would have similar metabolic rates as their shallow living relatives (at the same temperature) and hadal animals relying on vision would have lower rates than similar animals in the photic zone but would be similar to bathyal and abyssal inhabitants which inhabit the same dark environment.

Adaptations at a metabolic level are therefore not unique to the hadal environment, or indeed to areas of high pressure, but are, in fact, simply a mechanism for survival beyond the photic zone, within the deep sea. Like many aspects of hadal ecology it is currently very difficult to disentangle the co-varying effects of food supply, temperature, light, pressure and oxygen on any observed trends. The trench ecosystem provides a unique opportunity in which food supply may actually show an inverse relationship with depth (increases towards the axis), while light levels, temperature and oxygen do not vary very much, yet hydrostatic pressure continues to increase across the deepest 45% of the water column.

4.4   Light

In clear oceanic water, solar light decreases by an order of magnitude every ~75 m, until ceasing to be detectable at ~1000 m (Denton, 1990). Beyond 1000 m the only visually relevant source of light is of biological origin, bioluminescence. Bioluminescence can range from bacterial glow, gelatinous luminescent organisms or larger invertebrates and fish with light producing organs or the ability to eject luminescent material (reviewed in Herring, 2002; Haddock *et al.*, 2010). Pelagic bioluminescence has been detected at abyssal–pelagic depths (Priede *et al.*, 2006a; Gillibrand *et al.*, 2007a) and deep benthic communities are known to exhibit spectacular bioluminescence displays on the abyssal seafloor in the vicinity of food-falls (Gillibrand *et al.*, 2007b).

Bioluminescence has, so far, not been detected at hadal depths. However, most marine planktonic organisms are known to produce light (Herring, 2002) and plankton have been recorded in the hado-pelagic zones of four major trenches to at least 8000 m (Bougainville, Kermadec, Kuril–Kamchatka and Mariana Trenches; Vinogradov, 1962), albeit in low numbers. Although bioluminescence is typically realised using artificial stimulation such as bait (Priede *et al.*, 2006a; Gillibrand *et al.*, 2007b) or mechanical stimuli (Craig *et al.*, 2009, 2011a) it is known to naturally occur with some frequency. Craig *et al.* (2011b) showed that the visual environment of the deep sea is perhaps as heterogenic as the physical substrata. Planktonic organisms in the near-bottom waters produce light when they naturally collide with physical structures, such as rocks or sessile organisms (e.g. stalked crinoids). All these elements are present within the hadal trenches; rocky outcrops are found particularly on the fore arc (Lemche *et al.*, 1976), plankton survives to at least 8000 m (Vinogradov, 1962) and stalked crinoids to at least 9000 m (Oji *et al.*, 2009). Therefore, although it has not been observed directly so far, there is potential for bioluminescence at full ocean depth and thus it is a potentially heterogenic, visual environment (Craig *et al.*, 2011b).

4.5 Substrata

The nature of the seafloor substratum has a profound effect on the composition of the benthic communities (Thistle, 2003); surface-dwelling fauna dominate hard substrata, while burrowing fauna dominate sedimentary substrata.

Hard substrata in the deep sea are typically found on the continental slopes, seamounts and mid-ocean ridges, however, the nature of tectonic subduction is such that the slopes, particularly on the fore arc, are steep (often exceeding 45°). These slopes herald large areas of hard, steep projections characterised by numerous sections with base rock exposed outcroppings (e.g. Oji *et al.*, 2009), littered with copious bedrock fragments and stones beneath (Bruun, 1956a; Belyaev, 1989). Such geological debris is common on the trench floors due to seismically induced rockslides and turbidity currents, and thus redeposition of sediments and rocks via removal is apparently a typical phenomenon in trenches (Wolff, 1960; Itou, 2000; Otosaka and Noriki, 2000; Rathburn *et al.*, 2009).

The typical deep seafloor substratum type is soft sediments. Generally, grain size decreases with depth but whether this trend extends to hadal depths is unknown. The carbonate compensation depth (CCD), the depth beyond which calcium carbonate dissolves, results in seafloor >4000–5000 m comprising siliceous ooze, whereas above the CCD, the sediments are primarily carbonate ooze (Tyler, 1995). The continental margins of the Pacific Ocean, situated above most major trenches, are comprised of soft, terrigenous sediment (Smith and Demopoulous, 2003). These sediments consist of mineral particles of continental origin, integrated with various planktonic components (e.g. diatoms, Foraminifera shells) and many other dust and particle types (Berger, 1974; Angel, 1982). It is these sediments that ultimately sink towards the continental flank of the trenches and their transport to the great depths are likely to be accelerated by the steep, often bedrock-exposed slope of the fore arc, as well as occasional seismic activity.

At the abyssal depths beneath productive waters, siliceous mud composed of diatom and radiolarian tests are typical, with organic-carbon contents between 0.25% and 0.5% (Smith and Demopoulous, 2003). In contrast, beneath the central gyres of the North and South Pacific Ocean, the sediments are composed of fine-grained clay particles of continental origin, transported by wind and volcanic eruptions (Berger, 1974). It is the sediments of the surrounding abyssal plains that ultimately descend into the trench via the oceanic slopes as the oceanic plate is subducted.

The distribution of soft sediments differs on inter- and intra-trenches levels depending on the nature of the relief, internal topography, proximity to continental land masses and seismic activity. Sedimentation rates using ion-thorium methods are estimated at 0.5–6.3 mm per 1000 year in the northwest Pacific trenches, with some estimates as high as 5–10 to 50–1000 mm per 100 year (e.g. Kuril–Kamchatka

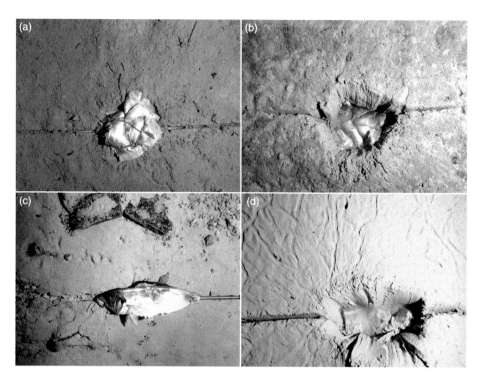

Figure 4.5 Examples of various substrata at hadal depths as photographed from the Hadal-Lander, where the field of view is 62×46.5 cm (0.29 m^{-2}). The metal bar holding the bait runs horizontally through the middle of the image and gives an indication of seafloor hardness. (a) Taken at 6173 m and shows little sinking into typical soft abyssal sediment; (b) is from 7050 m and shows some sinking into clay-like sediment littered with iron deposits; (c) is from 7561 m on the trench axis and shows a hard sediment surface with rock slabs in the top centre with stones and pebbles in the top right corner; (d) is from 9281 m and shows an extremely soft bottom littered with animal tracks, presumably made by holothurians. (a) and (b) are from the Peru–Chile Trench and (c) and (d) are from the Kermadec Trench. Images taken by Hadal-Lander B, courtesy of HADEEP.

Trench; Belyaev, 1989). These estimates are considerably greater than those for the surrounding abyssal plains (Jumars and Hessler, 1976).

The distribution of soft sedimentary substrata is thought to be driven by the physical topography of trenches. The steep slopes create a downward transport and subsequent accumulation of soft sediment along the trench axis (Danovaro *et al.*, 2003; Romankevich *et al.*, 2009), resulting in a heterogenic distribution of soft sediments within the trench that is different to that of the flat, neighbouring abyssal plains. It is therefore hypothesised that the trenches act as large sedimentation tanks that accumulate both surface-derived material and material transported from the surrounding abyssal plains (Jamieson *et al.*, 2010). The enclosed nature of trench topography, compared to other features such as plains, seamounts and submarine canyons, means that material that is transported into the trenches is physically captured at these depths, thus determining the high sedimentation rate within the trench environment.

The complex topography and the subsequent variation in sedimentation rate within areas of the same trench, coupled with the diverse types of substrate (from fine-grained silts to exposed rocky outcrops) potentially create diverse ecological niches that should promote the existence of diverse communities. Examples of different substrates are shown in Figure 4.5.

5 Hydrostatic pressure

Every life form on Earth experiences some degree of pressure, but none more so than those inhabiting the hadal zone. In the deep sea, pressure is a key physical parameter that has influenced the evolution and distribution of both microorganisms and macro-organisms (Bartlett, 2002). Pressure represents an absolute, continuous gradient from the sea surface to the deepest place on Earth, the Mariana Trench (~11 000 m) and it is one of the only parameters to maintain a linear relationship with depth. Adaptation to high hydrostatic pressure is an essential prerequisite for survival at hadal depths as this zone accounts for the deepest ~45% of the total depth range of the oceans (Jamieson et al., 2010). Hydrostatic pressure increases by 1 atmosphere (atm) every 10 m depth. Pressure is often expressed in Megapascal (MPa) or bar (bar) and all are relatively easy to covert; 1 atm = 1 bar = 0.1 MPa = ~10 m depth. The pressure found at hadal depths ranges from ~600 to 1100 atm (6000 m to 11 000 m).

The effects of hydrostatic pressure and survival in the deep sea have been reviewed several times (e.g. Somero, 1992; MacDonald, 1997; Pradillon and Gaill, 2007). It is not the intention of this chapter to review all aspects of high pressure and other adaptations, which could be another book in itself, but rather to highlight some of adaptations that are particularly relevant to or identified from hadal organisms.

5.1 Piezophiles

The discovery of piezophily (high-pressure adaptation) has been rooted in microbiology for over a century (Simonato et al., 2006). Arguably, its discovery came to fruition as a result of the hadal sampling undertaken on the *Galathea* expedition, most notably by Claude E. ZoBell and other former-Soviet contemporaries such as Anatolii Evseevich Kriss (Bartlett, 2009). It was ZoBell who obtained the first evidence of physiological adaptation to high hydrostatic pressure in trenches (from over 10 000 m in the Philippine Trench) and established the scientific discipline of high-pressure microbiology (ZoBell and Johnson, 1949; ZoBell, 1952). These pioneering experiments involved the rapid repressurisation of recently sampled deep-trench sediment back to its ambient pressure in laboratory-pressurised steel vessels, the fundamentals of which are mostly unchanged today (Bartlett, 2009). Although these early experiments laid the foundations for high-pressure biology research and provided evidence of piezophilic life, the first strains of piezophilic bacteria were not cultured until 1979 (Yayanos et al., 1982). This was soon followed by the discovery of a piezophilic microbial species that was extracted from an amphipod found in the Mariana Trench. This species could not be cultured at atmospheric pressure and thus was coined 'obligately piezophilic' (Yayanos et al., 1982) and assigned to a relatively new field of research that aimed to study organisms termed 'extremophiles' (MacElroy, 1974).

The definition of an extremophilic organism is one that thrives under extreme environmental conditions. It is mainly unicellular and prokaryotic organisms that are categorised under this definition, although not all extremophiles are unicellular, but most extremophiles are microorganisms (Horikoshi and Bull, 2011). The upper optimum growth temperature for Archea is 113°C, 95°C for bacteria and 62°C for single-celled eukaryotes. In contrast, multicellular eukaryotes rarely grow above 50°C.

Extremophiles that are specifically adapted to thrive at high hydrostatic pressure were originally termed 'barophiles' (*sensu* ZoBell and Johnson, 1949), however, in 1995 this name was replaced by the more appropriate term 'piezophiles' (Yayanos, 1995; Kato, 2011); 'baro' is derived from the Greek word meaning 'weight', while 'piezo' means 'pressure'. The terminology used in this research field comprises some of the following expressions: piezosensitive (not tolerant of high pressure), 'piezotolerant' (survival and growth at atmospheric pressure but can tolerate some degree of high pressure), piezophilic (survival and growth are optimised at high pressure) and obligatory piezophilic (requires high pressure for survival and growth). Other studies use definitions such as 'piezophilic' where optimal growth pressure exceeds 40 MPa, and 'moderately piezophilic' where optimal growth is above atmospheric but less than 40 MPa (Kato and Bartlett, 1997). The piezophilic relationship between optimal growth and depth of capture is illustrated in Table 5.1.

Many piezophilic bacteria have been recovered from hadal depths (reviewed in Eloe et al., 2011; Kato, 2011). The first pure culture of a piezophilic isolate was strain CNPT-3 which showed a rapid doubling rate at a pressure of 50 MPa but no colonies formed at atmospheric pressure, even after long incubations (Yayanos et al., 1979). It is now well known that many deep-sea heterotrophic bacteria grow optimally in the laboratory at pressures and temperatures close to those of their natural habitat (Yayanos et al., 1979). Thus, piezophilic bacteria are increasingly piezophilic with increasing depth of origin (Yayanos et al., 1982). The latter study also concluded that the pressure value which allows the maximal reproductive rate of the bacteria at 2°C to be achieved is always lower than the pressure at depth of capture. Many of the piezophilic strains are also psychrophilic ('psychropiezophilic'; both low-temperature and high-pressure adapted; Eloe et al., 2011) and cannot be cultured above 20°C (Kato and Qureshi, 1999). The effects of pressure and temperature on cell growth are similar, in that all strains become more piezophilic at higher temperatures (Kato et al., 1995a). This indicates that piezophilic isolates are obligately piezophilic above the temperature at which growth occurs at atmospheric pressure. Therefore, high pressure can extend the upper temperature limit for growth. Likewise, at lower temperatures (~2°C), piezophilic bacteria reproduce more rapidly when the pressure is less than that at its depth of capture (Kato and Qureshi, 1999). It is thought that the pressure at which the rate of reproduction at 2°C is maximal may indicate the true habitat depth of an isolate (Yayanos et al., 1982).

Table 5.1 Optimal growth pressure and temperature for various piezophilic, moderately piezophilic and piezotolerant bacterial strains. Modified from Kato and Bartlett (1997).

Bacterial strain	Optimal growth pressure [Temp.]	Depth of capture	Reference
Piezophilic bacteria			
DB5501	50 MPa [10°C]	2485 m	Kato *et al.*, 1995a
DB6101	50 MPa [10°C]	5110 m	Kato *et al.*, 1995a
DB6705	50 MPa [10°C]	6356 m	Kato *et al.*, 1995a
DB6906	50 MPa [10°C]	6269 m	Kato *et al.*, 1995a
DB172F	70 MPa [10°C]	6499 m	Kato *et al.*, 1995b
PT99	69 MPa [10°C]	8600 m	DeLong and Yayanos, 1986
Moderately piezophilic bacteria			
DSS12	30 MPa [8°C]	5110 m	Kato *et al.*, 1995a
S. benthica	30 MPa [4°C]	4575 m	MacDonell and Colwell, 1985
SC2A	20 MPa [20°C]	1957 m	Yayanos *et al.*, 1982
SS9	20 MPa [18°C]	2551 m	DeLong, 1986
DSJ4	10 MPa [10°C]	5110 m	Kato *et al.*, 1995a
Piezotolerant bacteria			
DSK1	0.1 MPa [10°C]	6356 m	Kato *et al.*, 1995a
DSK25	0.1 MPa [35°C]	6500 m	Kato *et al.*, 1995c
S. hanedai	0.1 MPa [14°C]	–	MacDonell and Colwell, 1985

5.2   Pressure and depth

The relationship between the rate of pressure change and depth of the ocean is linear. The relative rate of pressure change with depth is much higher in the shallower zones. For example, an organism descending from 500 to 1000 m will experience a pressure change of 101.2% but, in contrast, an organism moving from 9500 to 10 000 m experiences a pressure change of only 5.3%, although the absolute change is still ~500 dbar per 500 m (Fig. 5.1a). The same trend is seen with temperature and salinity and depth (Fig. 5.1b and c). For example, migration from 500 to 1000 m induces a 45% decrease in temperature, whereas any 500 m step at hadal depths equates to a decrease of only 5% or less. Similarly, although not as pronounced, a migration from 500 to 1000 m induces a salinity change of 9%, whereas beyond 4000 m, no such migration will result in a change in salinity.

These values for migration through the pressure gradient represent theoretical compression and decompression of an organism as it migrates vertically upwards and downwards through the water column. However, most current research into hadal fauna focuses on benthic fauna; animals associated with the seafloor. Therefore, vertical migrations do not necessarily account for any routine pressure changes that hadal benthic fauna experience. Hadal fauna presumably move around on the trench floor,

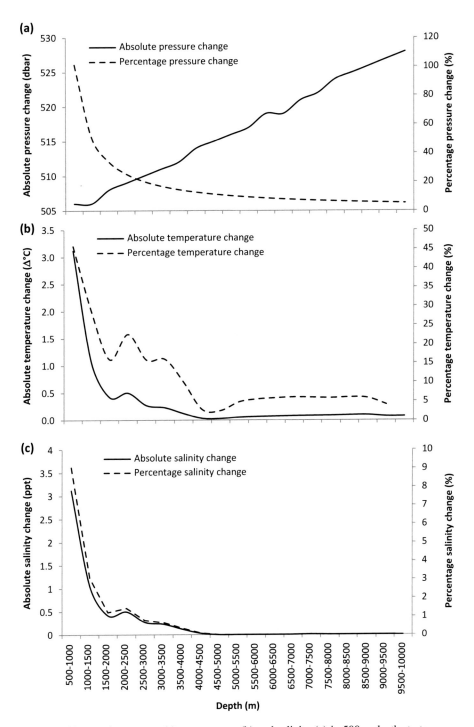

Figure 5.1 Changes in pressure (a), temperature (b) and salinity (c) in 500 m depth strata expressed in absolute values (solid lines) and percentage change per 500 m (dashed lines). Data taken from the Kermadec Trench.

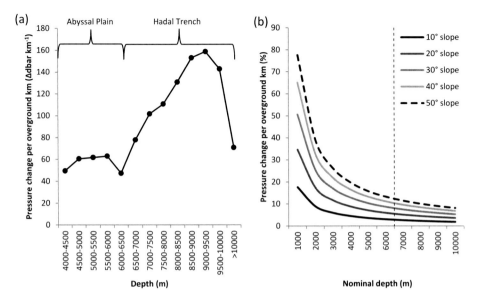

Figure 5.2 (a) Hydrostatic pressure change (%) experienced if theoretically traversing one kilometre across the seafloor perpendicular to the trench axis. Note increase in pressure change on the trench slopes relative to the abyssal plains. (b) Hydrostatic pressure change (%) if traversing one kilometre up or down a varying degree of slopes (10–50°) at different depths.

often traversing areas of extreme topography and many species are known to occur across the abyssal–hadal transition zone, which represents a flat-to-steep transition. Therefore, what are the spatial variations in hydrostatic pressure experienced by hadal fauna on a more natural timescale?

If an organism traverses the seafloor parallel to the trench axis (i.e. following a depth contour), then no significant change in pressure occurs. Likewise, if an organism traverses a flat and vast abyssal plain, the variation in pressure is also minimal. However, if an organism moves within the trench, perpendicular to the trench axis in either direction, then it experiences compression (if moving towards the axis) or decompression (if moving away from the axis).

Changes in hydrostatic pressure can be calculated per overground kilometre perpendicular to the trench axis, using the topography of the Kermadec Trench as an example (Fig. 5.2a). Across the abyssal plains (4000–6000 m), the pressure changes by 30–60 dbar per km. This change steadily increases with depth down the trench slopes to ~158 dbar per overground km at 9000–9500 m (almost four times that of the abyssal plains).

In reality, there are varying degrees of slopes within the trench and, presumably, organisms must traverse these often steep gradients. To demonstrate the effects of pressure changes when traversing a steep slope, the absolute and percentage change in pressure, calculated for one overground kilometre traversed up (or down) a 10, 20, 30, 40 and 50° slope at 1000 m depth intervals, are shown in Figure 5.2b. Regardless of depth, travelling this kilometre at each of the gradients will result in the same change in depth (174, 342, 500, 643 and 766 m, respectively) and approximately the same change

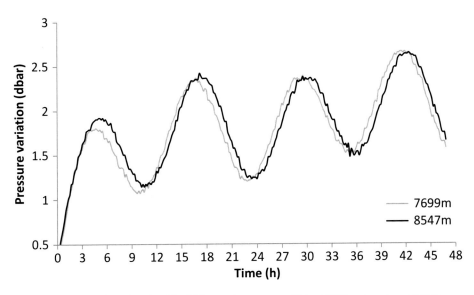

Figure 5.3 Pressure data showing M_2 tidal constituents over 48 h at 7699 m (grey) and 8547 m (black) in the Kermadec Trench.

in absolute pressure. However, the relative change in pressure at a given depth decreases dramatically. For example, travelling 1 km down a 10° and 50° slope at 1000 m will result in an 18 and 80% change in pressure, whereas beyond 6500 m, the relative pressure change is always less that 14%, regardless of the angle of the slope.

Irrespective of changes in depth or distance travelled overground, the ambient hydrostatic pressure is subject to tidal cycles as the sea surface rises and falls. Pressure data from 4329–8547 m from the Kermadec Trench, taken in both 2007 and 2009, show a combined mean tidal period of 12.42 h $\pm$ 0.64 S.D. (Fig. 5.3), suggesting the presence of an M_2 internal tidal period. The M_2 tidal period is one of the dominant semi-diurnal tides in the region and rotates anti-clockwise around New Zealand (Chiswell and Moore, 1999). It is also typical in bathyal and abyssal environments (Wagner *et al.*, 2007). The mean amplitude (peak to trough) of these pressure cycles was found to be 1.26 dbar $\pm$ 0.19 S.D., approximating to a 1 m 'swell'. These tidal cycles have been found as deep as 9900 m in the Kermadec Trench as well as in all other stations studied during the HADEEP project, irrespective of depth (Kermadec, Tonga, Izu-Bonin, Japan and Peru–Chile Trenches). In the North Pacific trenches the same signature of the cycle is seen as deep as 7700 m. It is therefore likely that hadal organisms may be able to detect slight tidal variations in pressure.

5.3   Carbonate compensation depth

The effects of high hydrostatic pressure can extend beyond the direct effects of pressure adaptation and toleration. It also has an effect on seawater chemistry, particularly at great depths. The oceans play an important primary role in the regulation and storage of carbon dioxide (CO_2) (e.g. Toggweiler *et al.*, 2006), through precipitation, burial and dissolution of carbonate sediment (Archer, 1996).

The oceanic absorption of atmospheric CO_2 in the last century has resulted in a decrease of the average pH of the surface oceans (Caldeira and Wickett, 2003) and a decrease in the carbonate ion concentration, leading to undersaturation of calcium carbonate ($CaCO_3$). Calcium carbonate becomes increasingly soluble at low temperatures and high hydrostatic pressures; the exact conditions that are found in the deep sea (Bostock *et al.*, 2011).

The carbonate compensation depth (CCD) is the depth, or more realistically, the gradual transition over which the calcium carbonate supply (calcite and aragonite) equals the rate of solvation, such that no solid $CaCO_3$ can accumulate (Pytkowicz, 1970; Takahashi and Broecker, 1977). As the CCD is not a sharply defined boundary, it is often defined as the depth at which the sediment $CaCO_3$ content decreases below 20% by weight due to dissolution (Broecker and Peng, 1982). The exact depth of the CCD varies between oceans and the largest regions therein, as a result of variations in temperature, circulation, $CaCO_3$ and organic carbon flux to the deep sea (Broecker and Peng, 1982; Archer, 1996; Feely *et al.*, 2004; Ridgwell and Zeebe, 2005). Generally, the CCD occurs within the abyssal depths between 3000 and 5000 m. It is therefore not an exclusively hadal feature, but rather, one of the wider deep sea. The CCD of the North Atlantic is ~5100–5200 m, in the South Atlantic it is ~4300–4400 m, in the Indian Ocean it is ~4500–4700 m, in the central Pacific Ocean it is 4500–4600 m and in the South Pacific Ocean it is 4000–4600 m (Bostock *et al.*, 2011).

In terms of deep-sea fauna, the CCD is significant because it has been proposed as a physiological barrier to deep-ocean colonisation that has promoted the adaptation of some taxa (Angel, 1982). Calcium carbonate is widely used as a structural component by foraminiferans, corals, crustaceans and molluscs. As carbonate solubility increases with increasing depth, ossification becomes harder and ultimately impossible to achieve (McClain *et al.*, 2004). This may explain why ossified groups (e.g. ophiuroids, echinoids) tend to be replaced by soft-bodied organisms (e.g. holothurians, soft and organic-walled foraminiferans) with increasing depth (Sabbatini *et al.*, 2002; Gooday *et al.*, 2004; Todo *et al.*, 2005).

5.4   Adaptations to high pressure

Many deep-sea fauna are constrained within species-specific depth limits (Vinogradova, 1997). Although, the relative role of ever-increasing pressure with depth versus other environmental correlates, such as food supply, isolation, hydrography and life history, is likely to be extremely complex. Nevertheless, hydrostatic pressure in the hadal zone represents the extreme end of one of the most significant environmental parameters in the ocean.

High-pressure adaptation is evident in many deep-sea organisms, from the lower eukaryotes (Simonato *et al.*, 2006), invertebrates and fishes (Kelly and Yancey, 1999), to deep diving marine mammals (Castellini *et al.*, 2001).

Pressure and temperature are two of the most fundamental abiotic factors (Danovaro *et al.*, 2004; Carney, 2005) and their effects on organismal physiology are complex because these variables are inversely related (Pradillon and Gaill, 2007). An increase in pressure at constant temperature leads to a compression and an ordering of molecules, whereas a loss of ordering occurs with an increase in temperature (Brehan *et al.*, 1992). Temperature and pressure also act slightly differently from each other, in that pressure effects only result from volume changes, whereas temperature affects both volume and energy. Furthermore, the increase in pressure with depth is linear, while temperature varies non-uniformly with depth and also varies depending on geographic location. In the polar regions, many shallower species inhabit depths with similar low temperatures to other deep-sea biozones, however, adaptation to low temperatures does not pre-adapt these organisms to function at high pressure (Somero, 1992). Given that most deep-sea species' physiological systems are improved under high pressures, they are piezophiles, or even obligate piezophiles, since they actually require high pressure for survival (Yayanos, 1986).

The high and low depth limits at which an organism can survive vary depending on the species, from tens to thousands of metres (Pradillon and Gaill, 2007). Furthermore, the maximum and minimum depth of a species may also vary depending on geographic location, ontogenic stage (Tyler and Young, 1998) and other environmental correlates such as food availability, oxygen and temperature (Tyler, 1995). The bathymetric distribution of marine organisms may be explained, in part, by interspecific variation in pressure resistance. High pressures may limit the maximum depth to which shallow-water organisms can extend, and conversely may limit the minimum depth into which the barophilic, deep-water organisms can penetrate (Somero, 1992).

Hydrostatic pressure can have several perturbing effects on biological structures and processes following Le Chatelier's principle which states that at equilibrium, a system tends to minimise the effects of disturbing external factors (Pradillon and Gaill, 2007). This essentially means that a reduction in a system's volume is favoured during an increase in pressure and vice versa. Therefore, the fundamental effect of pressure on a biological system is the change in volume that accompanies a process, whether this is compression or decompression (Somero, 1992). For example, when a biological process requires an increase in volume, it is inhibited by pressure and, conversely, when a decrease in volume is required, the pressure augments the process.

Pressure can affect biological processes in many ways and in response a number of adaptations have evolved to counter these effects in the deep sea. For example, there are pronounced pressure adaptations in the neural functions of deep-sea invertebrates (Campenot, 1975) and fishes (Harper *et al.*, 1987). Studies have shown that subjecting shallow-living fishes to pressures of up to 409 atm causes a 50% decrease in the peak amplitude of the compound action potential (c.a.p.) of the vagus nerve, an effect that is not detected in deep-sea fishes from 4000–4200 m (*Coryphaenoides armatus* and *Bathysaurus mollis*; Harper *et al.*, 1987). Similar differences between shallow- and deep-water fishes have been documented in relation to depolarisation rates (Harper *et al.*, 1987), where the apparent activation volumes for the rate of depolarisation is higher in shallow-water species than deep-sea species. Furthermore, the nerves of some deep-sea fish can be restored to an apparent normal function after recompression to the species' ambient pressures (Harper *et al.*, 1987). Likewise, the heart function of deep-sea fish can also be partially restored if pressurised to the species' ambient pressure (Pennec *et al.*, 1988). On a microbial level, high pressure has been found to affect the growth and viability of bacteria. Research is currently focusing on the isolation of pressure-resistant mutants, regulation of gene expression by high pressure, the role of membrane lipids and proteins in determining growth ability at high pressure, pressure effects on DNA replication, topology and cell division and the modulation of enzyme activity at high pressure (reviewed in Bartlett, 2002 and Simonato *et al.*, 2006).

5.4.1   Effects and tolerance of pressure

The vertical (bathymetric) zonation of marine species implies that each has adapted to a specific pressure range in which the species has evolved or adapted to function most efficiently. To fully understand the perturbing effects of hydrostatic pressure on an organism, simple experiments have demonstrated the effects of exposing organisms to pressures that normally lie outside their typical environment, whether higher or lower than normal. Although the susceptibility of shallow-living species to high pressures was demonstrated in the 1870s, it was not until the 1970s that experiments were conducted to ascertain the effects of higher and lower pressure on deep-sea organisms (reviewed in MacDonald, 1997).

  MacDonald (1997) gives a detailed account of what happens to shallow-water animals when they are exposed to ever-increasing pressure. He states that organisms respond to this perturbation through a series of changes in their motor activity. During the first few atmospheres of compression there is an initial increase in normal activity, followed by a period of impaired coordination and otherwise normal behaviour. As the pressure increases further, excitability increases culminating in spasms or convulsions at 100 atm. In decapods, tail-flip escape responses are elicited early on and at higher pressures they merge into similar, but more violent, convulsive responses. In amphipods, relatively slow dorsally directed spasms of the longitudinal musculature are elicited at high pressure. At higher pressure still, a progressive immobilisation sets in, albeit initially reversible, but it eventually culminates in mortality. Some deep-water species, however, have been shown to exhibit no hyperexcitability when exposed to pressure higher than at their depth of capture (MacDonald, 1997). MacDonald and Gilchrist (1982) recovered several species of amphipods in a pressure-retaining trap at 394–442 atm and pressurised the specimens to 700 atm. They noted that the amphipods did not convulse at the higher pressure, although they did exhibit mild hyperexcitability. The failure to convulse at high pressure means that the specimens from 4000 m differed radically in their pressure tolerance from those captured at the shallower depth of 2700 m. These studies suggest that for some amphipod species at least, the deeper the normal range of the species, the greater their tolerance to higher pressures.

  Conversely, the effects of decompression on deep-sea organisms are not straightforward. Due to the ease of capture in pressure-retaining traps, amphipods have been the test species of most of these types of study (e.g. Yayanos, 1978, 1981; MacDonald and

Gilchrist, 1980, 1982). During decompression, amphipods are considered to be 'relatively hardy in this respect' and the extent to which deep-sea amphipods can tolerate decompression is species-specific (MacDonald, 1997). Some amphipods can be recovered from moderate depths without any drastic impairment of the locomotor activity (Brown and Thatje, 2011). In some instances, they have also been shown to exhibit no hyperexcitability during a step-wise change in pressure. Many crustaceans appear dead after collection from the deep sea but can be reactivated if restored to their ambient pressure, although pressure resuscitation may take several minutes to hours (MacDonald, 1997). Likewise, the chances of observable activity from deep-sea organisms on the surface are increased if they are kept at their ambient low temperature (Truede *et al.*, 2002). Many amphipod species, such as *Eurythenes gryllus* and *Paralicella caperesca*, have been readily resuscitated by recompression following decompression paralysis, from as deep as 4000 and 5900 m (MacDonald and Gilchrist, 1980; Yayanos, 1981). The *P. caperesca* from 5900 m at 2°C were recovered at ambient pressure (~600 atm) and decompressed to atmospheric pressure. Loss of locomotor activity occurred at 215 atm but was regained following recompression, suggesting that this species is capable of 3000 m of vertical migration (Yayanos, 1981).

By maintaining the cold ambient temperature in thermally insulated baited traps, Truede *et al.* (2002) recovered hundreds of amphipod specimens from ~4400 m. *Abyssorchemene abyssorum* and specimens of the genus *Paralicella* from 4400 m only survived decompressions of up to 300 atm and were, therefore, considered stenobathic (Truede *et al.*, 2002). However, the same study showed that A*byssorchemene distinctus* and *Eurythenes gryllus* exhibited a high decompression tolerance since both species were recovered from abyssal depths without any detectable problems. They were subsequently classified as eurybathic. This is not surprising given that *E. gryllus* is known for its pronounced vertical migration behaviour (Ingram and Hessler, 1987; Christiansen *et al.*, 1990) and can tolerate pressure changes between 1 and at least 526 atm (George, 1979). Similarly, the high decompression tolerance of *A. distinctus* may also be related to large vertical migrations since it has been recovered from several mid-water trawls at ~2500 m above bottom (Thurston, 1990).

Experiments using amphipods from the Mariana and Philippine Trenches (from approximately 10 000 m) collected in pressure-retaining traps (Yayanos, 2009) showed that they were capable of a significant vertical migration (i.e. decompression) to as shallow as 3800 m. This decompression tolerance was greater than had been shown for species living at 5800 m. Although these experiments demonstrated that amphipods in particular herald impressive decompression tolerances, all other samples recovered from hadal depths died upon reaching atmospheric pressure and this was verified by a lack of observable activity after recompression (Yayanos, 2009). A range of comparative pressure tolerance data is shown in Table 5.2.

The pressure tolerance of the Amphipoda is perhaps demonstrated well by analysis of each amphipod species' bathymetric range compared with other full ocean depth taxa (Bivalvia, Gastropoda, Polychaeta, Holothuroidea and Isopoda). The average bathymetric range of trench amphipods, excluding species with ranges of <100 m to eliminate single finds and rarities, is 1562 m compared to the other taxons which range from

Table 5.2 Overview of the pressure tolerances of deep-sea amphipods in both compression and decompression.

Species	Capture depth (m)	Absolute pressure tolerance (atm)	Tolerance range (atm)	Reference
Compression				
Paralicella caperesca, *Orchomene* sp., plus others	4000	400–700	300[a]	MacDonald and Gilchrist, 1982
Decompression				
Abyssorchemene abyssorum, Paralicella spp.	3950–4420	442–140[b]	302	Truede *et al.*, 2002
Paralicella caperesca, *Eurythenes gryllus*	4000–4300	400[c]	400	MacDonald and Gilchrist, 1980
Abyssorchemene distinctus, *Eurythenes gryllus*	3950–4420	442–0[b]	442	Truede *et al.*, 2002
Eurythenes gryllus	5260	526–1	525	George, 1979
Paralicella caperesca	5900	601–0 (215[d])	601 (386)	Yayanos, 1981
Cf. *Hirondellea* sp.	10 000	1000–380	620	Yayanos, 2009

[a] This increase of 200 atm was when progressive inhibition was starting and not necessarily mortality.
[b] The success of live capture was aided by thermally insulated traps.
[c] 50% of the 155 individuals were resuscitated following recompression to 400 atm.
[d] Loss of locomotor activity occurred at 215 atm, but individuals were resuscitated following recompression.

730 to 852 m, therefore almost twice the range of the other full ocean depth groups (Fig. 5.4). Furthermore, these high values are consistent across the species with a far greater percentage of species with bathymetric ranges of 2000 m or more (Table 5.3).

Pressure tolerances have also been investigated in bacteria, where mortality was determined by the loss of colony-forming ability (CFA; Yayanos and Dietz, 1983). These studies demonstrated that a decompression sensitive (obligately piezophilic) bacterium from 10 476 m in the Mariana Trench (strain MT-41) lost CFA after 150 h, following decompression from 103.5 MPa to 0.101 MPa (atmospheric pressure) (Chastain and Yayanos, 1991). They found severe changes in cell shape and ultimately cell death when this pychropiezophilic bacterium was decompressed, but death did not occur immediately. This is interesting, as decompression mortality has not been observed in abyssal bacteria from 5900 m in the Pacific Ocean (Yayanos *et al.*, 1982). These abyssal bacteria are piezophilic, not obligately piezophilic and can be retrieved without pressure-retaining equipment, recompressed to high pressure and grown in the laboratory for five months (Yayanos *et al.*, 1979).

Conversely, the exposure of a mesophilic bacterium, *Escherichia coli*, to elevated pressures results in the induction of 55 proteins, including 11 heat-shock proteins and four cold-shock proteins (Welch *et al.*, 1993). Pressure appears to be the only stressor known to induce both heat-shock and cold-shock proteins simultaneously, as they are otherwise induced following exposure to opposing thermal regimes (Bartlett, 2002).

Table 5.3 The bathymetric ranges of six full ocean depth-dwelling taxons, expressed as the percentage of the total number of species in each 1000 m range. Only species with range >100 m were used to omit single findings and rarities.

		Amphipoda	Isopoda	Bivalvia	Gastropoda	Holothuroidea	Polychaeta
Total number of species		69	76	75	42	76	99
Percentage of total number of species within each bathymetric range (m)	0–1000	43	72	72	74	72	67
	1000–2000	23	20	15	14	18	25
	2000–3000	19	7	8	7	5	4
	3000–4000	10	1	4	2	1	1
	4000–5000	4	0	0	0	1	0
	5000–6000	0	0	0	0	1	0

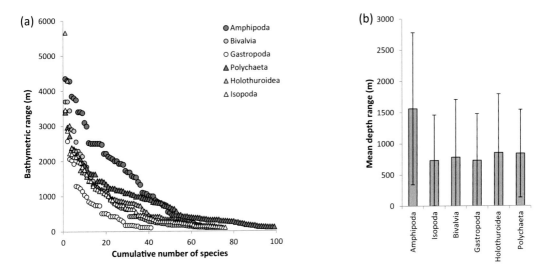

Figure 5.4 Bathymetric ranges of six full ocean depth groups showing that (a) the Amphipoda have consistently larger bathymetric ranges than the other groups and (b) the average is almost twice that of the others.

Heat-shock proteins or similar stress proteins can also be induced in piezophiles, such as the bacterium *Thermoccus barophilus*, during decompression (Marteinsson *et al.*, 1999). High pressure and low temperature exert similar effects on protein synthesis and membrane structure, and, therefore, the simultaneous induction of both heat- and cold-shock proteins may be an attempt to counteract the perturbing effects of high pressure on membrane integrity, translation processes and the stability of macromolecules (Bartlett, 2002).

Conspicuous examples of adaptation by organisms to high pressure in the deep sea include (1) the increased use of unsaturated fatty acids in cell membrane phospholipids in order to maintain their fluidity and cellular function (Hazel and Williams, 1990) and (2) the use of intracellular protein-stabilising osmolytes, such as trimethylamine

N-oxide (TMAO; Samerotte *et al.*, 2007), that act to maintain enzyme function by increasing cell volume to counteract the adverse effects of pressure.

5.4.2   Lipids

Another adaptation in some marine organisms is the accumulation of metabolic energy stores in the form of lipid reserves (Lee *et al.*, 2006) that comprise four major classes: triacylglycerols, wax esters, phospholipids and diacylglycerol ethers. All these groups contain long-chain fatty acids but differ in the ways in which the fatty acids are bound together. All classes may be found within an organism, but can vary in composition between taxa, for example, wax esters are the predominant storage lipid in many deep-living and cold water crustaceans (Bühring and Christiansen, 2001; Lee *et al.*, 2006).

Lipids provide solutions to several biological problems in the deep sea, one of which is membrane fluidity regulation at high pressure. Lipids are particularly sensitive to hydrostatic pressure (Bartlett, 2002) and are, on average, an order of magnitude more compressible than proteins (Weber and Drickamer, 1999).

Hydrostatic pressure, as well as temperature, exerts an evolutionary force on the composition of biological membranes, reducing their fluidity by increasing the packing of fatty acyl chains (Simonato *et al.*, 2006). Healthy membrane functioning is dependent upon their maintenance in a liquid crystal-like state that allows the appropriate movement of enzymes and transmembrane proteins. In a temperature–pressure-phase diagram, for many membranes, the gradient that represents the transition of gel to a liquid crystalline state increases about 20°C for every 100 MPa at pressures <100 MPa. Therefore, the combined effect of temperature and pressure on the membrane phase state of a deep-sea organism at full ocean depth (100 MPa and 2°C) is similar to a membrane at atmospheric pressure and at a temperature of −18°C (MacDonald, 1984a; Cossins and MacDonald, 1989; Bartlett, 2002).

The phospholipid bilayers of biological membranes have been the focus of many pressure effect studies (e.g. Wann and MacDonald, 1980; Somero, 1992; MacDonald, 1997), since cell membrane-based processes are, perhaps, the most susceptible to perturbation by high pressure and consequently exhibit extreme sensitivities in shallow-water organisms (MacDonald, 1984; MacDonald and Cossins, 1985). Decreasing temperature and increasing pressure will reduce membrane fluidity in shallow-living organisms (Brehan *et al.*, 1992). Modifications in membrane proteins, or the phospholipid bilayer (or indeed both) could prompt pressure adaptation in membrane-based processes. Bilayer fluidity offers a potential for adaptive regulation in deep-sea organisms as a result of strong interacting effects of high pressure and low temperature on the properties of lipids. Deep-sea organisms' membranes exhibit changes in lipid composition with an increase in the fraction of unsaturated fatty acids (UFAs). These UFAs permit the maintenance of fluidity under pressure (maintained within a narrow range of viscosity), a process sometimes referred to as 'homeoviscous adaptation' (Cossins and MacDonald, 1984; DeLong and Yayanos, 1985; Fang *et al.*, 2000). This is well illustrated by comparison with butter: a high to low temperature change causes otherwise soft butter to harden, whereas if the unsaturated fatty acid content is increased, it can remain soft at low temperatures (like spreadable butter; Herring, 2002).

The phospholipid bilayer in a deep-sea organism found at 2–4°C, at a depth of 4000 m, has an effective temperature of −3 to −6°C, while the effective temperature for a 2°C membrane in hadal organisms from the Mariana Trench is −11 to −19°C (MacDonald and Cossins, 1985), suggesting that phospholipid bilayers of deep-sea species differ from those of shallow-living species. Deep-sea species have an intrinsically higher degree of fluidity, such that under high hydrostatic pressure and low temperature, membrane fluidity is conserved within the optimal range for function. Therefore, homeoviscous adaptation may be an imperative component of adaptation to high pressure (MacDonald and Cossins, 1985; Cossins and MacDonald, 1989). Such pressure acclimations are also of particular importance in organisms with vertical ontogenetic structures spanning large bathymetric ranges (e.g. hadal amphipods; Yayanos, 1978, 2009; MacDonald and Gilchrist, 1980).

The fatty acid composition of piezophilic bacterial strains changes as a function of pressure (Kato, 2011). Generally, a greater amount of polyunsaturated fatty acids (PUFAs) are synthesised at higher pressure conditions for their growth. A piezophilic bacterium (strain CNPT3, probably of the genus *Vibrio*) cultured at pressures of 1, 172, 345, 517 and 690 atm (at 2°C) showed that the ratio of total unsaturated fatty acids (TUFAs) to total saturated fatty acids (TSFAs) increased with increasing pressure (DeLong and Yayanos, 1985), a trend found in other piezophilic bacteria (DeLong and Yayanos, 1986). These studies revealed that physiological acclimation to high pressure by a deep-living organism are possible, although homeoviscous adaptation has not been found in all studies of pressure acclimation. For example, contrary to the predictions of homeoviscous theory, MacDonald (1984b) did not detect homeoviscous adaptation in *Tetrahymena* cells that were grown at different pressures and found that the fatty acid changes observed were opposite from those expected. This suggested that the enzymatic and genetic mechanisms important in regulating membrane fluidity in shallow-water species during temperature changes are not pre-adapted to respond suitably to increasing pressure (Somero, 1992).

Psychrophilic and piezophilic bacteria are believed to produce one of the long-chain PUFAs, either eicosapentaenoic acid (EPA) or docosahexaenoic acid (DHA), but this does not appear to be obligatory (Kato, 2011). Of the bacteria strains found at hadal depths, the piezophilic *Shewanella* strains produce EPA (Nogi *et al.*, 1998b), *Moritella* strains produce DHA (Nogi *et al.*, 1998a; Nogi and Kato, 1999) and *Psychromonas kaikoae* produces both EPA and DHA (Nogi *et al.*, 2002). Yano *et al.* (1998) examined the lipid composition of barophilic bacterial strains that contained docosahexaenoic acid (DHA) and their adaptations in response to growth pressure. They also concluded that the general shift from saturated to unsaturated fatty acids is one of the adaptive changes in response to increasing hydrostatic pressure and suggested that DHA may also play an important role in maintaining optimal membrane lipid fluidity at high pressure.

5.4.3 Piezolytes

High hydrostatic pressure can have large perturbing effects on biological molecules. Membranes and proteins are known to have structural adaptations that provide pressure resistance (Hochachka and Somero, 1984). In recent years, another adaptation to pressure has been suggested involving 'piezolytes' (*sensu* Martin *et al.*, 2002); small, organic solutes first discovered as organic osmolytes. Solutes that are accumulated by most marine organisms prevent osmotic shrinkage of their cells by osmoconforming to the ambient environmental conditions (osmotic pressure of about 1000 mOsm). Among others present in marine organisms, one of the main osmolytes is trimethylamine oxide (TMAO; the source of the 'fishy smelling' trimethylamine) and this is found in deep-sea fish. Most marine Osteichthyes (bony fish) are osmoregulators, which means that they maintain a relatively high internal osmotic pressure of about 300–400 mOsm compared to shallower bony fish (~40–50 mOsm).

Organic solutes such as TMAO are thought to be selected as osmolytes over inorganic solutes, since the latter can perturb macromolecules while the former usually do not; i.e. they are not only compatible with cellular functions (Brown and Simpson, 1972) but can both stabilise macromolecules and counteract perturbants such as hydrostatic pressure (Yancey, 2005). Laboratory studies have shown that TMAO counteracts the perturbing effects of hydrostatic pressure on enzyme kinetics and protein stability and assembly (Yancey and Siebenaller, 1999; Yancey *et al.*, 2001, 2004; Yancey, 2005).

Analyses of TMAO in deep-sea bony fishes, elasmobranchs (sharks and rays) and decapods (shrimp and crabs) have shown an increase in this osmolyte with depth (Kelly and Yancey, 1999; Samerotte *et al.*, 2007). In bony fish (teleosts), internal osmotic pressure increases with depth as a result of increasing TMAO content. In osmoconformers, on the other hand, other osmolytes decrease in osmotic compensation as TMAO increases with hydrostatic pressure, e.g. urea in elasmobranchs and glycine in decapods (Kelly and Yancey, 1999).

There are currently no TMAO content data for any major hadal fauna, except for one species of fish, *Notoliparis kermadecensis* (Lipardiae; Nielson, 1964). TMAO content analyses of other deep-sea fishes revealed a linear relationship with depth, down to 4900 m (Gillett *et al.*, 1997; Kelly and Yancey, 1999; Samerotte *et al.*, 2007). Extrapolation of these data suggested that fishes would become isosmotic with seawater at about 8000–8500 m, roughly the depth of the deepest fishes observed or captured (Nielsen, 1977; Jamieson *et al.*, 2009a; Fujii *et al.*, 2010). To test this hypothesis, muscle, plasma and gel tissues were sampled from the *Notoliparis kermadecensis* specimens recovered by the *Latis* fish trap from 7000 m in the Kermadec Trench and analysed for TMAO content (Yancey *et al.*, in press). The results showed that the liparids had values of 386 ± 18 mmol kg^{-1}, which is in keeping with the extrapolation from the shallower dataset (teleosts between 0 and 4850 m are between 40 mmol kg^{-1} and 261 mmol kg^{-1}). Furthermore, the osmolality of the hadal samples were 991 ± 22 mosmol kg^{-1}, which again sits very close to the extrapolation of isosmosis (1100 mosmol kg^{-1}) at approximately 8200 m (Fig. 5.5). Therefore, the Yancey *et al.*

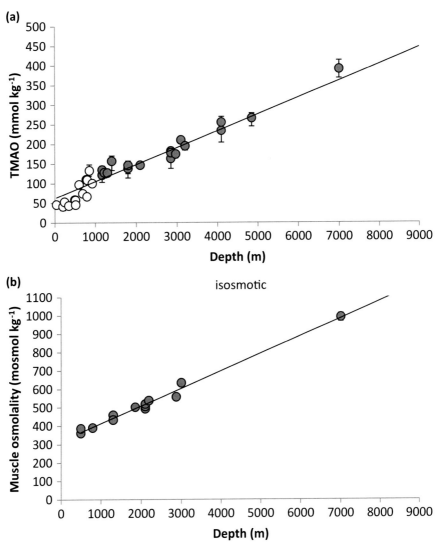

Figure 5.5 (a) Muscle TMAO contents (mmol kg^{-1} wet mass) versus depth for teleosts. Open circles represent shallow-water fishes, and solid dots show deep-sea fish with linear fit (TMAO $= 62.1 + 0.429 \times$ depth, p < 0.001). (b). Osmolarities (mosmol kg^{-1}) of muscle fluid versus depth for teleosts. The linear fit extrapolates to isosmosis at ~8200 m (mosmol kg$^{-1} = 320 + 0.0953 \times$ depth, p < 0.001). Modified from Yancey *et al.* (in press).

(in press) study supports the hypothesis that TMAO accumulation defines the depth limit for teleosts and is the first study to demonstrate that hydrostatic pressure can biochemically constrain an entire taxonomic group with the diversity and complexity of the teleosts.

Similarly, it has also been shown that Chondrichthyes (sharks, rays and chimaeras) are limited to bathyal depths, as predicted by linear regression of global database records for fish (Priede *et al.*, 2006b). This study proposed that chondrichthyans are excluded from abyssal and hadal depths as a result of their high-energy demands. One such demand is their oil-rich liver that is essential for buoyancy and which cannot be sustained in extreme oligotrophic conditions. However, Laxson *et al.* (2011) analysed the major organic osmolytes of 13 chondrichthyan species, caught between 50 and 2850 m. While their urea concentrations declined with depth, TMAO concentrations increased from 85–168 mmol kg^{-1} in the shallowest group, to 250–289 mmol kg^{-1} in the deeper group and indicated a plateau at the greatest depths, suggesting that the deepest chondrichthyans may be unable to accumulate sufficient TMAO to counteract the perturbing effects of hydrostatic pressure at upper abyssal depths (3000–4000 m; Laxson *et al.*, 2011). Whilst Chondrichthyes are not present at hadal depths, the findings provide further support for the TMAO hypothesis that fish are physiologically restrained by the high pressures found at depths of 8000–8500 m.

With further sampling across other taxa at hadal depths, the apparent inability to penetrate the trenches by many of the organisms may be accounted for by TMAO or other osmolyte concentrations. One such example are the Decapoda. In these organisms, extrapolations of the TMAO concentration data (Kelly and Yancey, 1999) and observational data (Jamieson *et al.*, 2009a, 2011a; Fujii *et al.*, 2010) of shallower-water specimens indicated that they share a similar bathymetric range to that of the fish; limited to depths of less than 8000–8500 m.

Although the effects of piezolytes, such as TMAO, may explain the depth limit for teleosts, many other hadal taxa exhibit no such depth limitations and in fact thrive at full ocean depth (e.g. bivalves, gastropods, holthurians), some maintaining extraordinary bathymetric ranges (e.g. amphipods). Exactly how these animals survive at higher pressures is still unknown, but through examining the upper end of the pressure gradient, such as that experienced in the trenches, future investigations into depth zonation and pressure adaptation should provide great insight.

6 Food supply to the trenches

In addition to hydrological and geological factors, food supply has an incredibly strong spatial and temporal influence on species composition and abundance (Gage, 2003). The deep sea, including the hadal zone, is essentially a heterotrophic environment where most of the available food descends from the ocean surface waters (Tyler, 1995). Surface-derived food (allochthonous) ranges in size, from particulate organic matter (POM), the carcasses of gelatinous organisms, fish and even dead whales (e.g. Britton and Morton, 1994; Beaulieu, 2002; Smith and Baco, 2003; Robison *et al.*, 2005, respectively) to terrestrial plant and wood debris (Turner, 1973; Wolff, 1976). Supplementary energy is available, however, in the form of autochthonous production from the hydrothermal vents and cold seeps found in the deep sea (Childress and Fisher, 1992).

Although the food sources and types of food found in the hadal zone are very similar to those of the general deep sea, 'hadal' food differs in that the distance travelled by surface-derived nutrition to hadal depths is formidable and, as a result, both the quantity and quality of the food is likely to be reduced. Furthermore, the trenches are of a unique topographical setting where food distribution is different to that of the flat neighbouring abyssal plains. Moreover, when coupled with the periodic seismic-induced sediment slides, the food distribution factor is central to the creation of this unique setting in which the deepest of deep-sea organisms survive. The other key factor to point out when considering food supply to the trenches is that, due to isolation, there is little resource gradient between different trenches; they are, after all, a disjunct collection of deep trenches with often very different biogeographical settings. Thus, the food availability, in terms of both type and quantity, may differ from one trench to another.

6.1 Particulate organic matter (POM)

POM comprises a vast range of sizes and types, ranging from phytodetritus (Rice *et al.*, 1986), mucilaginous aggregates (Martin and Miquel, 2010), larvacean houses (Robison *et al.*, 2005) and faecal pellets (Turner, 2002). In regions close to continental margins (which includes most trenches), increasing amounts of particulate and dissolved phytodetritus are derived from coastal and terrestrial material (Gage, 2003). It is the sinking POM that represents the main energy input to the deep-sea benthic community. Surface primary production varies greatly depending on geographic location (Romankevich *et al.*, 2009) or, indeed, biogeographic province (Longhurst *et al.*, 1995). Therefore, each trench or trench system underlies a gradient of eutrophic to oligotrophic surface productivity regimes, which presumably results in inter-trench heterogeneity in POM supply. However, regardless of trench location, the reduction that occurs in the POM supply as it falls from surface waters is drastic, resulting in less than 1% reaching abyssal depths (Tyler, 1995). This reduction occurs as sinking POM is intercepted by heterotrophic bacteria and zooplankton that either solubilise or mineralise it, prior to it reaching the deep sea (De La Rocha and Passow, 2007; Buesseler and Boyd, 2009). The trench communities are, therefore, generally considered to be energy- (organic carbon) limited systems (Smith *et al.*, 2008). However, despite the decrease in the quantity of POM, it is believed that qualitative aspects of the POM, such as the phytopigments, proteins and essential fatty acids (EFAs) may play a significant role (Danovaro *et al.*, 2002, 2003; Wigham *et al.*, 2003). However, heterotrophic pelagic organisms can selectively remove these highly labile compounds as the POM sinks through the water column, reducing both the quantity and quality of POM with increasing depth (Wakeham *et al.*, 1984, 1997).

As a result of the regular but ever-decreasing POM supply with depth, the deep sea is often considered nutrient poor. However, periodically large-scale, significant pulses of relatively 'fresh' phytodetritus are delivered to abyssal depths in many parts of the oceans (Deuser and Ross, 1980; Billett *et al.*, 1983; Lampitt, 1985). These pulses of POM are a result of seasonal surface blooms and they contribute substantially to the export of both organic carbon and nutritious compounds into the ocean's interior (Fabiano *et al.*, 2001, Beaulieu, 2002; De La Rocha and Passow, 2007). These seasonal pulses can elicit significant responses by benthic organisms, in particular in the abundance of deposit feeders (Bett *et al.*, 2001; Billett *et al.*, 2010) and in the stimulation of seasonal growth and reproduction (Starr *et al.*, 1994). However, both the routine input of POM and the magnitude of seasonal pulses can vary interannually and in response to climate variation (Billett *et al.*, 2001; Ruhl and Smith, 2004; Smith *et al.*, 2006; Vardaro *et al.*, 2009).

It can be speculated with some certainty that a sufficient quantity of POM reaches hadal depths. Furthermore, sufficient POM input to the deep abyssal plains, including significant seasonal pulses, is well documented, therefore, to have reached lower abyssal depths suggests that in areas overlying hadal depths, this food would continue to sink to the trench floor relatively untouched. The rate at which heterotrophic bacteria and zooplankton can solubilise or mineralise POM in the hado-pelagic is unknown and whether this would be higher or lower than in the abyssopelagic remains unresolved. Only further studies involving direct measurements of seafloor POM and POC flux in the hado-pelagic will resolve the issues of quantity, quality and temporal variability of surface-derived input to the trenches. In the meantime, most information must be obtained from models.

Longhurst *et al.* (1995) estimated the annual primary production in 57 distinct biogeochemical ocean provinces, based on monthly mean near-surface chlorophyll fields for 1979–86. The provinces were specified from regional oceanography and through examining the chlorophyll fields. The net primary production rate ($g\ C\ m^{-2}\ y^{-1}$) can be used as a proxy for relative food supply in each of the trenches,

Table 6.1 The biogeochemical provinces of each of the major subduction trenches with the annual primary production rate in the overlying surface waters. Data taken from Longhurst *et al.* (1995).

Trench	Biogeochemical province	Province primary production rate $(g\,C\,m^{-2}\,yr^{-1})$	Mean POC flux $(g\,C\,m^{-2}\,yr^{-1}\pm S.D.)$	Total trench POC flux $(g\,C\,yr^{-1})$
Banda	Sunda–Arafura Seas Coastal (SUND)	328	1.60 ± 0.44	160.46
Java	Sunda–Arafura Seas Coastal (SUND)	328	1.06 ± 0.62	252.45
Peru–Chile	Chile–Peru Current Coastal (CHIL)	269	3.17 ± 1.43	997.24
Kuril–Kamchatka	Pacific Subarctic Gyre (PSAG)	264	2.26 ± 0.77	3118.86
Aleutian	Pacific Subarctic Gyre (PSAG)	232*	1.76 ± 0.87	1827.27
Philippine	Kuroshio Current (KURO)	193	0.69 ± 0.21	395.12
Ryukyu	Kuroshio Current (KURO)	193	0.90 ± 0.54	145.05
Japan	Kuroshio Current (KURO)	193	3.05 ± 0.91	909.87
Cayman	Caribbean (CARB)	190	0.77 ± 0.46	40.65
South Sandwich	Antarctic (ANTA)	165	0.66 ± 0.28	384.57
Izu-Bonin	North Pacific Subtropical Gyre (NPST)	110	1.69 ± 0.54	1595.41
Puerto-Rico	North Atlantic Tropical Gyre (NATR)	106	0.85 ± 0.33	505.09
New Britain	Western Pacific Archipelagic Deep Basins (ARCH)	100	1.07 ± 0.55	110.29
San Cristobal	Western Pacific Archipelagic Deep Basins (ARCH)	100	0.82 ± 0.47	37.15
New Hebrides	Western Pacific Archipelagic Deep Basins (ARCH)	100	0.86 ± 0.75	18.15
Tonga	South Pacific Subtropical Gyre (SPSG)	87	0.99 ± 0.30	711.16
Kermadec	South Pacific Subtropical Gyre (SPSG)	87	1.64 ± 0.45	1270.33
Yap	Western Pacific Warm Pool (WARM)	82	0.56 ± 0.29	67.99
Palau	Western Pacific Warm Pool (WARM)	82	0.61 ± 0.36	11.51
Mariana	North Pacific Tropical Gyre (NPTG)	59	0.55 ± 0.20	606.59

* indicates an average rate between PSAG East and West. The mean POC flux per unit area and the total POC flux per trench are derived from the Lutz *et al.* (2007) model by M.C. Ichino (NOCS, UK).

in their respective biogeochemical provinces. These values are given in Table 6.1. The trenches that underlie the most productive province, the Sunda–Arafura Sea Coastal (SUND), are the Banda and Java Trenches. The most oligotrophic provinces overlie the Mariana and Volcano Trenches in the North Pacific Tropical Gyre (NPTG). By comparison, the annual primary production overlying the trenches of the SUND province is ~5.5 times higher than those underlying the NPTG province. This is, of course, a rather rough estimate of the relative food supply to each trench. The use of POC flux models provides a more accurate representation of both POC flux and total POC input, accounting for the size of the trench.

Using GIS software and the Lutz et al. (2007) POC flux model, a mean POC flux value in grams of carbon per square metre per year can be calculated for each of the trenches (Fig. 6.1a). This shows that trenches such as the Japan, Kuril–Kamchatka and Peru–Chile Trenches receive a relatively high amount of POC compared to the other trenches. By converting these data to the total POC input to the trenches (by accounting for total trench area), the larger trenches underlying areas of high surface productivity predictably produce high total POC input values (Fig. 6.1b). In both instances, the trenches underlying regions of low productivity are predictably oligotrophic, as inferred from the biogeochemical province values (Table 6.1). The indication that some of the trenches are likely to receive the highest POC input based on the biogeochemical province (e.g. Banda and Java Trenches) is not supported by the Lutz et al. (2007) model. Interestingly, the trench with the highest POC flux is the Peru–Chile Trench, which was also described as a 'depocenter of organic matter' by Danovaro et al. (2003). The POC flux modelling described here and shown in Figure 6.1 was calculated by M.C. Ichino (NOCS, UK).

The delivery of POM to the trenches differs to that on the surrounding abyssal plains in the way that it is distributed on the seafloor. The distribution of settled particles is thought to be affected by the physical topography of the trenches. The steep slopes create a gravity- (and subduction) driven downward transport and therefore a subsequent accumulation of POM along the trench axis (Otosaka and Noriki, 2000; Danovaro et al., 2003; Romankevich et al., 2009). This is further driven by occasional seismic-induced mass transport of sediments towards the axis (Itou, 2000; Rathburn et al., 2009); an effect that is not possible on the flat abyssal plains. This topography-driven accumulation of food resource is evident in continental shelf submarine canyons (Duinevald et al., 2001) and the increase in trench deposit feeders (e.g. holothurians; Belyaev, 1989). This increased availability of food along the trench axis, or the 'trench resource accumulation depth' (TRAD; sensu Jamieson et al., 2010) results in the quantity of food on the trench axis and slopes being respectively higher and lower than what would have otherwise fallen on the flat plains (Fig. 6.2). The relatively impoverished zones above the TRAD may serve as a biological barrier, impeding exploitation of the accumulated food resources at the deeper depths. The steeper slopes, often comprised of bare rocky outcrops, are likely to support relatively low levels of settled POM compared to the opposing oceanic slope. Observations of high abundances of deposit feeders and facultative scavengers (Belyaev, 1989; Blankenship and Levin, 2007) as well as dense aggregations of filter-feeders (Oji et al., 2009) at the deepest parts of the trench axes, regardless of actual depth, provide anecdotal support for the TRAD.

Unfortunately, at hadal depths, there have been no direct studies using conventional sampling such as moored sediment traps undertaken to quantify the amount of surface-derived food that reaches the trench floor, nor a comprehensive survey to investigate whether the TRAD truly exists. Direct analysis of sediment from multiple depths within

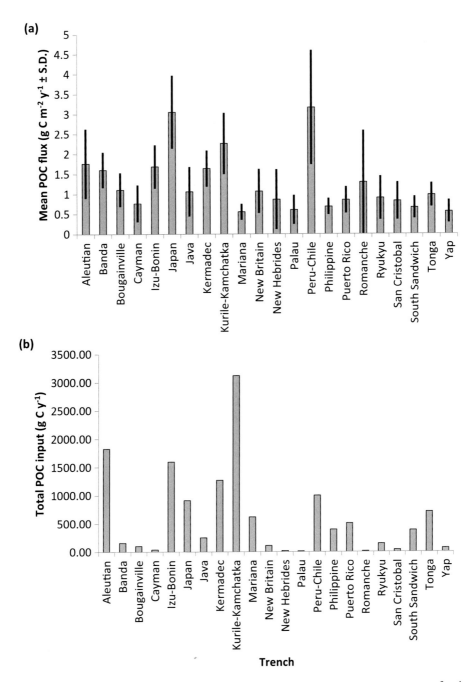

Figure 6.1 Surface-derived POC flux to the trenches, where (a) is the mean POC flux $(g\,C\,m^{-2}\,y^{-1})$ and (b) is the total POC flux $(g\,C\,y^{-1})$ accounting for the size of the trench. POC flux derived from Lutz et al. (2007) by M.C. Ichino, NOCS, UK, unpubl.

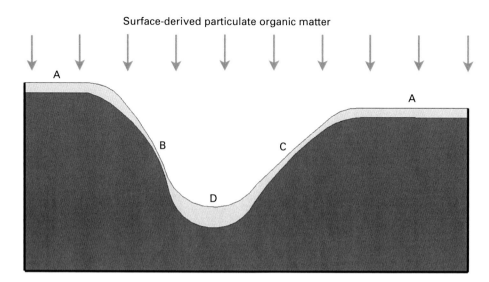

Figure 6.2 Graphical representation of the Trench Resource Accumulation Depth (TRAD) across a cross-section of a trench, where the distribution of surface-derived particulate organic matter raining from the surface settles across the flat abyssal plains (A). On the steel side of the trench the quantity of settled POM should in theory be lower (C) and lower still on the steeper fore arc (B) as more material is accumulated at the deeper trench axis (D).

a trench are few and studies that have been undertaken report highly variable results depending on the trench examined. For example, Richardson *et al.* (1995) found that the Puerto-Rico Trench sediments were low in nutritional value and were of oceanic origin, despite its close proximity to land. They found a low percentage of organic carbon (OC) and high C:N ratio which is characteristic of older, more refractory organic material from abyssal sources, as opposed to the high OC and low C:N ratio characteristic of more labile material derived from shallower-water sources. Richardson *et al.* (1995) concluded that the island of Puerto Rico had not recently been a significant supplier of sediment to the trench and the trench had accumulated low-nutrient, terrigenous clay. However, they did find that organic-rich turbidite sedimentation had once occurred but had since been buried under nutrient-depleted oligotrophic sediment and led to a present day 'depauperate' benthic community.

Conversely, large quantities of labile, phytoplankton-derived compounds have been found in trench sediments, albeit from a very few samples (Danovaro *et al.*, 2003). Such studies confirm that pulses of fresh POM are received occasionally, at least in certain trenches. The study by Danovaro *et al.* (2003) from the Atacama sector of the Peru–Chile Trench found that the functional chlorophyll-a ($18.0 \pm 0.10 \, \mathrm{mg \, m^{-2}}$), phytodetritus ($322.2 \, \mathrm{mg \, m^{-2}}$) and labile organic carbon ($16.9 \pm 4.3 \, \mathrm{g \, C \, m^{-2}}$) deposited on surface sediments at 7800 m reached concentrations similar to those encountered in highly productive shallow coastal areas. They concluded that the trench behaves like a deep oceanic trap for organic matter and that despite the extreme conditions, it represents an 'ultra-abyssal [hadal] eutrophic environment'.

A more recent study in the Mariana Trench by Glud *et al.* (2013) showed that sediment from nearly 11 000 m deep had higher concentrations of microbial cells than at neighbouring abyssal localities, as a result of enhanced deposition of organic material. The sedimentary organic material in the trench was found to be, on average, younger, more labile and possibly more nutritious than at the 6000 m abyssal reference site where the sediment represents a much longer deposition period. This was thought to be indicative of relatively rapid sediment deposition and burial, potentially as a result of erratic down-slope sediment transport aided by occasional seismic activity. Furthermore, a study arising from the aftermath of the 2011 Tōhoku-Oki Earthquake in Japan found ^{134}Cs-containing particles on the trench floor 4 months after the earthquake (Oguri *et al.*, 2013). Scientists realised that these radioactive particles must have been transported rapidly from the surface and speculated that this transport was facilitated by a concurrent springtime phytoplankton bloom that occurred 2 weeks after the earthquake and lasted for 2 weeks. The deposition and aggregation of phytodetritus in the Japan Trench was confirmed by seafloor images down to 5800 m. This led to the conclusion that the transport time for the ^{134}Cs-contaminated phytodetritus to reach the trench from the surface layers was a minimum of 78–64 m day^{-1}.

The relative importance of surface-derived organic matter and chemosynthetic sources to the hadal community is not yet fully understood but is likely to vary spatially as the number, size and extent of chemosynthetic resources are not well known and the accumulation of POM at the trench axis is still speculative based on various observations which are largely shallow; ~7800 m (Danovaro *et al.*, 2003) and ~7550 m (Oguri *et al.*, 2013) and both sites underlie relatively productive surface waters (Watling *et al.*, 2013) and are near sources of terrestrial input. The relatively fast settling of particulate matter may likely explain the accumulation of fresh sedimentary material in the trench axes of these trenches (Peru–Chile and Japan, respectively). However, the Mariana Trench differs from these other trenches in that it is located in the central Pacific Ocean away from continental land masses and underlies oligotrophic surface waters. Despite this, and the much greater depth of the Mariana Trench, increased amounts of fresh organic matter in its axis sediments relative to the shallower trench-rim sediments were found by Glud *et al.* (2013).

Surface-ocean productivity is the likely driver of the *absolute* amounts of pelagic sediments reaching trench axis but there may be additional factors that control the *relative* efficiency in the way sedimenting material is transferred through the water column and into the underlying trench-axis sediments. This led Turnewitsch *et al.* (in prep.) to speculate that the comparatively abundant fresh sedimentary materials found in sediments of the trench axis relative to sediments from shallower adjacent areas suggests that, in addition to the material inputted directly from the surface waters, there is perhaps input through quasi-lateral transport. Occasional mass-wasting events in the form of either turbidity currents transporting sediments down the slopes (Jumars and Hessler, 1976; Nozaki and Ohta, 1993; Fryer *et al.*, 2002) or earthquake-induced lateral transport (Itou et al., 2000; Oguri et al., 2013) are becoming ever more evident, but events such as these are likely to be on timescales of decades or longer (Nozaki and

Ohta, 1993) and are perhaps not sufficiently frequent to maintain the more routine supply of food to the heterotrophic hadal community.

In light of this, Turnewitsch et al. (in prep.) theorised that perhaps internal tides also play a prominent role in the transport of surface-derived POM. They examined ratios of measured (sediment-derived) and expected (water column-derived) sedimentary inventories of the naturally occurring and radioactive particulate-matter tracer $^{210}Pb_{xs}$ in trench-axis sediments of the northwest Pacific Ocean (Japan, Izu-Bonin and Mariana Trenches). They detected some evidence for a positive relationship between inventory ratios and POC fluxes albeit not a particularly strong relationship. In fact, they concluded that the more continuous supply of surface-derived food may not only be influenced by the magnitude of the productivity of the overlying surface-ocean waters but also by higher-frequency (tidal, near-inertial) fluid dynamics both in the water column and near the seafloor, with propagating internal-tide beams potentially playing a prominent role. The speculative conceptual mechanism of this is underpinned by the notion that the larger, aggregated and comparatively rapidly settling particles are converted into suspended (very slowly settling) particles and, consequently, reduced axis-accumulation of pelagic sediments into trench-axis sediments. Turnewitsch et al. (in prep.) also propose that this conversion takes place along internal-tide beams, particularly at locations where beams are reflected at the seafloor and where beams interfere with each other in the water column. Further studies are, however, required to fully test this hypothesis.

6.2   Carrion-falls

The spatial distribution and temporal variation of POM at hadal depths is punctuated with the input of carrion-falls, for example, from fishes and birds (mesocarrion ~1 kg), to seals and dolphins (macrocarrion ~100 kg) and the largest cetaceans (megacarrion >100 000 kg) (Stockton and De Laca, 1982; Britton and Morton, 1994; Bailey et al., 2007).

Carrion-falls should, in theory, occur irrespective of depth, as sinking rates are relatively high and there are little pelagic processes in place to impede descent. In the hado-pelagic, the average size of nutritious particles descending from shallower or surface waters must be significantly larger than at bathyal or abyssal depths. This is because the longer a particle takes to descend, the lower its nutritive value becomes as a result of continuous degradation during descent. For example, as smaller particles settle slower, those which may have reached the abyssal seafloor with some nutritional value may well be spent before arriving in the trenches. Therefore, if a greater proportion of the food-falls that occur at hadal depths consist of larger parcels that are more widely scattered across the seafloor, then it may well be the case that mobile scavengers play a proportionately greater role in such communities.

Mesocarrion-falls are extremely important as pelagic fish are known for their high essential fatty acid (EFA) content (Litzov et al., 2006), and could represent a highly concentrated, energy- and nutrient-rich resource for trench communities. Predicting the quantity and quality of naturally occurring carrion-falls to the deep sea is complex and speculative (Stockton and DeLaca, 1982) and direct observations of these events are rare (Klages *et al.*, 2001; Soltwedel *et al.*, 2003; Yamamato *et al.*, 2009; Aguzzi *et al.*, 2012). The significance of carrion-falls at hadal depths is evident as observed when using baited cameras and traps, which essentially simulate a carrion-fall. At bathyal and abyssal depths, baits are normally consumed by synaphobranchid eels, somniosid sharks or macrourid fishes (Priede *et al.*, 1991, 2003; Jamieson *et al.*, 2011c), however, direct carrion consumption in the trenches is typically the role of scavenging amphipods. The amphipods, ability to intercept and consume food-falls is often rapid and the numbers and speed with which they do so increases with depth (Blankenship *et al.*, 2006; Jamieson *et al.*, 2009a). Therefore, the role of scavenging amphipods in the dispersal of organic matter is significant, as the redistribution of organic matter by amphipods in the deep trenches may also provide nutrients to the wider hadal fauna via two pathways. (1) In the upper depths (<8000 m) there have been no other direct observations of scavenging, however, larger predators such as benthesicymid prawns and liparid fishes intercept the food-fall but rather than consume the carcass, they exploit the temporarily high densities of amphipods. (2) At depths greater than 8000 m where there are no apparent predators, amphipods eventually die and are scattered uniformly across the seafloor, thus themselves distributing food to the wider hadal fauna. In the upper depths (<8000 m), interception is fast but consumption is relatively low. Beyond 8000 m, where the bait attending community is almost exclusively amphipods, even large baits can be entirely consumed in less than 24 h (Hessler *et al.*, 1978; Angel, 1982; Jamieson *et al.*, 2009a; Fig. 6.3). The amphipods' ability to rapidly intercept and consume carrion-falls suggests that this does not directly contribute to the trench resource accumulation depth since the duration prior to consumption is short. The contribution to the food-webs at depths >8000 m appears to be highly significant when observed *in situ*, however, analysis of the gut contents of scavenging hadal amphipods has shown that scavenging is not the primary feeding strategy, but rather detritus feeding is dominant. Therefore, at these depths, punctuated carrion-falls may simply supplement the routine POM resources. The significance of carrion-falls in the upper trenches is perhaps greater. Despite low consumption rates and far fewer numbers of bait attending amphipods, the carrion-fall itself triggers a secondary response in larger predators. Natantian decapods (Benthesicmydae), snailfish (Liparidae) and cusk-eels (Ophidiidae) have consistently been observed to attend baited cameras and traps, not to feed directly on the bait itself but rather to prey upon the scavenging amphipods (Jamieson *et al.*, 2009a, b, 2011a; Fujii *et al.*, 2010). These larger species tend to approach the bait from down current, suggesting that they are using olfactory senses to detect the bait and follow the odour plume presumably to exploit the temporarily high densities of prey in the vicinity of the bait. Therefore, the energy pathway of carrion-falls in the hadal zone differs to that in the bathyal and abyssal zones. In the hadal zone the energy is directly consumed by amphipods that, in turn, provide a temporary feeding opportunity for larger predatory species in the upper depths, as opposed to the bulk of carrion being consumed directly by large scavenging fishes.

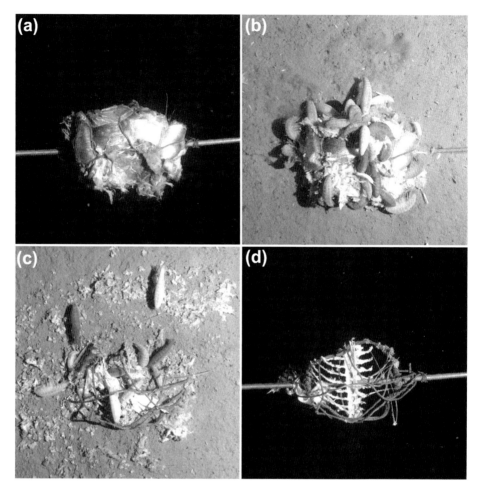

Figure 6.3 Rapid consumption of simulated carrion-fall at 8074 m in the Peru–Chile Trench, where (a) shows ~1 kg of tuna just before reaching the seafloor; (b) shows an aggregation of scavenging amphipods (*Eurythese gryllus*) 2 h later; (c) shows the remnants of the bait scattered around the seafloor after 18 h; and (d) shows the skeletal remains after leaving the seafloor.

Larger macro- and megacarrion-falls such as cetacean carcasses are also evident in the deep sea (Smith *et al.*, 1989; Smith and Baco, 2003; Dahlgren *et al.*, 2004; Kemp *et al.*, 2006). These large food-falls provide temporary elevated biodiversity and provide hard substrates and organic enrichment in the otherwise nutrient-poor deep sea. As these large injections of organic matter should reach the deep-sea floor irrespective of depth, it is likely that they also occur in the trenches, however, no direct observations have ever been reported. The likelihood of a hadal whale-fall or other cetacean carcasses is high given that most trenches are located close to continental margins underlying surface waters known to host such communities. The duration of these ephemeral habitats may also be prolonged at hadal depths, particularly at depths >8000 m where there is a distinct lack of large scavenging fish and elasmobranchs to remove the bulk of the tissues.

The input of gelatinous zooplankton carcasses to the deep sea is a newly emerging phenomenon known as 'jelly-falls' (Billett *et al.*, 2006). The occurrence and significance of jelly-falls are perhaps only recently being realised due to the spatiotemporal challenges of observation (Lebrato *et al.*, 2012). Jellyfish blooms in the surface waters occur with similar seasonal patterns to conventional surface blooms, particularly in areas of upwelling and/or in temperate or subpolar zones. As the bloom diminishes, gelatinous carcasses are deposited across the seafloor, often in low densities (Sweetman and Chapman, 2011) or they form thick and vast, gelatinous 'jelly-lakes' (Billett *et al.*, 2006; Lebrato and Jones, 2009) comprising Cnidaria (Scyphozoa) and Thaliacea (Pyrosomida, Doliolida and Salpida; Lebrato *et al.*, 2012). Although jelly-falls are emerging as a potentially significant input of remineralised organic/inorganic material to the deep sea, it is not known if, or what quality and condition of such material would reach hadal depths. These questions will hopefully be answered one day, when further, more extensive exploration of trenches is undertaken using ROV technology.

6.3 Plant and wood debris

Supplementary to the aforementioned food sources, terrestrial and coastal plant debris has been reported in some of the deep trenches. Pratt (1962) reported blades of coastal turtle grass *Thalassia testudinum* from 7860 m depth in the Puerto-Rico Trench. Later, George and Higgins (1979) trawled turtle grass, eel grass and red algae from depths over 8000 m. Wolff (1976) found that on the blades and rhizomes of *Thalassia* there was extensive evidence of consumption and within the samples were many invertebrate species (e.g. gastropods, tanaids, isopods). More recently, Fluery and Drazen (2013) deployed pieces of seagrass on a lander to ~5000 m in the Sargasso Sea and observed a different, mostly invertebrate, assemblage of scavengers attracted to the seagrass compared to conventional fish bait. These samples were taken from the Sargasso Sea, Cayman and Puerto-Rico Trenches in the vicinity of the Caribbean Sea where there are extensive seagrass meadows and hurricanes are frequent. It has been suggested that the amount of seagrass transported into the deep sea after removal by hurricanes may equal the amount of material washed ashore (Moore, 1963).

The input of seagrass to a trench is not entirely limited to the trenches around the Caribbean Sea. Such inputs have also been observed in the Palau Trench in the central Pacific Ocean and in the New Britain and New Hebrides Trenches in the Indo-Pacific (Lemche *et al.*, 1976). The photographs analysed by Lemche *et al.* (1976) revealed blades of seagrass as the most commonly identified objects in the Palau Trench between 8021 and 8042 m, with densities estimated at 1 blade per $30 \, \text{m}^{-2}$. They also reported larger pieces of wood (sticks, twigs, branches, fragments of tree trucks) and even coconut shells. Patches of unidentified plant remains, albeit also probably seagrass blades, were observed in densities of approximately 1 blade per $100 \, \text{m}^{-2}$ in the New Britain and New Hebrides Trenches (Lemche *et al.*, 1976). The input of seagrass appears to be highly significant at these depths as it is known to increase species diversity through the provision of hard substrate, shelter and as food (Wolff, 1976). Plant- and wood-falls can also host specialised plant/wood-dwelling species, such as

Table 6.2 Details of the plant debris trawled by the *Galathea* at depths exceeding 6000 m. HOT= Herring Otter Trawl, ST200, ST300 and ST600 are sledge trawls which were 2, 3 and 6 m wide, respectively (Bruun, 1957).

Station	Location	Depth (m)	Gear	Area swept (km^{-2})	Plant debris volume (ccm)	Plant debris biomass (g)
466	Java Trench	7160	HOT	0.178	240	82
497	Banda Trench	6490–6650	HOT	0.178	312	194
494	Banda Trench	7280	ST300	0.017	62	37
661	Kermadec Trench	5230–5340	ST600	0.039	75	135
649	Kermadec Trench	8210–8300	ST600	0.033	32	21
521	New Britain Trench	8830–8780	ST200	0.013	3	2
517	New Britain Trench	8940	ST300	0.017	121	55
418	Philippine Trench	10150–10190	ST300	0.008	25	20
419	Philippine Trench	10150–10210	ST300	0.011	75	40

cocculiniform limpets, which are prevalent in both number and diversity in the Cayman Trench (Leal and Harasewych, 1999; Strong and Harasewych, 1999). A recent and perhaps surprising study by Kobayashi *et al.* (2012) showed that the hadal amphipod *Hirondellea gigas* from 10 897 m in the Mariana Trench possesses a unique digestive enzyme capable of digesting sunken wood debris. This suggests that even in the deepest place on Earth, plant and wood debris of terrestrial origin are sufficiently frequent and important enough that species have evolved to utilise it.

More recent research on seagrass meadows has revealed that seagrass meadows, mangroves and salt marshes have a great capacity to trap carbon (Duarte *et al.*, 2005), potentially storing 50 times more carbon than tropical forests, per hectare (Kennedy *et al.*, 2010). Seagrass meadows are thus natural hotspots for carbon sequestration, accounting for 10–18% of the total carbon burial in the ocean despite only accounting for 0.1% of the coastal ocean (Kennedy *et al.*, 2010). Seagrass, therefore, offers the potential for rapid transport of carbon from the photic zone to the hadal zone where it is either recycled by the benthic community or buried, in which case, there is potential for this carbon to be eventually subducted back into the Earth's mantle.

The quantity of plant and wood debris in the trenches is specific to individual trenches in close proximity to continental land masses. Furthermore, if the occasional mass input of material such as seagrass is largely instigated by hurricanes, then the introduction and community responses are likely to be seasonal and highly subject to interannual variation. Quantitative data on plant debris at hadal depths are limited but the *Galathea* reports do, in fact, record a detailed description from their trawl data (Bruun, 1957), shown in Table 6.2.

6.4   Chemosynthesis

Autochthonous production occurs as a result of chemosynthesis by free-living or symbiotic bacteria (Childress and Fisher, 1992). The seismic activity synonymous with trenches can trigger turbidity flows and massive slope instability and it should not be surprising if methane emissions and exposed sulfidic sediments are found along slopes of trenches in the future (Blankenship-Williams and Levin, 2009). In these areas, methane seeps can support high densities of symbiont-bearing clams that rely on autochthonous chemosynthesis rather than the surface-derived (allochthomous) input of organic matter. The Aleutian Trench is known to host chemosynthetic communities at seep sites, mostly large vesicomyid clams, albeit, so far only found at abyssal depths (Rathburn *et al.*, 2009). Chemosynthetic bacterial communities such as cold seeps have been found at hadal depths at 6437 and 7326 m, in the Japan Trench. These habitats provide localised resources for a host of specialised organisms, in particular, mass aggregations of bivalves, which, as with all seeps, appear to occur in localised high densities (Boulègue *et al.*, 1987; Fujikura *et al.*, 1999; Fujiwara *et al.*, 2001). Few chemosynthetic sites at hadal depths have been found to date, but their close association with subduction zones and other geological features (Suess *et al.*, 1998; Rathburn *et al.*, 2009) suggest that more will be discovered as the sampling effort increases. Indeed, Blankenship-Williams and Levin (2009) speculated that seep communities, which are widespread along Pacific Ocean margins at shallower depths (Levin, 2005), are likely to be quite common in the tectonically active hadal trenches and may well host new species.

Chemosynthetic primary production in the deep sea is thought to be an almost insignificant component of the availability of organic carbon on a global scale (Tyler, 1995), yet its significance in the hadal zone is yet to be resolved. Seep environments at any depths are extremely localised and patchy habitats and are, therefore, difficult to find without the AUVs that can detect methane plumes followed by the usage of ROVs to survey the seafloor once evidence of methane is located (e.g. Newman *et al.*, 2008).

6.5   Heterogeneity

The concept of habitat heterogeneity developed between the first review of conditions on the deep-sea floor (Menzies, 1965) and the more recent Tyler review (1995). Within a given trench it appears that temperature and hydrostatic pressure (with the exception of tidal cycles) vary only with depth, and salinity does not vary at all. There is, however, evidence to suggest that oxygen may vary spatially and temporally, but this evidence is still only anecdotal (Belyaev, 1989). Current flow varies slightly between opposing flanks and is likely to vary on temporal and tidal scales. The habitat substrata is one key feature that exhibits great heterogeneity, perhaps even more so within the relatively small area of a trench than on the surrounding abyssal plains. The heterogeneity of substrata can be characterised on the large scale; steeper projection, rocky outcrops and debris on the continental flank (fore arc), more gradual, softer, sedimentary seafloor on the oceanic flanks and a deep but soft accumulation of sediment along the trench axis, particularly at the greatest depths. On the smaller scale, the seafloor is likely to be as heterogenic as most deep-sea settings; areas of flat soft sediment partitioned by steeper and/or outcrops with different degrees of sediments depending on underlying topography.

Given the size of each trench, there is still no comprehensive review of spatial heterogeneity in seafloor habitat for any trench. Asides from a perhaps similar spatial variation to that of abyssal plains, underlying often steep continental slopes and rises, punctuated with submarine canyons and seamounts, the trenches are exposed to somewhat unique environmental conditions of seismic-induced land-slides and turbidity currents. These presumably catastrophic events must reshape vast areas of seafloor in relatively short periods of time (Itou *et al.*, 2000; Fujiwara *et al.*, 2011).

The input of POM to the trenches varies depending on the biogeochemical province and upwelling. Furthermore, the presence of seasonal surface blooms provides occa-sional mass depositions of POM, as also occur on the abyssal plains. The punctuation of carrion-falls is likely to occur in all trenches, but the frequency and magnitude is, once more, subject to geographic location. While the mesocarrion-falls may represent a relatively frequent but short-term source of food, the macro- and megacarrion-falls introduce packages of food with much longer timescales. The input of plant and wood debris, utilised as both food and substrata/shelter, are also likely to be trench-specific and cases where material, such as seagrass, is deposited on mass following adverse atmospheric events, such as hurricanes and typhoons, are also very much trench-specific and exhibit spatial, temporal and seasonal variation.

Is summary, many of the environmental conditions found in the hadal zone are not significantly different from those of the abyssal plains; temperature, salinity, oxygen, absence of solar light, tidal cycles, current velocities, food supply and perhaps seasonal variation therein, but it is the complex topography and seismic activity that are truly unique. While many of these environmental conditions may appear intrinsically inhibit-ing to trench colonisation by deep-sea fauna, the fact remains that many taxa have, indeed, colonised these habitats and many successfully thrive.

6.6   Adaptation to low food availability

Adaptation to a low food environment can be achieved by the efficient storage of energy, in order to survive long periods of food deprivation. Lipids are the most widespread, long-term, energy storage strategy in marine organisms (Lehtonen, 1996). Furthermore, some of the fatty acids ingested by marine organisms are incorpor-ated unmodified into their lipid reserves and thus provide information about their dietary composition (Dalsgaard *et al.*, 2003; Stowasser *et al.*, 2009); certain accumu-lated fatty acid classes serve as biomarkers for different trophic levels and feeding strategies, i.e. herbivory, omnivory or carnivory (Graeve *et al.*, 1994, 1997; Kirsch *et al.*, 2000).

Deep-sea benthic organisms, such as amphipods, have shearing mandibles and capacious guts that allow them to take advantage of sporadic and potentially infre-quent food-falls (Dahl, 1979; Sainte-Marine, 1992). They are, therefore, adapted for bursts of feeding activity, followed by lengthy periods of digestion and fasting. This lifestyle is supported by the presence of wax esters in their tissues (Bühring and

Christiansen, 2001), which serve as energy reserves during prolonged periods of food deprivation (Lee *et al.*, 2006).

Data on lipid concentrations in hadal fauna are currently few. However, Perrone *et al.* (2003) found that lipid concentrations of *Eurythenes gryllus*, recovered from 7800 m in the Peru–Chile Trench, accounted for 7–18% of the dry weight (D.W.), much lower than those reported for deposit-feeding amphipods of the Baltic Sea (15–45% of D.W.; Lehtonen, 1996) and for *E. gryllus* from the Southern Ocean (>40% of D.W.; Reinhardt and Van-Vleet, 1985), but close to that of another hadal amphipod, *Hirondellea gigas* from 9800 m in the Philippine Trench (26.1% of D.W.; Yayanos and Nevenzel, 1978).

The differences in lipid content that were observed between the Peru–Chile Trench amphipods and those from the other deep-sea localities were deemed unlikely to be dependent on thermal variation, given that the temperatures were similar (~2°C). The concentrations of *E. gryllus* lipids were also found to decrease significantly with increasing body size. Other studies suggested that *E. gryllus* accumulates large amounts of lipids during sexual maturation to cope with reproductive effort (Ingram and Hessler, 1987), therefore, the decrease in lipid concentration in these hadal specimens could be explained by the fact that they were all immature females (Perrone *et al.*, 2003).

The lipid composition of these hadal amphipods was dominated by monounsaturated fatty acids with a very small fraction of polyunsaturated fatty acids. Conversely, shallower amphipods (*Anonyx nugax* and *Stegocephalus inflatus*) from 150–250 m in the Barents Sea had much higher amounts of polyunsaturated fatty acids (Graeve *et al.*, 1997). The low values reported for *E. gryllus* in Perrone's 2003 study suggest that hadal amphipods may be more dependent upon lipid reserves than species living at shallow-water depths and the data were thought to be indicative of a scavenging lifestyle, interspersed with long periods of starvation (Perrone *et al.*, 2003).

Part III

The hadal community

Introduction

This part details all that is currently known about the different forms of life in the hadal zone. In an ideal world, the taxa should be split into easily navigable sections in taxonomic order, but the hadal community is not as well known as many other environments and there are several taxa on which a disproportionate amount of information is available. Therefore, the next four chapters are divided into sections that cover the type of organisms found using different sampling methods. Chapter 7, *Microbes, protists and worms*, includes the types of organisms largely found in sediment cores. Chapter 8, *Porifera, Mollusca and Echinodermata*, incorporates the types of invertebrate epifauna that are encountered using trawls and sledges. Chapter 9, *Crustacea*, are largely represented by amphipods caught by baited trap, but other crustaceans have also been included for consistency. Chapter 10 details the current knowledge concerning *Cnidaria and fish*, most of which is derived from baited camera observations or plankton net hauls. For completeness, the Cnidaria section also includes the sessile Anthozoa, which are often recovered via trawl or dredge, or filmed by ROV.

Representatives of most marine organisms are found at hadal depths and each class heralds its own interesting groups. For example, the most diverse and commonly sampled groups are the polychaetes (Annelidea, ~164 species), bivalves (Mollusca, ~101 species), gastropods (Mollusca, ~85 species) and holothurians (Echinodermata, ~59 species). All of these groups are found at full ocean depth and in most cases are found in large aggregations, particularly the holothurians. Other groups such as the Bivalvia differ in that they do not appear to be uniformly distributed within the trenches, but rather they aggregate at chemosynthetic habitats (Fujikura *et al.*, 1999; Fujiwara *et al.*, 2001). Other groups such as the Crinoidea are rather conspicuous when found, forming 'crinoid meadows' as deep as 9700 m (Oji *et al.*, 2009), although identification of such organisms to species level has not yet been possible. Other groups, such as the Sipunculids and the Echinoids, are less well known and appear to be limited to the upper hadal depths. Likewise, the Asteroidea appear limited to upper depths, with the exception of one or two deeper findings. Other groups such as the Byrozoa are not well known at all, some are lacking species identification, partly due to the poor condition in which samples have been recovered. The microorganisms such as bacteria and Foraminfera are, of course, present at all depths and have been the focus of several detailed studies (e.g. Kato, 2011 and Gooday *et al.*, 2008, respectively), such as high-pressure adaptation (Bartlett, 2002; also discussed in Chapter 5) and connectivity with shallow waters (e.g. Eloe *et al.*, 2010). Despite the depth of detail reported in studies, information concerning the organisms is often derived from very few samples. In contrast, amphipods have recently emerged as one of the best sampled hadal fauna in recent decades, due to their ease of capture with baited traps. Consequently, they represent one of the few taxa that can be readily captured in sufficient numbers and diversity to provide statistically meaningful data. Hadal members of the Decapoda have also been identified and are included in this book in some detail. This is, perhaps, surprising given that most literature referring to the hadal zone cites the fact: 'decapods are not represented in the hadal zone' (e.g. Herring, 2002). On the contrary, hadal decapods do exist and have been found frequently over the last few years, in several trenches (Jamieson *et al.*, 2009b). Their previously reported absence highlights the 'wrong tool for the job' scenario and emphasises the fact that trawling at great depths is simply insufficient for catching fast-moving prawns.

7 Microbes, protists and worms

7.1 Bacteria

Scientists have proposed that life may have originated in the deep sea and that early forms of life may have possessed high-pressure-adapted mechanisms of gene expression (Kato and Horikoshi, 1996). It has been proposed that the primary chemical reactions involved in the polymerisation of organic materials (i.e. amino acids) could have occurred in high-pressure environments (Imai *et al.*, 1999). Thus, the study of high-pressure-adapted microorganisms, such as those found at abyssal and hadal depths, may enhance our understanding of the deep sea and offer new perspectives on the origin and evolution of life (Kato, 2011).

Most of the research concerning hadal bacteria has originated from the northwest Pacific trenches (Japan, Izu-Bonin and Mariana Trenches; Table 7.1), mainly due to the activities of Japanese researchers (e.g. Kato *et al.*, 1995a, b, c, 1996, 1998; Nogi *et al.*, 2002, 2004, 2007). While most bacterial studies focus on seafloor sediment samples, including the Puerto-Rico Trench sediments (Deming *et al.*, 1988), other strains have been isolated from hado-pelagic waters (Fig. 7.1; Eloe *et al.*, 2011) and even from hadal amphipods in the Mariana Trench (Yayanos *et al.*, 1981) and the Kermadec Trench (Lauro *et al.*, 2007).

Many of the deep-sea isolates are novel, psychrophilic and piezophilic bacteria: *Photobacterium profundum*, *Shewanella violacea*, *Moritella japonica*, *Moritella yayanosii*, *Psychromonas kaikoi* and *Colwellia piezophila*. These piezophilic strains belong to five genera in the *Gammaproteobacteria* subgroup and possess high-pressure adaptations such as the inclusion of significant amounts of unsaturated fatty acids in their cell membranes in order to maintain fluidity at low temperature and high pressures. It is believed that these piezophilic microorganisms are well distributed on our planet (Kato, 2011). Table 7.1 details the most common genera of piezophilic bacteria currently known from hadal depths (*Colwellia*, *Psychromonas*, *Moritella* and *Swellanella*).

Species of the *Gammaproteobacteria* genus *Colwellia* are defined as facultatively anaerobic and psychrophilic bacteria (Deming *et al.*, 1988). There are three piezophilic species known from hadal trenches; *C. peizophila* Y223G^T (Nogi *et al.*, 2004), *C. hadaliensis* BNL-1^T (Deming *et al.*, 1988) and *C.* sp. strain MT41 (Yayanos *et al.*, 1981). These have been found at 6278 m in the Japan Trench, 7410 m in the Puerto-Rico Trench and 10 476 m in the Mariana Trench (Table 7.1), the latter was discovered from a decaying amphipod (*Hirondellea gigas*). Cells of strain Y223G^T are Gram-negative

Table 7.1 A summary of known hadal bacteria including optimal growth temperature (T_{opt}) and pressure (P_{opt}). Modified from Eloe *et al.* (2011).

Isolate	Trench	Depth (m)	T_{opt} (°C)	P_{opt} (MPa)	Reference
Colwelliacae					
Colwellia peizophila Y223G^T	Japan	6 278	10	60	Nogi *et al.*, 2004
Colwellia hadaliensis BNL-1^T	Puerto-Rico	7 410	10	90	Deming *et al.*, 1988
Colwellia sp. strain MT41	Mariana	10 476	8	103	Yayanos *et al.*, 1981
Psychromonadaceae					
Psychromonas kaikoae JT7304^T	Japan	7 434	10	50	Nogi *et al.*, 2002
Psychromonas hadalis K41G	Japan	7 542	6	60	Nogi *et al.*, 2007
Moritellaceae					
Moritella japonica DSK1	Japan	6 356	15	50	Kato *et al.*, 1995a
Moritella yayanosii DB21MT-5	Mariana	10 898	10	80	Nogi and Kato, 1999
Shewanellaceae					
Shewanella benthica DB6705	Japan	6 356	15	60	Kato *et al.*, 1995a
Shewanella benthica DB6906	Japan	6 269	15	60	Kato *et al.*, 1995a
Shewanella benthica DB172R	Izu-Bonin	6 499	10	60	Kato *et al.*, 1996
Shewanella benthica DB172F	Izu-Bonin	6 499	10	70	Kato *et al.*, 1996
Shewanella benthica DB21MT-2	Mariana	10 898	10	70	Kato *et al.*, 1998
Shewanella sp. strain KT99	Kermadec	9 856	~2	~98	Lauro *et al.*, 2007
Non-*Gammaproteobacteria*					
Dermacoccus abyssi MT1.1^T	Mariana	10 898	28	40	Pathom-aree *et al.*, 2006
Rhodobacterales bacterium PRT1	Puerto-Rico	8 350	10	80	Eloe *et al.*, 2011

rods, 2.0–3.0 μm long by 0.8–1.0 μm wide and motile by means of a single unsheathed polar flagellum (Nogi *et al.*, 2004). *Colwellia* species produce the long-chain PUFA, docosahexaenoic acid (DHA; Bowman *et al.*, 1998), however, *C. piezophila* do not produce eicosapentaenoic acid (EPA) or DHA in the membrane layer but, rather, high levels of unsaturated fatty acids (16:1 fatty acids) are produced (Nogi *et al.*, 2004). This suggests that the possession of long-chain PUFA is not a prerequisite for piezophilic classification in bacterium, although unsaturated fatty acid production is likely to be common in piezophiles (Kato, 2011).

The *Gammaproteobacteria* genus *Psychromonas* is a psychrophilic bacterium. Based on 16S rRNA gene sequence data, they are closely related to the genera *Shewanella* and *Moritella*. *Psychromonas kaikoae* was isolated from sediment retrieved from the deepest known cold seep environment: 7434 m in the Japan Trench (Nogi *et al.*, 2002). *Psychromonas kaikoae* has an optimal growth temperature of 10°C and optimal growth pressure of 50 MPa. *Psychromonas kaikoae* JT7304 are an unusual strain as they produce both EPA and DHA in the membrane layer, whereas piezophilic and psychrophilic bacteria, such as *Shewanella* and *Photobacterium,* typically produce only one or the other (Nogi *et al.*, 2002). Interestingly, one of the shallower-water counterparts in Antarctica, *P. Antarctica,* does not produce either (Kato, 2011). The obligately piezophilic species *Psychromonas hadalis* K41G was also isolated from sediment from 7542 m in the Japan Trench and was found to have an optimal growth temperature and pressure of 6°C and 60 MPa, respectively (Nogi *et al.*, 2007). No growth was found

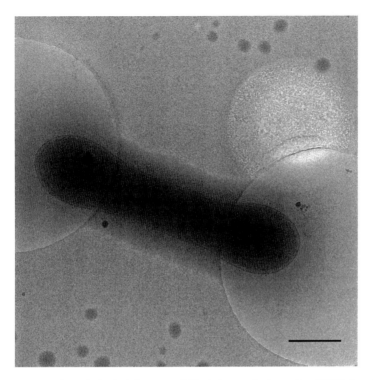

Figure 7.1 Morphological features of *Rhodobacterales* bacterium PRT1 from 8350 m in the Puerto-Rico Trench, imaged using cryo-transmission electron microscopy (Cryo-TEM) at 12 000× magnification. Scale bars = 0.5 μm. Modified from Eloe *et al.* (2011), ©American Society for Microbiology.

to occur at atmospheric pressure. Cells of strain K41G are Gram-negative rods, 1.5–2 μm long by 0.8–1.0 μm wide and motile by means of a single unsheathed polar flagellum.

Psychromonas antarctica has been isolated from sediment in the Antarctic (Mountfort *et al.*, 1998) and was found to be very similar to piezophilic *Psychromonas* strains from the Japan Trench (Nogi *et al.*, 2002). Similarly, *Psychrobacter pacificensis*, isolated from water from the Japan Trench at 5000–6000 m, was taxonomically similar to the Antarctic isolates *Psychrobacter immobilis*, *P. gracincola* and *P. frigidicola* (Maruyama *et al.*, 2000). From these findings, it was concluded that the global deep-ocean circulation linked to the sinking of cooled seawater in polar regions may have influenced bacterial habitation of the deep sea and their evolution.

Moritella marina is the type strain of the genus *Moritella* and is one of the most common psychrophilic marine organisms but is not a piezophilic bacterium (Kato, 2011). The first piezophilic species of the genus *Moritella* was *Moritella japonica* DSK1, a moderately piezophilic bacterium isolated from 6456 m in the Japan Trench (Kato *et al.*, 1995a). Production of the long-chain PUFA, DHA, is a characteristic property of the genus *Moritella* and is characteristic of *M. japonica* DSK1 (Nogi *et al.*,

1998). Cells of the DSK1 strain were found to be Gram-negative rods, 2–4 µm long by 0.8–1.0 µm wide, motile by means of a single, unsheathed, polar flagellum. The optimal growth temperature and pressure for this species is 15°C and 50 MPa, respectively, making it moderately piezophilic. The extremely piezophilic bacterium *Moritella yaya-nosii* DB21MT-5 was isolated from 10 989 m in the Mariana Trench (Nogi and Kato, 1999). The optimal growth pressure of *M. yayanosii* strain DB21MT-5 is 80 MPa and it can grow well at pressures as high as 100 MPa. It is, however, unable to grow at pressures <50 MPa and for this reason it is an extremely piezophilic strain (Kato *et al.*, 1998). *M. yayanosii* strain DB21MT-5 also represented the first evidence of the existence of an obligately piezophilic *Moritella* sp. The cells of DB21MT-5 are rods, 2.5–~3 µm long by 1 µm wide with a single polar flagellum for motility (Nogi and Kato, 1999). The membrane lipids of this strain are approximately 70% unsaturated fatty acids, consistent with its adaptation to very high pressures (Fang *et al.*, 2000).

Shewanella are not exclusively found in marine environments. Gram-negative, aerobic and facultatively anaerobic *Gammaproteobacteria* (MacDonell and Colwell, 1985) and several novel marine *Shewanella* species are not piezophilic (Kato, 2011). *Shewanella* strains PT-99, DB5501, DB6101, DB6705, DB6906, DB172F, DB172R and DB21MT-2 are all piezophilic members of the same species, *S. benthica* (Nogi *et al.*, 1998b; Kato and Nogi, 2001). The extensively studied *S. violacea* strain DSS12 (Kato *et al.*, 1995a; Nakasone *et al.*, 1998, 2002) is moderately piezophilic, with a reasonably constant doubling time at pressures between 0.1 and 70 MPa, as opposed to most piezophilic *S. benthica* strains that have doubling rates that change significantly with increasing pressure (Kato, 2011). Strain DSS12 exhibits few differences in the growth characteristics under varying pressure conditions and thus represents a convenient deep-sea bacterium for research focused on high-pressure adaptations. In fact, the genome analysis on strain DSS12 has been used as a model deep-sea piezophilic bacterium (Aono *et al.*, 2010). Both *S. benthica* and *S. violacea* are also considered psychrophilic at atmospheric pressure (Nogi *et al.*, 1998b). *Shewanella benthica* DB21MT-2 represents one of the deepest bacteria ever sampled, having been isolated from sediment from Challenger Deep at 10 898 m (Kato *et al.*, 1998). The optimal growth pressure for this strain is 70 MPa, with no growth occurring at pressures <50 MPa, indicating that it is extremely piezophilic. Furthermore, this strain was able to grow well at pressures up to 100 MPa. The cells of DB21MT-2 are rods, 2 µm long by 0.8–~1 µm wide with a single polar flagellum for motility.

Kato (2011) identified two distinct groups within the *Shewanella* genus from the *Gammaproteobacteria* phylogenetic tree. Most psychrophilic or psychrotrophic *Shewa-nella* species belong to the same group. Also within this group are *S. benthica* and *S. violacea*, which show piezophilic or piezotolerant growth properties under high-pressure conditions. The second group comprises species that exhibit no growth at a pressure of 50 MPa, and are therefore piezosensitive (Kato and Nogi, 2001). Generally, members of *Shewanella* group 1 are characterised as cold adapted and pressure tolerant, whereas the members of *Shewanella* group 2 are mostly mesophilic and piezosensitive. This is supported by evidence that members of the former group produce substantial amounts of the PUFA EPA (11–16% of total fatty acids), whereas members of the latter group produce no or limited amounts of EPA, thus defining the two taxonomic groups as psychro- and piezophilic (group 1) and mesophilic and piezosensitive (group 2). Indeed, the piezophilic strains have been isolated from sediments in the Japan Trench (6269–6356 m; Kato *et al.*, 1995a), the Izu-Bonin Trench (6499 m; Kato *et al.*, 1996) and the Mariana Trench (10 898 m; Kato *et al.*, 1998). Another strain, KT99, was isolated from 9856 m in the Kermadec from the amphipod, *Hirondellea dubia*, homogenate (Lauro *et al.*, 2007).

On a larger scale, Takami *et al.* (1997) characterised the microbial flora community of sediments collected from 10 897 m in the Mariana Trench. The community comprised actinobacteria, fungi, non-extremophile bacteria and various extremophilic bacteria such as alkaphiles, thermophiles and psychrophiles. The non-extremophilic bacteria were found in frequencies of 2.2×10^4–2.3×10^5 colonies per gram of dry sediment. The filamentous fungi and actinobacteria were found at the same frequency as facultative psychrophilic bacteria (2.0×10^2 per gram of dry sediment). They did not, however, find any of the piezophilic bacteria in these samples that were reported shortly after by Kato *et al.* (1998) and previously reported by Yayanos *et al.* (1981). Takami *et al.* (1997) concluded that the sediments of the Challenger Deep are a depository of active or dormant microbes usually found in particulate matter constituting 'marine snow'. With estimated sinking rates of 1.0 to 0.1 m per day (or 5000 m in 1–50 per year; Jannasch and Taylor, 1984), many microbes are being deposited over the course of long periods of time. However, Glud *et al.* (2013) found higher organic carbon concentrations and more disturbed and inconsistent sediment profiles at Challenger Deep than at an abyssal reference site (6018 m). This was thought to be indicative of relatively rapid sediment deposition and burial, potentially as a result of erratic down-slope sediment transport, aided by occasional seismic activity. The same study concluded that the sediment of Challenger Deep exhibited intensified mineralisation mediated by the prokaryote community (combined bacteria and Archea), which was sustained by the enhanced deposition of organic material. The average prokaryote density at 10 813 and 10 817 m was $0.97 \times 10^7\,\mathrm{cm}^{-3}$, compared to just $0.14 \times 10^7\,\mathrm{cm}^{-3}$ at the abyssal site. The intensified heterotrophic microbial activity resulted in sediment O_2 concentrations attenuating faster at full ocean depth than at abyssal depths (Glud *et al.*, 2013).

The hado-pelagic zone represents one of the least studied environments on Earth as very little research has been undertaken. Of the few studies available, Eloe *et al.* (2010) examined the compositional differences in particle-associated ($>3\,\mu$m) and free-living (3–$0.22\,\mu$m) microbial assemblages (bacterial, archaeal and eukaryal) from pelagic waters at 6000 m in the Puerto-Rico Trench. This study obtained 541 bacterial and 675 archaeal gene sequences from both free-living and particle-associated fractions, and 339 eukaryal sequences from the former fraction. No significant differences were found in the archaeal libraries, whereas the bacterial libraries were statistically different. The particle-associated bacterial fraction hosted a greater diversity compared to the free-living fraction which was consistent with other, albeit shallower studies. Bacterial sequences recovered from the two size fractions were comprised of ~40% of six named orders of Alphaproteobacteria: Rhodobacterales, Rhizobiales, Sphingomondales, Caulobacterales, Rhodospirillales and Rickettsiales, with the latter representing the greatest number of sequences. The order Rhizobiales accounted for a large fraction of the particle-associated fraction, which expanded this order from soil and sediment environments to the deep sea. Relatively few *Gammaproteobacteria* sequences were recovered from either fraction but included Legionellales, Xanthomonadales, Alteromonadales, Chromatiales and Oceanospirillales, of which the order Xanthomonadales accounted for half of the *Gammaproteobacteria* sequences in the free-living fraction. *Betaproteobacteria* sequences accounted for 2.7% of the particle-associated fraction and 26.5% of the free-living fraction. The majority were from the order Burkholderiales, within the three families Alcaligenaceae, Comamonadaceae and Burkholderiaceae. This high proportion of *Betaproteobacteria* was surprising given that they had not been found in significant abundance in other marine bacterioplankton communities. Deltaproteobacteria sequences accounted for 13% of the particle-associated fraction and 4.4% of the free-living fraction with more than half of the total 49 sequences being within the same clade, SAR324. The remaining sequences were comprised of non-*Proteobacteria*, which accounted for 38 and 107 sequences from the free-living and particle-associated fractions, respectively. The majority of the particle-associated sequences grouped within the phyla Bacteroidetes and Planctomycetes. Four cyanobacterial-like and 12 plastid-related sequences were identified exclusively from the particle-associated fraction, suggesting that some of the particle-associated fraction is surface-derived. Of the free-living fraction, the most dominant non-*Proteobacteria* was the well-documented deep-dwelling SAR406.

The study by Eloe *et al.* (2010) indicated that the phylotypes recovered from the Puerto-Rico Trench were closely related to sequences known from other deep-sea habitats. However, some of the phylotypes from the particle-associated fraction may have originated from the overlying surface waters and have sunk and survived at high pressure. They concluded that the particle-associated microorganisms present at depth may reflect the magnitude and duration of exposure to elevated pressures and the relative pressure-resistance of the associated microbes. The presence of both eukaryotic and bacterial phytoplankton in the particle-associated library appears to be evidence of a shallow-water microbial community connection. However, given the high overall similarity of this community with other deep communities and the free-living fraction, the portion of surface-water microorganisms present must be relatively small.

7.2   Foraminifera

Within the meiofaunal size fraction (small eukaryotes), most of the biomass is made up by nematodes and foraminiferans. Benthic Foraminifera (Protista, Rhizaria) are represented at hadal depths by more species than from any other taxon, apart from crustaceans (Belyaev, 1989). The first hadal Foraminifera were retrieved by the *Challenger* expedition collected in the 1870s, from a depth of 7224 m in the Japan Trench (Brady, 1884), although most of the 14 species found were known from shallower sites. The majority of hadal Foraminifera samples have emerged from the Soviet expeditions of the 1950s onwards and were reviewed by Belyaev (1989). At that time, 103 foraminiferal species (five orders, 15 families) were known between 6000 and 10 687 m. These

Table 7.2 List of Foraminfera families sampled from depths exceeding 6000 m. The list is derived from Belyaev (1989) and updated with Gooday *et al.* (2008), Kitazato *et al.* (2009), Akimoto *et al.* (2001). All records have been updated using the World Register of Marine Species (WORMS).

Order	Family	Genera	Species	Depth range (m)
Allogromiida	Allogromiidae	4	5	2140–10 896
	Allogromiidae incertae sedis	1	1	10 896
Astrorhizida	Ammodiscidae	5	6	68–9220
	Astorhizidae	2	4	1760–6980
	Botellinidae	1	1	2000–8430
	Dendrophryidae	1	2	5510–10 002
	Hyperamminidae	2	8	1739–10 002
	Normaninidae(Komikiacea)	1	3	2890–10 687
	Polysaccamminidae	1	1	3360–8006
	Psamminidae	1	1	6860–7320
	Psammaophaeridae	2	2*	2532–10 687
	Rhabdamminidae	3	7**	500–10 924
	Rhizamminidae (Komikiacea)	1	3	1015–6520
	Saccamminidae	3	6	1724–10 924
	Stannomidae (Xenophyophorea)	1	2	6116–6675
	Syringamminidae (Xenophyophorea)	1	3	2760–8950
Litoulida	Reophacidae	1	1	10 896
	Ammosphearoidinidae	6	11**	252–8380
	Discamminidae	1	1	7 225
	Haplophragmoidae	4	4*	1450–6740
	Hormosinellidae	2	3	1134–7660
	Hormosinidae	3	4	1620–10 924
	Lituolidae	2	5	640–7316
	Prolixoplectidae	1	1	2862–7225
	Reophacidae	3	8*	1739–9580
	Spiroplectamminidae	3	3**	1887–9540
	Trochamminoidae	1	1	750–6250
Loftusiida	Cyclamminidae	1	3	2750–6240
	Globotextularidae	2	2*	1550–6070
Miliolida	Cornuspiroidinae	1	1	2197–6240
	Hauerinidae	2	2	2 048–7 225
	Sprioloculinidae	1	1	4930–6927
Textulariida	Eggerellidae	1	1	1748–6250
	Textulariidae	1	1	10 896
Trochamminida	Conotrochamminidae	1	1	2507–7300
	Trochamminidae	1	4	713–9220

* Indicates the number of species listed in Belyaev (1989) that are not listed by WORMS.

numbers have now increased to seven orders and 36 families after cross-checking and updating this list using the World Register of Marine Species (WORMS) database (Table 7.2). This is still a relatively low number compared to an estimated 2140 known species worldwide (Murray, 2007). This low number is not a reflection of low diversity but rather a chronic undersampling of hadal sediment, compared to shallower zones.

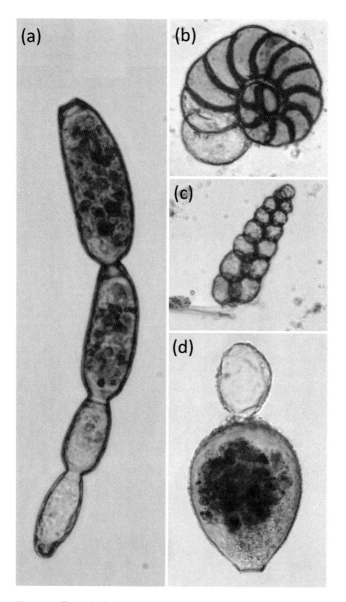

Figure 7.2 Foraminifera from the Challenger Deep, Mariana Trench. (a) *Resigella laevis*, (b) unidentified coiled foram, (c) *Textularia* sp. and (d) *Resigella bilocularis*. Images courtesy of Andrew J. Gooday, National Oceanography Centre, Southampton (b–d), and (a) copyright of the Lennean Society of London.

Of the 103 species listed by Belyaev (1989), there were two with organic-walled tests: *Nodellum membranacea* (from the *Challenger* expedition; Brady, 1884), and *Xenothekella elongata*, from over 9000 m in the Kuril–Kamchatka Trench (Saidova, 1970). Since the Belyaev (1989) review there have been more findings, primarily from two trenches: the Mariana Trench (Akimoto *et al.*, 2001; Todo *et al.*, 2005; Gooday *et al.*, 2008) and the Peru–Chile Trench (Sabbatini et al., 2002). Of the remaining species, six were calcareous, six were ammodiscaceans, 49 were multilocular, agglutinated species and 46 (including three komokiaceans) belonged to monothalamous (single-chambered), agglutinated taxa. Other than the two organic-walled species ('allogromiids') listed by Belyaev (1989), records of other 'allogromiids' from hadal depths were sparse.

Allogromiids did, however, account for 41% of meiofaunal taxa in a box core from 7298 m in the Aleutian Trench (Jumars and Hessler, 1976). Sabbatini *et al.* (2002) reported numerous soft-shelled Foraminifera (organic-walled allogromiids and agglutinated saccamminids) from the Atacama sector of the Peru–Chile Trench from 7800 m and these accounted for 82% of 'live' individuals in one core sample. They reported a total of 546 Rose-Bengal stained, and therefore 'living', soft-walled specimens in the >20 μm fraction of the 0–6 cm layer of the core. Most specimens were allogromiids (82.0%), followed by saccamminids (11.0%) and psammosphaerids (6.0%). In the size distribution, a distinct peak around 120–160 μm was observed in the allogromiids, particularly *Nodellum*- and *Resigella*-like forms, while a spherical allogromiid species dominated the larger-size classes. This study provided further evidence for the widespread occurrence of soft-walled, monothalamous Foraminifera in marine habitats (Sabbatini *et al.*, 2002).

The majority of recent hadal Formaminifera have been sampled from the Mariana Trench, particularly Challenger Deep, as a result of the research interests of JAMSTEC. Akimoto *et al.* (2001) examined a 54 ml sediment sample from 10 897 m. Of the 91 individual agglutinated foraminiferans found in the sample *Lagenammina difflugiformis* (46 individuals) and *Rhabdammina abyssorum* (27 individuals) dominated, but also included *Hormosina globulifera* (9 individuals), *Hormosinella* (*Reophax*) *guttifer* (3 individuals) and 6 miscellaneous species. Only 4 out of 91 individuals were considered 'live', all of which were *L. difflugiformis*.

Following this study, Todo *et al.* (2005) reported 432 'living' benthic Foraminifera in the top 1 cm of a sediment core from 10 896 m in Challenger Deep. This equated to a density of ~449 individuals per 10 cm^{-2} which was similar to other shallower, but still hadal, samples (7088 m in the Japan Trench and 7761 m in the Kuril–Kamchatka Trench) and greater than at many abyssal sites, including their own reference site (550 m; Table 7.3). The assemblage was very similar to shallower samples from the same trench at 7123 m. The assemblage was dominated by delicate, soft-walled species except for four individuals of the multichambered agglutinated genera *Leptohalysis* and *Reophax*. Of these 428 specimens, 85% were organic-walled allogromiids, resembling the genera *Chitinosiphon*, *Nodellum* and *Resigella*. The organic-walled species accounted for 99% of the total foraminiferal assemblage at Challenger Deep. The high percentage of organic-walled allogromiids was thought to be very unusual because in most deep-sea environments they constitute 5–20% of the living Foraminifera assemblage.

Gooday *et al.* (2008) later described four new species and a new genus of Foraminifera from 10 896 m, again from Challenger Deep, which were first mentioned in Todo *et al.* (2005). The species were *Nodellum aculeate*, *Resigella laevis*, *R. bilocularis* and

Table 7.3 Number of 'live' specimens (N) and number of species (S) of major foraminiferal taxa at trench and abyssal sites in the northwest Pacific Ocean. MAF = multilocular agglutinated Foraminifera. Modified from Todo *et al.* (2005).

Major taxonomic group	Other Trenches				Mariana Trench Region					
	Kuril–Kamchatka, 7661 m		Japan, 7088 m		Mariana Stn 40 7123 m		Challenger Deep 10 896 m		Abyssal Stn 64 5507 m	
	N	S	N	S	N	S	N	S	N	S
Nod/Res-like allogromiids	12	3	16	2	199	11	363	5	24	9
Other allogromiids	45	4	22	5	20	11	24	1	32	13
Psammosphaeridae	17	5	17	5	165	11	32	2	89	18
Saccamminidae	318	6	54	9	19	11	9	3	39	18
Other soft-shelled taxa	0	0	0	0	0	0	0	0	23	5
Lagenammina	1	1	0	0	10	3	0	0	52	9
Other Astrorhizacea	0	0	0	0	0	0	0	0	3	2
Ammodiscacea	1	1	0	0	3	2	0	0	2	1
Hormosinacea	9	3	26	3	21	5	4	2	28	6
Trochamminacea	126	8	8	2	2	1	0	0	9	3
Other MAF	3	2	1	1	1	1	0	0	9	3
Rotaliina	161	2	1	1	5	1	0	0	19	3
Total numbers	**693**	**35**	**145**	**31**	**445**	**57**	**432**	**13**	**329**	**89**
% soft-shelled monothalamous	56		75		91		99		63.5	

Conicotheca nigrans (also new genus). Previously, Akimoto *et al.* (2001) had reported several agglutinated foraminiferal species, identified as *Lagenammina difflugiformis*, *Hormosina globulifera*, *Reophax guttifera* and *Rhabdammina abyssorum*, from Challenger Deep. The findings from Akimoto *et al.* (2001) and those of Todo *et al.* (2005) and Gooday *et al.* (2008) differed somewhat, despite being from the same area and depth. This was attributed to the different methods used to analyse the samples since many of the very small, elongate morphotypes would have passed through the 125 mm mesh sieve used in the Akimoto *et al.* (2001) study. Furthermore, the earlier study examined dried residues in which delicate, organic-walled Foraminifera would have been destroyed or unrecognisable (Gooday *et al.*, 2008).

Gooday *et al.* (2010) detailed four small foraminiferan tests, three trochamminaceans and one species of *Textularia*, from Challenger Deep and found that some agglutinated Foraminifera living close to full ocean depth construct a test from biogenic and detrital particles, which subsequently dissolve. The incorporation of dissolved coccoliths and fragments of planktonic foraminiferal shells into two tests indicates that these delicate calcareous particles can reach full ocean depth intact. Presumably, they are conveyed by rapidly sinking particles, such as phytodetritus aggregates or faecal pellets.

The predominance of organic-walled or soft-shelled species at hadal depths is likely to be an effect of survival well beyond the CCD. Todo *et al.* (2005) also concluded that the very distinctive yet primitive Foraminifera found at Challenger Deep probably represented the remnants of an abyssal assemblage. It seems likely that this abyssal assemblage was able to adapt to the steady increase in hydrostatic pressure over the 6–9 million years that it is thought the Challenger Deep has taken to develop (Fujioka *et al.*, 2002).

7.2.1   Komokiacea

The superfamily Komokiacea (Tendal and Hessler, 1977) is one of the most widely distributed deep-sea foraminiferal groups and also the most controversial one (Lecroq *et al.*, 2009a). Prior to their classification as Foraminifera, the komokiaceans were placed within the sponges and xenophyophores (Hessler and Jumars, 1974). Komokiaceans (informally termed 'komoki') are particularly common in oligotrophic, abyssal regions. They sometimes have broad bathymetric ranges and have been found in the hadal zone. Tendal and Hessler (1977) and Kamenskaya (1989) reported a variety of komoki from the Pacific Ocean, including from eight trenches. The former study, from the *Galathea* expedition, found Komokiacea at seven stations >6000 m, the deepest at 9605 m. The deepest trawl sample examined was recovered from a reported depth of 10 915 m in the Tonga Trench and yielded komoki belonging to the genus *Edgertonia* (Kamenskaya, 1989). Tendal and Hessler (1977) stated that the greatest relative abundance of komoki have been found in the oligotrophic abyssal plains and hadal trenches, where their volume exceeds that of all the metazoans combined and equals that of the remains of other Foraminifera. More than 80% of the specimens came from the surface layer of sediment and only 2% were found below 2 cm. Gooday *et al.* (2007) re-examined several abyssal and hadal specimens from the earlier Soviet work reported by Saidova (1975), who described them as new species of foraminferans in the genera *Dendrophyra* and *Normania*. Gooday *et al.* (2007) concluded that *D. kermadecensis* from 8928–9174 m was of the genus *Reticulum*, while the *N. fruticosa* specimen from 6126 m was not only lost, but it was concluded from the examination of photographs that this specimen was probably not *Normanina*, but rather *Komokia* and from 9995–10 002 m. *N. ultrabyssalica* was probably a new komokiacean. At present, the Komokiacea are still classified within Foraminifera, based on morphology. Even molecular attempts to establish their phylogentic position have failed to unequivocally confirm that they are indeed Foraminifera (Lecroq *et al.*, 2009a). This study did, however, suggest that komokiaceans, and probably many other large testate protists, provide a habitat structure for a large spectrum of eukaryotes and significantly contribute to maintaining the biodiversity of micro- and meiofaunal communities in the deep sea.

7.2.2   Xenophyophorea

The somewhat enigmatic xenophyophores are a group of giant unicellular Foraminifera that construct fragile, agglutinated tests which provide habitat structure for a wide range of small metazoans (Levin, 1991). They themselves are therefore, already known to be 'hotspots' of deep-sea meiofaunal and macrofaunal diversity (Lecroq *et al.*, 2009a). Like the komokiaceans, their taxonomic position remained undetermined for almost a century, with various studies classifying them as Spongia, or different groups of Protozoa. However, they now reside in the Foraminifera (class Polythalamea, super-family Xenophyophorea).

They are deposit- or suspension-feeding benthic organisms, often greater than 10 cm in diameter. They consist mainly of foreign particles (xenophyae) and are found with a wide variety of morphologies (Lecroq *et al.*, 2009b). Despite their wide occurrence in the abyssal deep sea in particular, at times reaching densities as high as 1000 per 100 m^2 (Tendal and Gooday, 1981), xenophyophores are poorly understood since they are notoriously fragile and easily fragmented. In total, 14 genera and ~60 species of xenophyophores have now been described.

The first records of hadal xenophyophores emerged from the Russian and Danish literature following the *Vitjaz* and *Galathea* expeditions (Belyaev, 1989). Tendal (1972) established two orders in the subclass Xenophyophoria: Psamminida and Stannomida. These are now two families (Psammettidae and Stannomidae) in the superfamily Xenophyophoroidea. Species from the predominantly abyssal genus *Stannophyllum* (Stannomidae) were found at hadal depths. *Stannophyllum granularium* was known from various abyssal sites and several hadal sites in the Kuril–Kamchatka Trench, albeit <6900 m (6272–6282 m, 6710–6675 m and 6215–6205 m). The same species was also found in the Japan Trench at 6116 m, where another species, *S. mollum*, was found at 6380 m. Xenophyophores have also been responsible for extraordinarily high biomass estimates in core sampling. Shirayama (1984) sampled 12 stations in the northwest Pacific Ocean (2090–8260 m), in the region of Japan and found the highest biomass at the 8260 m site, in the Izu-Bonin Trench. This was the result of an elevated abundance of the xenophyophore *Occultammina profunda* (Syringamminidae).

Apparent xenophyophores also featured on many bottom photographs from the PROA expedition in the Palau, New Britain, Bougainville and New Hebrides Trenches, to depths of 8662 m (Lemche *et al.*, 1976). Scientists reported a mean density of the genus *Stannpohyllum* of 1 ind. 3 m^{-2} in the New Britain Trench at 8260 m, and 1 ind. 10 m^{-2} in the New Hebrides at 6770 m. The density of an unidentified *Psammetta* sp. was 1 ind. m^{-2} at 7875–7931 m in the New Britain Trench.

Belyaev (1989) hypothesised that Xenophyophorea may play a very significant role in the primary use and reprocessing of the organic matter contained in the benthic sediment at hadal depths. Given their extreme fragility, only further sampling and more delicate sampling methodologies (such as the use of ROVs) will the diversity and role of the xenophyophores be truly understood.

7.3   Nematoda

Nematoda (roundworms) are an extraordinarily diverse animal phylum, inhabiting a range of environments (Lambshead, 2003). They were known to exist at depths exceeding 4570 m until the mid-1950s (Wieser, 1956). However, in the 1960s, extraordinary large population densities in the region of 20 000–80 000 ind. m^{-2} (Thiel, 1966) and 156 000–278 000 ind. m^{-2} (Thiel, 1972) were found at lower abyssal depths,

Table 7.4 Nematode species diversity, richness and evenness from three stations in the Puerto-Rico Trench and corresponding abyssal and bathyal sites.

Site	Depth (m)	Species diversity (H')	Species richness (SR)	Evenness (J')
Hatteras Plain[a]	5411	4.10	19.14	0.87
Puerto-Rico Trench[a]	2217	3.97	15.89	0.89
Puerto-Rico Trench[a]	7460	3.58	11.47	0.87
Puerto-Rico Trench[a]	8189	3.58	11.05	0.92
Puerto-Rico Trench[a]	8380	3.33	8.60	0.86
Atacama Slope[b]	1050	3.1	14.1	0.878
Atacama Slope[b]	1140	3.2	14.5	0.897
Atacama Slope[b]	1355	3.2	13.0	0.897
Atacama Trench[b]	7800	2.7	7.9	0.862

Modified from [a] Teitjen (1989) and [b] Gambi *et al.* (2003).

suggesting that hadal populations may be as numerous. Indeed, they were subsequently found at all depths to 10 415–10 687 m in the Tonga Trench during the *Vitjaz* expeditions (Belyaev, 1989). Having been found at more than 60 stations in 18 trenches of all three oceans (Wolff, 1960), nematoda are now considered one of the most characteristic fauna of the meio- and microbenthos in the hadal zone and many other deep-sea settings. Nematode diversity is often very high, even at hadal depths, for example, Jumars and Hessler (1976) recovered 194 species from a single box core sample, from 7298 m in the Aleutian Trench. This corresponds to a density of 776 ind. m^{-2}.

Teitjen *et al.* (1989) investigated the meiofauna abundance and biomass at three sites in the Puerto-Rico Trench (5411, 7460 and 8189 m) and found that nematode abundance and biomass did not correlate to depth and that the deepest station supported four times the biomass. Fewer known families were found at the trench depths than at comparative abyssal (5411 m) and bathyal (2217 m) sites (Teitjen, 1989). The trench nematodes were overwhelmingly dominated by the families Oxystominidae, Chromadoridae and Xyalidae (also observed at the abyssal site). These three families accounted for 65.9 and 55.9% of the individuals and species at the two trench depths, respectively. Several other families were of local significance (Sphaerolaimidae at 7460 and 8189 m, Siphonolaimidae and Desmoscolecidae at 8189 m, Microlaimidae at 7460 m). Of the 110 genera found across all stations, 52% were known and the rest were new, undescribed genera. The unknown genera accounted for 6.4% (7460 m) and 19.5% (8189 m) of the total number of nematodes identified. The highest diversity was found at the abyssal site and the lowest at the deepest site. Teitjen (1989) concluded that, generally, species diversity was in agreement with many other abyssal localities, where it was found to be a function of species richness and where the nematode species tended to be evenly distributed. Nematode species diversity, richness and evenness are shown in Table 7.4.

The Puerto-Rico Trench, however, is thought to be a nutrient-depleted, oligotrophic environment (Richardson *et al.*, 1995). In contrast, a study similar to that of Teitjen *et al.* (1989) was undertaken in the relatively eutrophic Peru–Chile Trench, in the Atacama sector (Gambi *et al.*, 2003). This study examined nematode assemblages at bathyal and hadal depths (1050–7800 m) in an area of very high concentrations of nutritionally rich organic matter, at 7800 m. The assemblages at these depths displayed characteristics typical of eutrophic systems and revealed high nematode densities (>6000 ind. 10 cm^{-2}) that were different in composition from those found at bathyal depths.

The low affinity index between the bathyal and hadal communities of the Puerto-Rico Trench was thought not to be due to the presence or absence of certain families/genera, but rather to the different per cent compositions of the assemblages (Tietjen, 1989). Although Gambi *et al.* (2003) also reported a low affinity index, it was due to an entirely different composition between bathyal and hadal depths. The most abundant families in the bathyal Atacama Slope were Comesomatidae, Cyatholaimidae, Microlaimidae, Desmodoridae and Xyalidae, whereas the most abundant in the trench area were Monhysteridae (<1% at bathyal depths, 24% at hadal depths), Chromadoridae, Microlaimidae, Oxystominidae and Xyalidae.

In a comparison of the Peru–Chile Trench and the Puerto-Rico Trench (Tietjen, 1989), Gambi *et al.* (2003) found that the dominant families were the same: Monhysteridae, Chromadoridae, Oxystominidae and Xyalidae; but the relative importance of Monhysteridae (5% in the Puerto-Rico, 24% in the Atacama Trench) and Xyalidae (17.4% in the Puerto-Rico, 7.4% in the Atacama Trench) changed. Another peculiarity of the nematode assemblage from the Peru–Chile Trench was that the family Cyatholaimidae, particularly abundant in both trench and slope sediments (6.3% and 13.1%, respectively), accounted only for a minor fraction of nematode assemblages in all other deep-sea studies.

The two regions are quite contrasting, where the Peru–Chile Trench is characteristically typical of eutrophic systems with very high concentrations of nutritionally rich organic matter (Danovaro, 2002) and the Puerto-Rico is nutrient poor (Richardson *et al.*, 1995). The different trophic conditions observed between these trenches are thought to have contributed to the differences in the structure of nematode assemblages (Gambi *et al.*, 2003).

Based on buccal morphology, Teitjen (1989) classified the nematode assemblages into feeding guild (selective deposit feeders, non-selective deposit feeders, epistrate feeders and predators/omnivores). At all stations, deposit feeders were the dominant type, whereas at the two deepest sites non-selective deposit feeders comprised 55% of those observed. Generally, the relative abundance of epistrate feeders was lower in the trench than at the bathyal or abyssal sites. This was thought to be the result of increased resource partitioning by worms inhabiting a more heterogeneous sedimentary environment at bathyal depths than at hadal depths.

7.4   Polychaeta

In the hadal zone, Polychaeta ('bristle worms' of the Annelidea phylum) are one of the most abundant and diverse groups of benthic invertebrates (Table 7.5; Fig. 7.3). Polychaeta are the most frequently encountered among all benthic invertebrates, accounting

Table 7.5. Maximum depth of each order and family of Polychaeta and the number of genera and species therein. * indicates one undescribed species.

Order	Family	Genera	Species	Maximum depth (m)
Eunicida	Dorvilleidae	1	1	7298–7398
Eunicida	Lumbrineridae	2	4	6156–8100
Eunicida	Onuphiidae	2	2	6090–6330
Phyllodocida	Pilargidae	1	1	6580
Phyllodocida	Hesionidae	1	1	8980–9043
Phyllodocida	Nephtyidae	3	3	6180–9174
Phyllodocida	Nereididae	3	5	5800–8400
Phyllodocida	Aphroditidae	1	1	6766–6875
Phyllodocida	Goniadidae	1	1	7218–7934
Phyllodocida	Phyllodocidae	2	3	6052–8100
Phyllodocida	Polynoidae	10	21	6052–10730
Phyllodocida	Sigalionidae	1	1	6050–6150
Drilomorpha	Capitallidae	2*	3	6410–8660
Terebellida	Faveliopsidae	2	2	6052–6835
Terebellida	Flabelligeridae	2	5	5650–7934
Terebellida	Maldanidae	4	6	6156–7290
Terebellida	Opheliidae	3	6	6052–9734
Terebellida	Cirratulidae	4	5	6487–10015
Terebellida	Poecilochaetidae	1	1	10415–10687
Terebellida	Amparetidae	5	7	6150–8430
Terebellida	Terebellidae	1	2	6040–6328
Terebellida	Trichobranchiidae	1	1	6660–7587
Spionida	Chaetopteridae	1	1	6860
Sabellida	Siboglinidae	11	28	6156–9735
Sabellida	Oweniidae	3	3	6180–8300
Sabellida	Scalibregmatidae	2	3	5650–9174
Sabellida	Sabellida	3	4	6156–9735
Sabellida	Serpulidae	1	1	6410–9735

for 90% in both trawl and grab samples from the *Vitjaz* expeditions (Belyaev, 1989). Hadal polychaetes have been found in all the studied trenches, including at depths exceeding 10 000 m (10 160–10 730 m in the Philippine, Mariana and Tonga Trenches; Kirkgaard, 1956; Fig. 7.4). Their mean abundance and biomass at hadal depths is second only to the Holothurioidea and Bivalvia, which are perhaps the most diverse hadal fauna known. They are, therefore, one of the few classes where there is relative wealth of information available in terms of diversity and distribution compared to other faunal classes. The composition of Polychaeta dwelling at depths of more than 6000 m is extremely complex and urgently requires either new comprehensive sampling or major revision. Based on the *Vitjaz* and *Galathea* data, Belyaev (1989) lists 7 orders, 26 families, 50 genera and 75 species (one with two subspecies). Of these, 30 species (40%) are thought to be endemic to hadal depths, but 14 of these species are known from a single finding only and the other 16 by very few findings. Furthermore, many species have only been described to genus or family level.

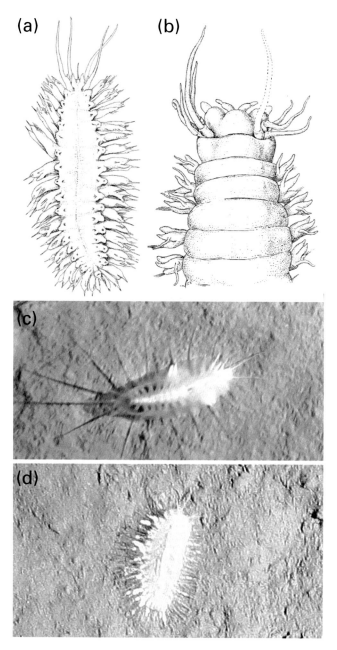

Figure 7.3 Examples of hadal polychaetes. (a) *Macellicephala hadalis* from 10 190 m in the Philippine Trench; (b) anterior part of *Nereis profundi* from 7250–7290 m in the Banda Trench; (c) polynoidae from 9300 m and (d) 6979 m in the Kermadec Trench. (a, b) taken from Kirkgaard (1956) reproduced with the permission of *Galathea Report* and (c, d) are courtesy of HADEEP.

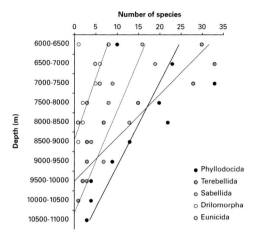

Figure 7.4 Number of species per 500 m depth stratum for the Polychaeta orders: Phyllodocida, Terebellida, Sabellida, Drilomorpha and Eunicida (Spionida not shown).

A recent census of marine polychaetes has been produced to evaluate whether there is indeed a distinct hadal fauna (Paterson *et al.*, 2009). The study was part of the 10-year Census of Marine Life programme (CoML; Snelgrove, 2010) of which one of the objectives was to assess global biodiversity in the oceans. To evaluate marine polychaetes, Paterson *et al.* (2009) assembled 3633 records, most of which are from depths >2000 m. These records were used to test for the degree of geographical similarity of polychaete found at >6000 m between 20 trenches using Parsimonious Analyses of Endemism; Rosen, 1988).

A total of 107 species of polychaetes were identified from trench depths and these comprised a mix of hadal endemic species, abyssal–hadal species and some species that extended from bathyal to hadal depths. The four most common families were Polynoidae, Ampharetidae, Maldanidae and Onuphidae. Each family contained several species that appear to be restricted to the abyss or to hadal depths: 17, 6, 3 and 2, respectively (although there are instances where some records are <2000 m). The analysis excluded species known from a single finding.

Ecocladistic analysis revealed high levels of trench endemism, with only limited resolution supporting biogeographical similarity. These results suggest that if a distinct hadal fauna exists, it is represented by a low number of species, many that are possibly endemic to a particular trench (although sampling bias must be acknowledged). The number of hadal species that have a wide bathymetric range suggests that the trench polychaetes are at the deeper end of a shallower distribution and not necessarily exclusively 'deep' or indeed 'hadal'. Paterson *et al.* (2009) also suggested that a source–sink system may apply to hadal communities deriving from, and maintained by, dispersal from abyssal and bathyal assemblages (Rex *et al.*, 2005). However, the scarcity of information regarding polychaete reproduction and larvae dispersal makes this a difficult hypothesis to test.

The same dataset was used to examine the biogeography of polychaetes based on the physical properties of the trenches; size, productivity and seismic activity versus polychaete communities; Tilston, 2011). Statistical analysis of trench size identified two groupings: one with the longest trenches, Aleutian, Peru–Chile and Java, and one group comprising the rest. A similarity matrix of POC flux also resulted in two groups: one with Aleutian, Kuril–Kamchatka, Middle America, Peru–Chile and Java, and another comprising the rest. The biogeography of polychaetes was split into two broad groups (meaning there were many similarities between communities) with pronounced intra-group gradient in faunistic change, and an overarching gradient with both latitude and POC flux. Although the volume of polychaete records are more numerous than most hadal classes, they are still insufficient to unequivocally draw conclusions of biogeographical patterns.

Of all the hadal Polychaeta, the Polynoidae family (scale worms) are the most characteristic with 20 species, and of those, 17 (85%) are endemic for depths over 6000 m (Belyaev, 1989); these 20 species belong to 9 genera, of which 6 are endemic to the trenches. Lemche *et al.* (1976) estimated the abundance of two Polynoidae species in the New Hebrides Trench between 6758 and 6776 m at one individual per 100 m^2. In these images, up to four specimens of Polynoidae were occasionally present in an image and were photographed at six out of seven stations. The *in situ* photography of Lemche *et al.* (1976) also showed that many species were free-swimming and not exclusively benthic; behaviour that was also noted in the HADEEP expedition to the Kermadec Trench (unpublished). In fact, it was in the PROA images where Polynoidae were first found to be capable of swimming (Lemche *et al.*, 1976). Similar images were obtained during the HADEEP project from 7884, 8613 and 9281 m in the Kermadec Trench. Here groups of polynoids were seen both on the seafloor and swimming 50 cm above it (Fig. 7.3). These polychaetes were in the region of 4–6 cm long.

In the Polychaeta listings in Belyaev (1989), there are almost as many species with extraordinary eurybathic depth ranges (sublittoral to hadal; 26%), as there are species that are limited to the deeper areas (21%). Belyaev noted with curiosity a few immediate bathymetric groups that had 2.5 times fewer intermediate depth range species than the eurybathic ones. It was suggested that these eurybathic species may be erroneous given further examination or sampling, as it is difficult to assume that populations spanning over 6000 m of water could be classified as the same species. The bathymetric ranges of all known polychaetes are shown in Figure 7.4.

Also within the Polychaeta phylum is the more recently reclassified family Siboglinidae, comprised of the former tube-dwelling Pogonophora and large tube worms Vestimentifera (Rouse, 2001). There are 29 species of hadal Siboglinidae (all formerly Pogonophora) listed from the *Vitjaz* and *Galathea* collections, comprising 11 genera (Belyaev, 1989). Of these, only five species also occur at abyssal depths, one is known from bathyal depths and one from shallower depths still (22 m). The 22 endemic hadal species (76%) are confined to single trenches or to several neighbouring trenches. The deepest record of Siboglinidae is 9715–9735 m in the Izu-Bonin Trench (*Heptabrachia subtilis* Ivanov, 1957).

Ivanov (1963) suggested that as filter-feeders, these organisms depend on the quantity of detritus suspended in the near-bottom water and the bacterial flora developing on it. Therefore, the Siboglinidae are likely to be more frequently found where concentrations of near-bottom suspended organic matter are plentiful. This is supported by their absence in open-ocean areas far from continents and tentatively supported by their apparent absence in the open-ocean trenches (Volcano, Mariana, Yap, Palau, New Hebrides, Tonga, Kermadec, Romanche), although further exploratory research may detect them elsewhere (Ivanov, 1963; Belyaev, 1989). Dense aggregations have been found in the Kuril–Kamchatka Trench where Siboglinidae are both relatively diverse and abundant. There are 10 known species and the trawl catch frequency in this trench is 50% (in general it is 28%). The *Vitjaz* once hauled 1500 specimens comprising six different species from 9000 m, most of them were *Zenkevitchiana longissima* Ivanov, 1957, whose white leathery tubes reached up to 1.5 m long (Belyaev, 1989).

There are currently no records of any giant tube worms (formally Vestimentifera) from hadal depths, although they may well be discovered as more exploratory research is undertaken, particularly on the more seismically active fore arcs.

7.5   Miscellaneous worms

There are various other fauna found at hadal depths that comprise just a few species and were identified from a very low number of samples. Many of these organisms can loosely be categorised as 'worms' albeit from a variety of phyla and classes. In most instances their apparent scarcity or absence from the great depths may be a sampling artefact, since many of these species are highly fragile and often very small, and are easily damaged beyond recognition when trawled (such as the Turbellaris and Nemerea). Similarly, some species may not occur in the densities of other more conspicuous species and therefore have gone relatively unnoticed (e.g. Enteropneusta). There are, of course, others which although technically found at >6000 m, really only represent the deeper fringes of an abyssal population, such as the sipunculids. While there is very little information available by which to present a meaningful discussion on the 'miscellaneous worms', they are mentioned here for completeness.

Echiura (spoon worms) are known from 17 trenches at all depths (max depth = 10 150–10 210 m in the Philippine Trench). From 60 findings, accounting for a trawl catch frequency of 35%, all the hadal Echiura belong to the Bonelliidae family which is comprised of 10 genera and 15 species. Four of these species are thought to be endemic (but none at the genus level) while five and three species have been found in the abyssal and bathyal zones, respectively. The Echiura therefore comprise a characteristic fraction of the hadal community fauna.

Echiura are also known to utilise plant and wood debris. They were found in wood fragments from the Puerto-Rico Trench (5890–6000 m) and among seagrass rhizomes in the Cayman Trench (6740–6780 m). Wolff (1976) confirmed from the stomach contents that these worms feed directly on this debris.

Of the oligochaetes (Phylum: Annelida), there has only ever been one finding at hadal depths; four specimens of *Bathydrilus hadalis* (Tubificidae) from 7298 m in the Aleutian Trench.

Turbellaria (flat worms) from the phylum Platyhelminthes and the order Polycladida have been trawled from the Kuril–Kamchatka and Peru–Chile Trenches, albeit in extremely low numbers. In the Kuril–Kamchatka Trench, single specimens were captured from 7265–7295 m and 9170–9335 m (which may have been a pelagic bycatch; Belyaev, 1989) and 6000–6354 m in the Peru–Chile Trench (Frankenberg and Menzies, 1968). In the former case, a 500 μm mesh trawl net was used and proved that conventional trawling damages these extremely fragile animals. Such damage may have accounted for their absence in samples from the *Vitjaz* and *Galathea* expeditions. Trawling damage was later confirmed in the Aleutian Trench at 7298 m by Jumars and Hessler (1976) who, rather than using a trawl, collected 37 specimens of Turbellaria in a 0.25 m^2 box core (148 individuals m^{-2}).

Gastrotricha (hairy back worms) have only been described from 6000–6354 m in the Peru–Chile Trench (Frankenberg and Menzies, 1968).

Nemerea (ribbon worms) have been identified in the Kuril–Kamchatka, Aleutian, Peru–Chile and the South Sandwich Trenches to depths of 7230 m, although usually only in fragments (Belyaev, 1989).

Sipuncula (peanut worms) are known from the Aleutian, Kuril–Kamchatka, Japan and South Sandwich Trenches. Of the eight known species of the genera *Golfingia* and *Phascolion*, seven are eurybathic, extending from the bathyal zone or shallower. Despite a trawl catch frequency of 62%, they have only been found at <7000 m (the deepest being 7000 m in the Peru–Chile Trench; *Nephasoma* (*Golfingia*) *schuttei* (Augener, 1903; Golfingiidae). Although they do not represent a significant hadal fauna, they do, on occasion, occur in large numbers; in the northwest Pacific Trench the *Vitjaz* hauled in up to 615 specimens in a single trawl. This haul was comprised mainly of either *Phascolion lutense* Selenka 1885 or *Nephasoma* (*Golfingia*) *minuta* (Keferstein, 1862) (Belyaev, 1989).

Plankton net hauls by the *Vitjaz* in the Kuril–Kamchatka Trench recovered small Chaetognatha (arrow worms) from 7000–6000 m and 8700–7000 m, thought to be *Eukrohnia fowleri* of the family Eukrohniidae. An arrow worm was also observed in the Puerto-Rico Trench by Pérès (1965), although no images or samples were taken.

Enteropneusta (acorn worms) of the Hemichordata phylum are commonly found burrowing in soft sediment (Smith *et al.*, 2005) or in the deep sea in particular they live on the sediment surface leaving characteristic spiral and looped faecal trails of raised sediment behind them (Heezen and Hollister, 1971; Lemche *et al.*, 1976). Recently, they have also been found to swim or at least actively drift in near-bottom currents (Osborn *et al.*, 2012). They have been collected from several trenches: Kuril–Kamchatka (5615–8100 m), Aleutian (6520–7250 m) and the South Sandwich Trench (8004–8116 m) and observed in images from the New Britain and New Hebrides

Trenches (Lemche *et al.*, 1976). The density of acorn worms, as calculated from the PROA images, was ~1 ind. $100\,m^{-2}$ (Lemche *et al.*, 1976). After many years of convoluted taxonomy as a result of great morphological diversity, Osborn *et al.* (2012) re-diagnosed the family and established that most deep-sea enteropneusts are part of a single clade, the Torquaratoidae, which are also likely to be those observed at hadal depths.

8 Porifera, Mollusca and Echinodermata

8.1 Porifera

Porifera (sponges) are one of the simplest multicellular organisms. The occurrence and diversity of Porifera at hadal depths is somewhat low (Table 8.1) and many findings are only solitary specimens. Although the sponges are a seemingly diverse group of organisms that span multiple trenches, the species that are found in the trenches are also found at abyssal, bathyal and even shallower depths (<500 m). The Porifera classes are split evenly between the Demospongiae (most common in shallower waters) and the Hexactinella or glass sponges (more characteristic in the deep sea). The majority of the Porifera do not extend much deeper than 7000 m and Belyaev (1989) suggests that the hadal sponges are simply a depleted abyssal fauna. Evidence of the low densities of these organisms on the seafloor was provided by the PROA images; where only 3 out of 4000 images contained sponges (Cladorhizidae; Lemche *et al.*, 1976). Mass occurrences of sponges were, however, found on one occasion by the *Vitjaz*, on the east slope of the Emperor Trench fault at 6272–6282 m, where 207 specimens from five different species were trawled. Of these, 200 specimens were identified as *Hyalonema apertum* Schulze, 1886 (Koltun, 1970).

The absence of sponges from the Kuril–Kamchatka Trench is thought to be a result of the dominance of silt (which can clog sponge irrigation systems) and the lack of solid substrates for attachment to the seafloor (Belyaev, 1989). This finding suggests that in the other trenches, sponges may inhabit deeper depths where silt is less and where solid substrates are more plentiful; a hypothesis supported by the finding of several sponges from 8950–9020 m in the Tonga Trench and 9990 m in the Philippine Trench. Therefore, the dissemination of hadal Porifera is, perhaps, determined by more favourable substrata and not by depth, although supporting data are still circumstantial. Furthermore, the heterogeneity of substrata within any trench, particularly between the fore arc and the oceanic slopes, implies that sponges may well have an equally heterogenic distribution.

8.2 Mollusca

Within the phylum Mollusca (shellfish), the Gastropoda and Bivalvia classes are extremely characteristic hadal fauna in particular. Both diverse classes have been found in every trench sampled, often in extraordinarily high numbers and both are known to

Table 8.1 Maximum depth of each order and family of Porifera and the number of genera and species therein.

Class	Order	Family	Genera	Species	Depth range (m)
Hexactinellida	Amphidiscophora	Hyalonematidae	1	2	6090–6860
Hexactinellida	Haxasterophora	Caulphacidae	1	3	6090–6770
Hexactinellida	Haxasterophora	Euplectellidae	1	1	6296–6328
Hexactinellida	Haxasterophora	Rossellidae	2	2?	5650–8540
Demospongiae	Poecilosclerida	Chondrocladiidae	1	1	6090–8660
Demospongiae	Poecilosclerida	Cladorhizidae	3	5	6620–9990
Demospongiae	Poecilosclerida	Esperiopsidae	1	1	6860
Demospongiae	Poecilosclerida	Cladorhizidae	1	1	6920–7567

extend to full ocean depth (Table 8.2; Fig. 8.1). As well as these two classes, other Mollusca found at hadal depths include the Scaphopoda, Polyplacophora and Monoplacophora. While several records exist for these three classes in the trenches, their sitings are relatively few compared to the other molluscs and, generally, they appear to be restricted to upper depths (<7600 m).

8.2.1 Gastropoda

Gastropoda (sea snails and limpets; Fig. 8.2) are an important fraction of the hadal fauna and, to date, these organism have been found in all the trenches sampled and at full ocean depth; 10 687 m in the Tonga Trench and 10 730 m in the Mariana Trench. The Gastropoda were the seventh most dominant taxa collected from the *Galathea* expeditions in the 1950s (Wolff, 1970). The hadal Gastropoda comprise at least 40 putative genera, although the species lists presented by Belyaev (1989) include many undescribed species or specimens classified to genus or family-level only. Even so, the data indicated a total number of 60 hadal species. Belyaev (1989) also estimated that the number of known species of hadal gastropods will reach close to 100 once a systematic taxonomic exercise is undertaken.

The composition of the hadal Gastropoda is complex; there are 19 families, 40 genera and 58 species listed by Belyaev (1989) (Table 8.2). The families Cocculinidae, Turridae and Buccinidae dominate in diversity and trawl catch frequency compared to the others. The Turridae represent the most diverse gastropod family and the second most diverse family of the Mollusca (13 species, 8 genera). Endemism at the species level is estimated at 68%, but the majority of these endemics are known from single specimens. Of the non-endemic species, two-thirds are restricted to the abyssal zone.

The number of gastropod species decreases with depth while the percentage of endemism increases; a similar pattern in most groups (Fig. 8.1). Gastropoda, however, show unusually high endemism at the genus level; 26% of the genera are endemic to the hadal zone and another four genera are confined to a single trench and its surrounding abyssal plain. These values differ slightly from those calculated by Wolff (1970) who

Table 8.2 Hadal community composition of the Mollusca phylum (class, order, family, numbers of genera and species and maximum known depth).

Class	Order	Family	Genera	Species	Maximum depth (m)
Gastropoda	Docoglossa	Bathypeltidae	1	1	8560–8720
Gastropoda	Docoglossa	Bathysciadiidae	1	1	8240–9530
Gastropoda	Docoglossa	Propolidiidae	1	1	6090–6135
Gastropoda	Fissurellidae	Fissurellidae	1	1	6290–6300
Gastropoda	Alata	Seguenziidae	3	3	7000–7450
Gastropoda	Anisobranchia	Skeneidae	1	1	6290–6330
Gastropoda	Anisobranchia	Trochidae	3	4	6620–8035
Gastropoda	Aspidophore	Naticidae	1	1	6330–6430
Gastropoda	Hamiglossa	Buccinidae	4	8	5329–9050
Gastropoda	Hamiglossa	Cancellariidae	2	2	6660–7340
Gastropoda	Heterostropha	Aclididae	1	1	8210–8300
Gastropoda	Heterostropha	Piramidellidae	1	1	7000–7280
Gastropoda	Homeostropha	Eulimidae	1	1	6660–6770
Gastropoda	Planilabiata	Bathyphytophilidae	2	2	5800–8120
Gastropoda	Planilabiata	Cocculinidae	5	10	5179–8400
Gastropoda	Toxoglossa	Turridae	8	13	6052–10 730
Gastropoda	Tectibranchia	Phylinidae	1	4	6410–7587
Gastropoda	Tectibranchia	Retusidae	1	1	7974–8006
Gastropoda	Tectibranchia	Scaphandridae	2	2	5650–8035
Bivalvia	Nuculida	Ledellidae	5	20	5650–10190
Bivalvia	Nuculida	Malletiidae	1	3	5650–8035
Bivalvia	Nuculida	Nuculanidae	4	5	5650–10687
Bivalvia	Nuculida	Tindariidae	1	1	6296–7286
Bivalvia	Lucinida	Montacutidae	1	1	6290–8580
Bivalvia	Lucinida	Mytilidae	1	1	6050–6150
Bivalvia	Lucinida	Thyasiridae	5	9	6150–10,687
Bivalvia	Pectinida	Limarlidae	1	1	6135–9735
Bivalvia	Pectinida	Pectinidae	4	6	5650–8100
Bivalvia	Venerida	Pholadidae	1	2	6660–7290
Bivalvia	Venerida	Teredinidae	2	2	7250–7290
Bivalvia	Venerida	Vesicomyidae	3	7	6156–10730
Bivalvia	Cuspidariidae	Cuspidariidae	2	2	6290–9990
Bivalvia	Verticordiidae	Vertocordiidae	3	10	6040–9335
Scaphopoda	Galilida	Entalinidae	2	3	5900–6780
Scaphopoda	Galilida	Pulsellidae	-	1–5*	5650–7657
Polyplacophora	Cyclopoida	Chitonophilidae	2	4	6740–7657
Monoplacophora	Tryblidiida	Neopilinidae	3	4	1647–6354

* indicates up to five undescribed species.

reported that 16 species of gastropod were found at >6000 m, of which only 2 were also found at abyssal depths. Thus, he estimated gastropod endemism at hadal depths to be 87.5%.

Due to the relatively small size and colour of most hadal gastropods, they are often not visible or obvious from *in situ* photography (Fig. 8.2). For example, Lemche *et al.* (1976) did not report seeing any gastropods at all in over 4000 images from four

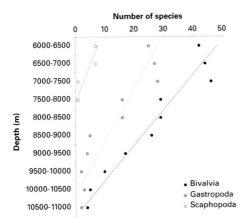

Figure 8.1 Number of species per 500 m depth stratum for the Mollusca: Bivalvia, Gastropoda and Scaphopoda.

Indo-Pacific trenches. There are, however, high quality images of the buccinid *Tacita zenkevitchi* from 5329 and 6173 m in the Peru–Chile Trench (Aguzzi *et al.*, 2012). These were photographed in time-lapse (every 1 min) over periods of 11 h 09 min and 18 h 40 min, respectively. These data showed that *T. zenkevitchi* is a gregarious scavenger, albeit a potentially facultative one. Several hours after the bait had been placed on the seafloor, the gastropods entered the field of view and positioned themselves directly on the bait where they stayed for several hours until being adversely removed by macrourid or ophidiid fish.

These data allowed each gastropod to be digitally tracked across the seafloor, providing estimates on their locomotion speed and area coverage. They traversed the seafloor at a mean absolute speed of 3.2 cm min^{-1} ± 1.5 S.D. (specific speed = 0.6 SL min^{-1} ± 0.3 S.D.) at 5329 m and 2.3 cm min^{-1} ± 1.2 S.D. (0.6 SL min^{-1} ± 0.2 S.-D.) at 6173 m. The tracks left in their wake were clearly visible as a mucus layer within the depressed sediment. The mean area of the gastropod tracks was estimated at 0.03 m^{-2} h^{-1} ± 0.02 S.D. The gastropods had a mean shell height of 5.33 cm at 5329 m and 4.7 cm at 6173 m. Three specimens were also recovered from the baited traps coupled to the camera system (mean shell height of 3.1 cm; Fig. 8.2).

The Gastropoda collected from the Kermadec Trench on the *Galathea* expeditions exhibited different feeding strategies (Knudsen, 1964). Based on morphological characteristics, Naticidae and Admete appear to be predators and the two species of Trochidae are 'scrapers' (feeding on surface film on rocks and stones) but may also be opportunistic scavengers or detrital feeders.

Cocculiniform limpets are a particularly complex group of gastropods. At hadal depths they have been described in six trenches but are most prevalent in number and diversity in the Cayman Trench (Leal and Harasewych, 1999). Cocculinids are exclusively associated with submerged plant or wood debris (Strong and Harasewych, 1999) and are often collected by a bottom trawl (e.g. George and Higgins, 1979). For example,

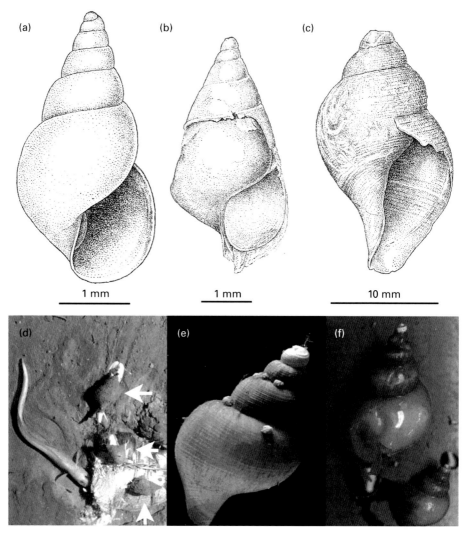

Figure 8.2 Hadal gastropods. (a) *Aclis kermadecensis* from 8210–8300 m in the Kermadec Trench. (b) *Melanella hadalis* from 6660–6770 m in the Kermadec Trench. (c) *Admete bruuni* from 6660–6770 m in the Kermadec Trench. (d) Three scavenging gastropods (*Tacita zenkevitchi*) at 5329 m in the Peru–Chile Trench (arrowed), which were later captured from 6173 m (e). (f) Unidentified Buccinidae from 7703 m in the Japan Trench, note the thumbprint and tear in the soft shell. (a–c) are taken from Knudsen (1964), reproduced with the permission of *Galathea Report* and (d–e) are courtesy of HADEEP.

Leal and Harasewych (1999) described several new species originally collected from the Cayman and Puerto-Rico Trenches in the 1950s and '60s, and reported that *Fedikovella caymanensis* Moskalev, 1976 and *Caymanabyssia spina* Moskalev, 1976 were recovered inhabiting wood debris, while *Amphiplica plutonica* Leal and Harasewych, 1999 was found attached to blades of turtle grass. Turtle grass (*Thalassia testudium*) is common within the seagrass meadows of the Caribbean and is often found on the floor of the neighbouring trenches amongst other plant debris (Moore, 1963; Wolff, 1976). Therefore, in the case of cocculinformes, their high diversity and abundance is likely to be due to their location within highly productive coastal regions containing vast seagrass meadows. When the seagrass is removed naturally or by seasonal adverse weather systems (to which the area is prone), it accumulates in the trench providing the ideal substrata for such organisms.

Coping mechanisms that enable organisms to deal with the environmental conditions at hadal depths are particularly pertinent in the gastropods, as beyond the CCD they are physically unable to ossify their shells. Nevertheless, gastropods prevail to full ocean depth. At 10 700 m in the Mariana Trench, remains were found of several small Gastropoda shells that only had a preserved periostracum (the thin organic coating of the shell), while the calcareous part was completely dissolved (Belyaev, 1989). In 2008, three specimens of unidentified Buccindae were recovered in baited traps from 7703 m in the Japan Trench (unpublished data, HADEEP; Fig. 8.2). Despite being perfectly formed, the shells were incredibly soft, so much so, they were easily torn upon careful removal from the trap. Ingeniously, gastropods have adapted to survive beyond the CCD by maintaining the structural formation of their shell but in a soft form; a strategy adopted by many calcified groups (e.g. Sabbatini *et al.*, 2002; Todo *et al.*, 2005).

8.2.2   Bivalvia

Bivalve molluscs (clams and mussels) are another dominant class in the hadal zone, often forming mass populations, particularly around chemosynthetic habitats (Boulègue *et al.*, 1987; Fujikura *et al.*, 1999; Fujiwara *et al.*, 2001). In terms of numbers recovered in trawls, Bivalvia are second only to the Holothurioidea (Wolff, 1960) accounting for catch rates of up to 82% (Belyaev, 1989). They have been found in all the studied trenches and at all depths including greater than 10 500 m (Table 8.2; Fig. 8.1) in the Mariana and Tonga Trenches.

The records of hadal Bivalvia listed in Belyaev (1989) are complex and include a lot of incomplete data and identifications (20 taxons only defined to the genus or family). The records list 33 taxons at genus level, 6 orders, 14 families and 47 species but Belyaev (1989) thought the numbers would be even greater once all specimens were examined in more detail. Indeed, on compiling an updated list for this study, a total number of approximately 70 species is perhaps a better estimate. The most diverse family of the Bivalvia is the Ledellidae (of the Nuculacea superfamily), which include 20 species from 5 genera. This is another family which is likely to herald many more species with further analysis. The remaining 13 families are much less diverse. Nuculaceans are known to exhibit extraordinary variation in species and morphology (Gage and Tyler, 1991), suggesting a variety of life history traits and exploitation of different habitat types.

Of the 47 known Nuculacean species found at hadal depths, 32 (68%) have not been found at abyssal depths. Endemic species are generally confined to either single trenches or clusters of adjoining trenches.

Table 8.3 Examples of mass findings of Bivalvia in the hadal zone. Table shows selected trawls from the Kuril–Kamchatka Trench. Modified from Belyaev (1989).

Depth (m)	Bivalve species	Number of specimens
7210–7230	*Bathyspinula vityazi*	184
	Tindaria sp.	189
7600–7710	*Parayoldiella mediana*	227
8185–8400	*Vesicomya sergeevi*	3496
8240–8345	*Vesicomya profundi*	119
9070–9345	*Vesicomya sergeevi*	186
9170–9335	*Yoldiella ultraabyssalis*	440
	Vesicomya sergeevi	191
9520–9530	*Yoldiella ultraabyssalis*	3380
	Vesicomya sergeevi	1935

On many occasions, even at the greatest depths, the Soviet expeditions discovered mass findings of Bivalvia. They sometimes recovered over 3000 specimens from a single trawl (Table 8.3).

Three of the Bivalvia species recovered during the Soviet expeditions belonged to the families Teredinidae and Pholadidae. Both are borer-molluscs that were retrieved from sunken plant debris recovered from the Banda Trench >7000 m. Although adults of these species inhabit hadal depths, it is not known if they reproduce there. Knudsen (1970) suggested they were 'guest' species whose trench populations may be maintained via recruitment from the shallower zones where these species typically live and reproduce.

Two common families of deep-sea bivalves are Thyarsiridae and Vesicomyidae. Both of these families are relatively diverse in the hadal zone (nine and seven species, respectively). They are often associated with hydrothermal-vent and cold seep habitats (e.g. Boulègue *et al.*, 1987; Fujikura *et al.*, 1999). Vesicomyid bivalves are a consistent component of the communities that inhabit sulphide-rich reducing environments and they have a global distribution, from shallow depths (100 m) to almost full ocean depth (9530 m; Krylova and Sahling, 2010), e.g. the genus *Calyptogena* (Boulègue *et al.*, 1987). Vesicomyids found in sulphide-rich reducing habitats are typically larger (up to 30 cm) than others. These larger vesicomyid clams live in symbiosis with the sulphur-oxidizing bacteria in their gills (Fisher, 1990). In these habitats, the clams have access to substantial concentrations of hydrogen sulphide in the pore water (Barry *et al.*, 1997). Conversely, far less is known about the smaller representatives of the family. The irregular mass occurrences of vesicomyid clams in the trenches may be indicative of chemosynthetic habitats within the trenches. A theory further supported by the discovery of the cold seeps at 7200 m in the Japan Trench that were dominated by Thyasirid clams of the genus *Maorithyas* (Fujikura *et al.*, 1999; Fujiwara *et al.*, 2001).

As the Bivalvia include molluscs with diverse habitat and feeding preferences, it could be construed that some of the mass findings were a result of localised density increases in the vicinity of sunken wood/plant debris, or the presence of chemosynthetic habitats.

8.2.3 Scaphopoda

Scaphopoda (tusk shells) are a small class of recent molluscs, with approximately 500 known living species dwelling in shallow and deep marine sediments (Gracia *et al.*, 2005). They have been recorded 20 times at hadal depths of which 19 reports were from Soviet expeditions (Belyaev, 1989). They have been found in many major trenches throughout the oceans but are limited to depths of <8000 m (Table 8.2; Fig. 8.1).

There are only 10 known records of hadal scaphopods, which include three species and one with two subspecies. Only one hadal scaphopod has ever been found at depths greater than 7000 m; this specimen is from one of five unidentified species belonging to the Pulsellidae family, one of which was caught between 6920 and 7657 m in the Bougainville Trench, the other four were from the Japan, Java, San Cristobal and Romanche Trenches. Of the described species, there are *Costentalina caymanica* Chistikov, 1892 (Cayman Trench; 5900–6780 m), *Costentalina tuscarorae* Chistikov, 1892 (Japan Trench; 6480–6640 m), *Striopulsellum* (*Siphonodentalium*) *galatheae* Knudsen, 1964 (Java Trench; 6900–7000 m) and *Entalinidae* sp. that have been found in the Kuril–Kamchatka Trench (6090–6675 m) and the South Sandwich Trench (6052–6150 m).

The hadal scaphopod (*S. galatheae*) examined by Knudsen (1964) was found to feed primarily on unicellular organisms and the microscopic larvae of benthic invertebrates. Other scaphopods are known to feed on other items such as Foraminifera tests (Morton, 1959), however, this was not evident in the *Galathea* specimens (Knudsen, 1964).

8.2.4 Polyplacophora and Monoplacophora

Specimens of hadal Polyplacophora (chitons), from the family Leptochionidae (order Lepidopleurida) have only been found on four separate occasions in trenches located within the tropical regions of the Pacific and Atlantic Oceans (Schwabe, 2008).

The first hadal sample discovered was *Leptochiton vitjazae* (Sirenko, 1977), which was found and described from the Bougainville Trench (6920–7657 m; Sirenko, 1977). The Soviet expeditions also retrieved 12 specimens of *Leptochiton* from the Palau Trench at 7000–7170 m, one specimen from the New Hebrides Trench at 6680–6830 m (Belyaev, 1989) and 39 specimens attached to sunken pieces of wood at 6740–6780 m in the Cayman Trench. The specimens found in the Cayman Trench were later found to be of a new genus and species: *Ferreiraella caribbea* (Sirenko, 1988).

The discovery of living Monoplacophora by the *Galathea* expedition was perhaps one of their greatest finds, since this order of Mollusca was previously known only from Paleozoic fossils; Cambrian–Devonian, *c.* 500–320 Ma (Lemche, 1957; Menzies *et al.*, 1959; Schwabe, 2008). While the name Monoplacophora should technically be Tryblidiida (Lemche, 1957; Wingstrand, 1985), it is still the most widely used (Schwabe, 2008).

Of the modern Monoplacophora, 35% of the 31 known species inhabit depths exceeding 2000 m with four species extending to hadal depths, one of which is endemic. These four species are *Veleropilina (Rokopella) oligotropha* (Rokop, 1972) (6065–6079 m; northwest Pacific Trough), *Vema bacescui* (Menzies, 1968) (5986–6134 m; Peru–Chile Trench), *Vema ewingi* (Clarke and Menzies, 1959) (5817–6002; Peru–Chile Trench) and *Neopilina* sp. which has large bathymetric range of >4000 m (1647–6354 m; Peru–Chile Trench and Scotia Ridge). All are members of the Neopilinidae family and are believed to be deposit feeders (either sediment ingesting or unselective detritus feeding). It is likely that they are preyed upon by ophiuroids, gastropods and fish (Menzies *et al.*, 1959).

Based on the current number of samples, which is admittedly low, the Monoplacophora appear in highest abundance in the vicinity of trenches, particularly up the eastern Pacific Ocean (0.04–0.7 ind. km^{-2}; Menzies *et al.*, 1959). Furthermore, of the abyssal species, none have been found in the western Pacific Ocean (Schwabe, 2008). The reasons for this remain unresolved.

8.3   Echinodermata

The phylum Echinodermata encompasses Crinoidea (sea lilies), Asteroidea (sea stars), Ophiuroidea (brittle stars), Echinoidea (urchins) and Holothuroidea (sea cucumbers). All of these are wholly represented in the hadal zone and in most trenches sampled (Gislén, 1956; Madsen, 1956; Hansen, 1957; Wolff, 1960, 1970; Table 8.4). These five groups all exhibit different trends within the trenches; some (e.g. Echinoids) are restricted to the shallower depths while others (e.g. Holothurians) thrive at full ocean depth (Fig. 8.3). Similarly, where some echinoderms prefer rocky outcrops and hard substrata (e.g. Crinoids), some prefer the softer sediments of the oceanic slopes (e.g. Asteroidea) and others the soft sedimentary accumulation at the trench axis (e.g. Holothurians).

8.3.1   Crinoidea

Crinoids (sea lilies) are reasonably well represented in most of the deep trenches trawled during the *Vitjaz* expeditions (Gislén, 1956). They have, so far, been recovered from the Aleutian, Kuril–Kamchatka, Japan, Izu-Bonin, Volcano, Palau, Bougainville, Kermadec, Java, New Hebrides and Peru–Chile Trenches (Lemche *et al.*, 1976; Belyaev, 1989; Oji *et al.*, 2009). They have been found at depths of 6000 to nearly 10 000 m (Belyaev, 1989). One species, in particular, that trawled from between 8175 and 9345 m in the Kuril–Kamchatka Trench is described as 'massive' (Belyaev, 1989). The general abundance of crinoids is also impressive; The *Vitjaz* trawled on occasion up to 100 and even 255 specimens in one hadal trawl. Such large numbers of crinoids per trawl is, in part, due to their tendency to grow in dense patches, utilising the rocky outcrops that are characteristic of trenches (Oji *et al.*, 2009). Video footage from the ROV *Kaikō* obtained by Oji *et al.* (2009) represents the deepest *in situ* observation

Table 8.4 Hadal community composition of the Echinodermata (class, order, family, numbers of genera and species and maximum known depth).

Class	Order	Family	Genera	Species	Depth range (m)
Holothuroidea	Apodida	Myriotrochidae	4	15	5650–10730
Holothuroidea	Aspidochirotida	Synalactidae	5	5	6490–8260
Holothuroidea	Elasipoda	Elpidiidae	7	30	2470–10000
Holothuroidea	Molpadonia	Gephyrothuriidae	1	2	6758–9530
Holothuroidea	Molpadonia	Molpadiidae	1	1	6490–6650
Ophiuroidea	Ophiurae	Ophiacanthidae	2	4	6065–7880
Ophiuroidea	Ophiurae	Ophiodermatidae	1	1	6052–6150
Ophiuroidea	Ophiurae	Ophioleucidae	1	1	6680–8006
Ophiuroidea	Ophiurae	Ophiuridae	9	19	5650–8662
Asteroidea	Brisingida	Freyellidae	2	4	5650–8662
Asteroidea	Paxillosida	Porcellansteridae	6	9	5650–7880
Asteroidea	Valvatida	Goniasteridae	1	1	8021–8042
Asteroidea	Valvatida	Caymanostellidae	1	1	6740–6780
Asteroidea	Valvatida	Pterasteridae	1	2	6052–9990
Echinioidea	Echinothuroida	Echinothuroida	1	1	6090–6235
Echinioidea	Spatangoida	Holasteridae	1	1	5800–6850
Echinioidea	Spatangoida	Pourtalesiidae	3	7	5650–7340
Echinioidea	Spatangoida	Urechinidae	1	1	5800–6780

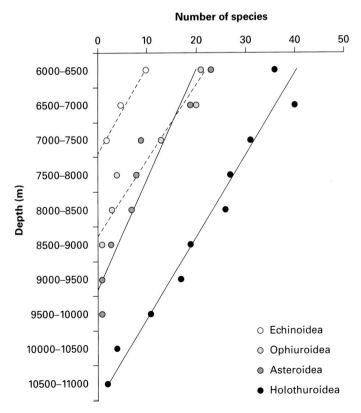

Figure 8.3 Number of species per 500 m depth stratum for the echinoderms: Holothuroidea, Asteroidea, Ophiuroidea and Echinoidea.

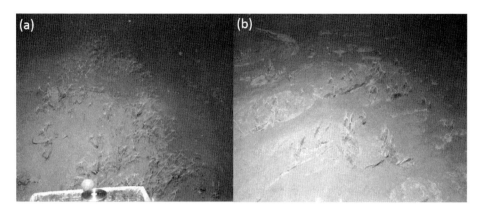

Figure 8.4 Examples of Crinoidea at hadal depth, where (a) is from 9092 m and (b) is from 9095 m in the Izu-Bonin Trench. Both images were taken with the ROV *Kaikō*. Images © JAMSTEC, Japan.

of stalked crinoids (Fig. 8.4). The footage clearly shows that they can occur in abundance at hadal depths and that they adopt the same feeding postures (mouths directed down-current) as many shallower-water counterparts. The dense aggregations on rocky outcrops are similar to the dense populations of stalked crinoids, or 'crinoid meadows', reported in shallower zones (Conan *et al.*, 1980; Messing *et al.*, 1990).

Based on earlier trawling expeditions, eight species of the most common genus are found in the hadal zone; the filter-feeding stalked crinoid *Bathycrinus* (family Bathycrinidae). This family is significantly represented in the hadal zone yet none of its species have been found at abyssal depths, suggesting a high degree of endemism. Specimens of *Bathycrinus* have also been recovered from geographically distant localities, from the Aleutian, Volcano and Palau Trenches in the North Pacific Ocean, to the Bougainville and Kermadec Trenches in the south (Mironov, 2000), across the Pacific Ocean in the Peru–Chile Trench (Menzies *et al.*, 1959; Menzies *et al.*, 1963) to the Southern Ocean's sub-zero South Sandwich Trench (Belyaev, 1989). The maximum depths at which crinoids have been recovered are between 9715 to 9735 m in the Izu-Bonin Trench. This is also the same location where they were filmed *in situ* by the ROV *Kaikō* (Oji *et al.*, 2009). Belyaev (1989) stated that the encounter frequency of crinoids between 6000 and 10 000 m is 22%, suggesting that they are indeed a major fraction of the hadal community.

Prior to the introduction of full ocean depth ROVs, photographs obtained by drop-camera revealed occasional stalked crinoids at >6000 m depth, with population density estimates of 1 ind. 100 m^{-2} (Lemche *et al.*, 1976). They reported photographing at least 25 *Bathycrinus* cf. *australis* Clark, 1907 (Bathycrinidae) from between 6758 and 6776 m in the New Hebrides Trench situated individually, in groups of three to six and mostly at the most exposed areas. They also noted seeing a further group of six individual crinoids between 8021 and 8042 m in the Palau Trench. This group was attached to what appeared to be a seagrass rhizome.

The significant abundance of crinoids at full ocean depth implies that there is sufficient food supply reaching the extreme depths to support these organisms. As they are filter-feeders, there must, therefore, be sufficient food suspended in the near-bottom currents to sustain these dense aggregations. The source of this food supply is still unknown, although Oji *et al.* (2009) hypothesised that conventional filter feeding may be supplemented by chemosynthetic sources, although evidence of such geological features in close proximity to the crinoids was lacking. They later concluded that perhaps the crinoids were exploiting the food resource accumulation thought to occur at the trench axis (as suggested by Danovaro *et al.*, 2003 and Jamieson *et al.*, 2010).

It seems likely, given the wide geographic distribution of stalked crinoids, that the hadal trenches provide an ideal setting for these organisms that are specialists when it comes to attaching to rocks and the other solid debris that is ever present in the trenches. The significance of crinoids in trenches will be further realised as more exploratory methods are undertaken.

8.3.2   Asteroidea

Asteroidea (sea stars) are known to inhabit at least 15 trenches to depths close to 10 000 m (Table 8.4), although they are more common at and generally restricted to depths of <8500 m. The frequency of encounter of Asteroidea in trawling hauls for depths from 6 to 10 km averaged 42% (Belyaev, 1989). Both the number of species and catch rate decrease linearly with depth (Fig. 8.3).

There are 41 known findings of 17 species belonging to five families of Asteroidea from depths exceeding 6000 m, many of which are undescribed. These families are Porcellansteridae (9 species, of which 4 are undescribed; 5650–7880 m), Pterasteridae (1 species, 11 undescribed; 6052 to 9990 m), Freyellidae (3 species and three findings of 1 undescribed species; 5650–8662 m), Goniastidae (1 undescribed species; 8021–8042 m) and Caymanostellidae (1 species; 6740–6780 m).

The deepest record (>8500 m) for a sea star is that of a single specimen of the genus *Hymenaster* (family Pterasteridae) from the Philippine Trench (Mironov, 1977, cited in Belyaev, 1989). While there are many species known from depths exceeding 6000 m, only one species has been described (*Hymenaster glegvadi*, from the Kermadec Trench; Madsen, 1956). *Hymenaster* spp. have been found in most of the trenches sampled by the *Galathea* and *Vitjaz* expeditions, from the North Pacific Ocean (Kuril–Kamchatka, Japan, Izu-Bonin, Volcano, Yap, Palau and Philippine Trenches), the South Pacific Ocean (Kermadec, New Hebrides and Bougainville Trenches) as well as the Atlantic Ocean trenches (South Sandwich and Romanche Trenches). The most numerous family in terms of species number is Porcellansteridae and this family is found in the same trenches as Pterasteridae, and also in the Aleutian, Ryukyu in the North Pacific Ocean and the Peru–Chile Trench in the southeast Pacific Ocean, the San Cristobal and New Britain Trenches in the southwest Pacific Ocean and the Cayman Trench in the Carribean Sea.

Asteroidea endemism in the hadal zone is 40% at the species level and of the 10 known genera from these depths, only one is endemic; *Lethmaster* (Belyaev,

1989). They were also found in bottom photographs from the New Hebrides and Palau Trenches (Lemche *et al.*, 1976) but have never been seen on any of the lander footage from the HADEEP projects. Lemche *et al.* (1976) do, however, provide a reasonably detailed description of the Asteroidea observed.

8.3.3   Ophiuroidea

Ophiuroidea (brittle stars) are a characteristic fauna at abyssal depths and are found in most major trenches, albeit mostly in the upper depths (Fig. 8.3). There are an estimated 24 known species belonging to four families; Ophiacanthidae (four species, three genera), Ophioleucidae (one species), Ophiodermatidae (one undescribed species) and the most representative family Ophiuridae (18 species, eight genera) (Table 8.4). The exact species number may vary as the records listed in Belyaev (1989) contain both undescribed and presumed species identifications.

The deepest known Ophiuroidea is *Perlophiura profundissima* Belyaev and Litvinova, 1972 (Ophiuridae) trawled from 8060–8135 m by the *Vitjaz* in the Kuril–Kamchatka Trench, although photographs also exist of an unidentified species from 8662 m in the Bougainville Trench (Lemche *et al.*, 1976). However, most species of ophiuroids are found at <7500 m with only four species found deeper (Fig. 8.3). Furthermore, these records span 17 of the world's trenches. The overall trawl catch frequency decreases from a mean of 50% between 6000 and 7500 m, to 9.6% at >7500 m. In general, the Ophiuroidea comprise a minimal component in most hadal trawl catches, however, there are instances of mass catches. For example, one trawl on the *Vitjaz* recovered 600 specimens of Ophiuroidea, accounting for 55% of the total number of samples. Similarly, findings of dense aggregation were found in the Palau Trench (55%) and the South Sandwich Trench (42% and 14%).

Ophiuroidea endemism at species level is estimated at 43%, with only one endemic genus in the trenches. The non-endemic species are apparently highly eurybathic, as they are known to inhabit both abyssal and bathyal depths suggesting bathymetric ranges of several thousand metres.

Ophiuroidea have been photographed several times at hadal depths in the HADEEP projects but these taxa are notoriously difficult to identify from images. At 7199 m in the Kermadec, 180 images of *Ophiura* aff. *loveni* (Lyman, 1878; Ophiuridae) were reported (Jamieson *et al.*, 2011a; Fig. 8.5a), two distinctly different species were observed in the Peru–Chile Trench at 4602 m and one further species at 6173 m (unpublished data, HADEEP; Fig. 8.5). The three Peru–Chile Trench species behaved slightly differently. Ophiuroid A was present periodically throughout the deployments, staying for either 3–8 min (40% of observations) or 30–90 min (60%). It is estimated that of the 10 observations made, seven of these were of individuals and only on one occasion were two individuals visible simultaneously. Ophiuroid B was observed three times; the first two occurrences saw solitary individuals traversing the seafloor seemingly indifferent to the bait and associated activity, while the third individual remained in the vicinity for over 3 h. Ophiuroid C aggregated to a maximum number of five

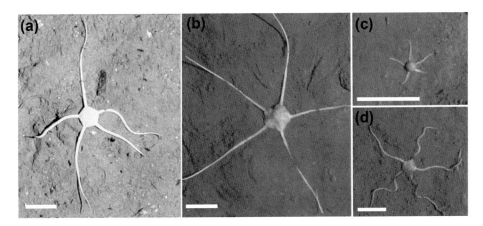

Figure 8.5 (a) *Ophiura* aff. *loveni* from 7199 m in the Kermadec Trench. (b, c, d) The ophiuroids, Ophiuroid A from 4602 m, Ophiuroid B from 4602 m and Ophiuroid C from 6173 m in the Peru–Chile Trench, respectively. Scale bars = 2 cm. Images courtesy of HADEEP.

individuals by 386 min into the deployment. As the numbers of the fish increased, these ophiuroids quickly exited the field of view. The behaviour of *O.* aff. *loveni* in the Kermadec Trench was similar to Ophiuroid A.

Lemche *et al.* (1976) made 350 observations of *Ophiura* sp. (and two specimens thought to be Ophiacanthidae) between 6758 and 6776 m in the New Hebrides Trench, using a drop-camera. The ophuiroids were frequently seen on the seafloor, situated between rocks and boulders. They observed up to 11 individuals in the same frame and estimated the average density to be at least 3 ind. 10 m^{-2}.

8.3.4   Holothuroidea

Holothuroidea (sea cucumbers) are thought to be one of the most significant hadal fauna (Hansen, 1957; Fig. 8.3), so much so that the hadal zone has been referred to as the 'kingdom of the Holothurians' (Belyaev, 1989). This is based on the results of deep trawls from the *Galathea* and *Vitjaz* expeditions where the trawl catch rate at >6000 m was 88%, only comparable to Polychaeta. The hadal Holothuroidea are, however, less diverse than the Crustacea and Polychaeta, but close to that of the Gastropoda and Bivalvia (Table 8.4). The holothurians are represented at hadal depths by seven families; Elpidiidae (30 species, 6 genera), Myriotrochidae (15 species, 4 genera), Synalactidae (5 species, 4 genera), Psychrolotidae (4 species, 2 genera), Gephyrothuridae (2 species, 1 genus), Leatmogonidae (1 species), Molpadiidae (1 species) and Palagothuriidae (1 species). Holothurians of the order Elasipoda, mainly the family Elpidiidae (Fig. 8.6), are thought to be one of the most important trench species based on genus diversity and density (Fig. 8.7). It also thought that Elasipoda are one of the most common animals in the deep sea (Hansen, 1972).

Holothuroidea species endemism is estimated at 69% in the trenches and most endemics are limited to a single trench or adjoining trenches. There is only one known

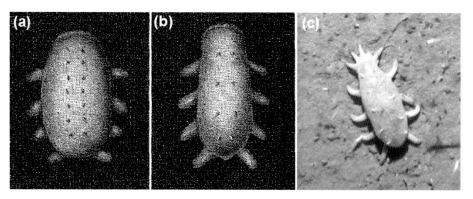

Figure 8.6 Two drawings of Elpidiidae from the *Galathea* expeditions; *Elpidia glacialis solomonensis* (a) and *E.g. kermadecensis* (b) and (c) *Elpidia atakama* photographed *in situ* at 8074 m in the Peru–Chile Trench. Images (a) and (b) taken from Hansen (1957), reproduced with the permission of *Galathea Report* and (c) is courtesy of HADEEP.

endemic genus, *Hadalothuria* Hansen, 1956. Of the non-endemic species, 6 are constrained to abyssal depths whereas the remaining 11 species are sufficiently eurybathic to extend as far as bathyal depths.

The number of trawl samples from hadal depths is still extremely low, even by comparison with abyssal depths. Therefore, there are some perhaps more abundant species where inter-trench trends can be observed, while there are others that still pose disparate trends that may change as further research in undertaken. An example of the disparate species is *Amperima naresi* (Théel, 1882; Elpidiidae) that has, so far, only been found in the Java Trench (Indian Ocean) and the Palau Trench in the North Pacific Ocean. It seems unlikely that a species would inhabit two such isolated trenches and not trenches in closer geographic proximity, such as the Yap, Mariana or Volcano Trenches. Similarly, *Amperima velacula* Agatep, 1967 and *Elpidia decapoda* (Belyaev, 1975) are recorded in the San Cristobal in the Indo-Pacific and the South Sandwich Trench in the Southern Ocean. Equally, there are many examples of holothurians that are endemic to hadal depths but that occupy a cluster of neighbouring trenches. For example, the *Hadalothuria wolffi* Hansen, 1956 species has been found in the closely located New Britain, New Hebrides and Bougainville Trenches and the *Elpidia glacialis kurilensis* species has been found to inhabit the Aleutian, Kuril–Kamchatka and Japan Trenches, all of which are linked. A good example of a species found to inhabit multiple trenches spanning the western Pacific Ocean is *Prototrochus bruuni* (Hansen, 1956; Myriotrochidae). This holothurian is found from the northern Izu-Bonin through the Palau and Philippine Trenches in the central Pacific Ocean, to the Bougainville, Kermadec and Tonga Trenches in the south and across into the Java Trench.

A greater faunal diversity exists between 6000 and 7500 m, however, holothurians still account for 25% of trawl catches at these depths. At depths exceeding 7500 m, their mean trawl catch rate exceeds 50%, with some as high as 75–98%. In the deepest parts of the trench, the holothurians represent one of the largest animals. Consequently, their

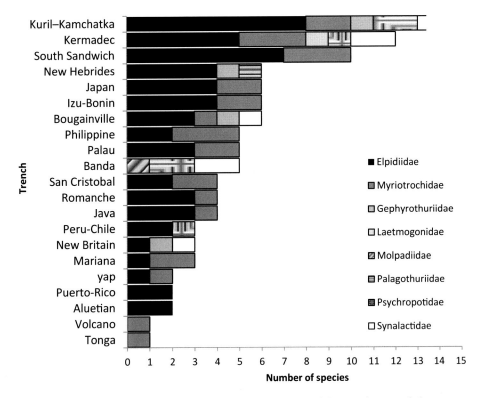

Figure 8.7 Number of species of holothurian per family in each of the trenches sampled.

percentage biomass from most deep trawls can exceed 90%. Wolff (1970) detailed the numbers of holothurian specimens belonging to the genus *Elpidia*, recovered by trawl in the Kuril–Kamchatka Trench that showed an exponential increase in numbers with depth (Fig. 8.8). Such mass aggregations have also been witnessed in the Japan Trench, by remotely operated vehicles (Fig. 8.9).

Most of the incidences of mass catches occur at highly productive regions in the North Pacific Ocean, mainly around the Kuril–Kamchatka, Japan and the Kermadec Trenches, and even at trenches in the Southern Ocean (Vinogradova *et al.*, 1993b). One similar result was obtained in the tropics, at the Java Trench, which is also characterised by high biological productivity. These mass catches are mainly confined to the deepest trench axis, thus supporting the trench resource accumulation hypothesis (Jamieson *et al.*, 2010); if the downward input of POM were to accumulate along the trench axis, particularly at the greatest depth of the trench, then this should be reflected in high numbers of deposit feeders, such as holothurians, as is observed at abyssal depths (Billett and Hansen, 1982; Bett *et al.*, 2001; Billett *et al.*, 2001). Mass catches are also an indication of the relationship between holothurians and shallower, complex topographies, since dense populations of elpidiids are a common feature of submarine canyons and other depressions known to accumulate organic matter (De Leo *et al.*, 2010).

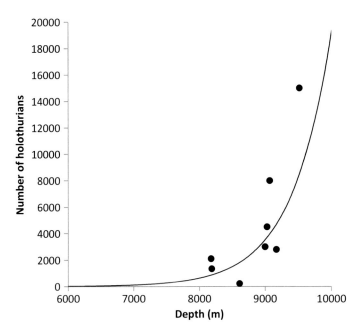

Figure 8.8 Number of holothurians of the genus *Elpidia* recovered by trawl from the Kuril–Kamchatka Trench. Data derived from Wolff (1970).

Figure 8.9 Mass aggregation of holothurians at 7323 m in the Japan Trench, taken by the ROV *Kaikō*. Image © JAMSTEC, Japan.

Based on the *Vitjaz* trawl data, Belyaev (1989) provided population estimates of other Elpidiidae in various other trenches. The density of *E. glacialis uschakovi* Belyaev, 1971 in the New Hebrides Trench and *Elpidia* sp. from the Palau Trench were both 0.1 ind. m^{-2} (= 1000 ind. ha^{-1}), while the density of *E. solomonensis* Hansen, 1956 from the New Britain and Bougainville Trenches ranged from 0.03 to 0.1 and 0.01 ind. m^{-2}, respectively (300–1000 ind. ha^{-1} and 100 ind. ha^{-1}). Summarising the whole family, Belyaev (1989) suggested that the density of Elpidiidae in trenches varies from 0.5 to 10 ind. m^{-2} (5000–100 000 ind. ha^{-1}). Other reports from the abyssal Orkney Trench have estimated the density of *E. decapoda* to be 15 ind. m^{-2} (150 000 ind. ha^{-1}) and 30 ind. m^{-2} (300 000 ind. ha^{-1}) at 6160 m and 5580 m, respectively (Gebruk, 1993). These estimates are somewhat higher than estimates from the abyssal North Pacific Ocean (15.5–193.3 ind. ha^{-1}; Kaufmann and Smith, 1997) and North Atlantic Ocean (8.77–337.92 ind. ha^{-1}; Billett *et al.*, 2001). In the western tropical Pacific trenches, various Holothuroidea were photographed at depths down to 9000 m (Lemche *et al.*, 1976). Here, population densities for the Elpidiidae were estimated at 0.5 to 10 m^{2}.

Aside from trawl catch statistics, holothurians have also been observed by drop-cameras in multiple trenches (Lemche *et al.*, 1976) and during ROV exploration (unpublished data, JAMSTEC, Japan and WHOI, USA). Jamieson *et al.* (2011b) serendipitously landed a Hadal-Lander in the vicinity of the holothurian *Elpidia atakama* (Belyaev, 1971) at 8074 m in the Peru–Chile Trench. Over a 20 h 25 min period, the lander recorded 1225 still images of the solitary holothurian traversing the seabed. The holothurian was tracked to provide information on locomotion rates and feeding behaviour. The results showed that the behaviour of *E. atakama* was not exceptional when compared to shallow-water counterparts. It exhibited the 'run and mill' foraging pattern also adopted by functionally analogous, abyssal species in the northeast Pacific Ocean (*Elpidia minutissima* Belyaev, 1971, *Abyssocucumis abyssorum* (Théel, 1886), *Synallactes profundi, Peniagone vitrea* Théel, 1882 and *Scotoplanes globosa* Théel, 1879; Smith *et al.*, 1993; Kaufmann and Smith, 1997; Fig. 8.10). Its locomotion and feeding rates were also comparable to these shallow-water holothurians. The 'run and mill' foraging behaviour observed suggested that these holothurians are a functionally important species capable of behavioural adaptation in response to localised resource heterogeneity (Godbold *et al.*, 2009, 2011). As unique as these data were, it is difficult to assess the ecological significance of hadal Elpidiidae in the absence of spatial and temporal information of the population, as there are no abundance estimates for *E. atakama* (or any other holothurian) in the Peru–Chile Trench. However, these data were used to calculate the area of seafloor bioturbated by a single holothurian with time (based on mean speed and animal width); *E. atakama* is capable of reworking 1 m^{2} of surface sediment every 5.1 days, or alternatively a population of 123 individuals could rework 1 m^{2} of surface sediment per hour.

Holothurians were also a dominating fauna in the trench images described by Lemche *et al.* (1976). These images provided further information regarding their behaviour rather than simply focusing on diversity and abundance. For example, they showed that *Peniagone purpurea*, a flat holothurian with 7–9 pairs of ventrolateral tubefeet, not only lie on the sediment surface but often partially bury themselves, with only the anterior visible. They discussed how the tubefeet are used for walking as well as milling (also observed in *E. atakama*; Jamieson *et al.*, 2011b). The illumination in some of the images was sufficient to ascertain that the tubefeet raise the ventral surface over the

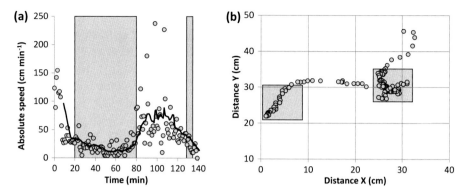

Figure 8.10 Locomotion and feeding behaviour of the hadal holothurian *Elpidia atakama*, where (a) is the time course of feeding and locomotion and (b) shows the X–Y coordinates as viewed from above. In both instances the shaded area indicated periods of feeding. Modified from Jamieson *et al.* (2011b).

seafloor during locomotion and that the veler papiliiae are as long as the body (not previously known to damaged trawl samples) and can move independently, seemingly to explore the waters and sediment immediately in front of the animal. Further observation of *Peniagone azorica* Marenzeller von, 1892 from the New Britain Trench, between ~7000 and 8000 m, also showed the tubefeet raising the body off the sediment surface, as well as traces of footprints. *P. azorica*, however, showed no evidence of burying behaviour. *Scotoplanes globosa* at 7057–7075 m in the New Britain Trench and at 6758–6776 m in the New Hebrides Trench was observed to be walking with every second foot moving at a time. In addition, no footprints were visible, suggesting a very minute weight in water. These observations also confirmed the observation of Hansen (1972), who reported that their bodies produce a wave of constrictions to aid in locomotion, i.e. peristaltic waves passing over the body, pressing fluid from the dermal ambulacral cavities to the tubefeet. The images also showed that *Pelgothuria*, hitherto considered a swimming holothurian, is also found resting on the seafloor, in this case at 6758–6776 m in the New Hebrides Trench. Interesting feeding behaviours were also documented describing how *Hadalothuria wolffi* is able to bend its anterior end at right angles to feed over a greater area without the need for further locomotion. It was also seen to leave shallow tracks in its wake and the scientists speculated that it moves in a gliding motion because its small tubefeet offer little resistance against the sediment.

8.3.5   Echinoidea

Echinoidea (urchins) are known to inhabit at least nine major trenches to depths of <7000 m (Madsen, 1956). They have, however, been found slightly deeper in two trenches; the Palau (7170 m) and Banda Trenches (7340 m). The Echinoidea are represented by four families: Pourtalesiidae (seven species, three genera), Holasteridae (one species), Echinothuroidea (one species) and Urechinidae (one species) (Table 8.4). With the exception of the Echinothuroidea, all the other families belong to the order Spatangoida.

Only two species and one subspecies of echinoid are endemic at hadal depths and of the non-endemic species, four are known to inhabit abyssal depths and just one eurybathic species extends to the bathyal zone.

Echinoidea are not especially common in upper hadal depth trawl catches and are typically represented only by fragments. The trawl catch frequency between 6000 and 7340 m is 31% (Belyaev, 1989). The Echinoidea, similar to the other Echinodermata, do appear to occur, on occasion, in mass aggregations. In the Banda Trench, *Pourtalesia heptneri* Mironov, 1978 (Pourtalesiidae) was found in all three trawls to a maximum depth of 7130 m. In total, 24 whole specimens were recovered along with the shell fragments of at least another 120 individuals. In the Java Trench, the endemic subspecies *Echinosigra amphora indica* Mironov, 1974 (Pourtalesiidae) was found in two out of six trawls undertaken between 6820 and 6850 m. One trawl included 32 specimens. Very little further information is known about the occurrence, bathymetric range and ecology of hadal echinoids.

8.4   Other benthic invertebrates

Ascidiae (sea squirts) are present in the hadal zone but are generally restricted to the upper half of the depth range. The hadal sea squirts fall into two orders (Phelebobranchia and Stolidobranchia) and five families (Corellidae and Octacnemidae, and Hexacrobylidae, Pyuridae and Styelidae, respectively). All records are taken from the trenches of the western Pacific Ocean where the trawl catch frequency is approximately 25% (Belyaev, 1989). Ascidens, possibly of the Corellidae family, were also visible in some of the PROA images in the New Britain (7875–7921 m) and New Hebrides Trenches (6758–6776 m; Lemche *et al.*, 1976). The New Britain ascidians were found in far greater densities than those in the New Hebrides, estimated at one specimen per 30 m^2 although whether this was an effect of trench, depth or substrata is unknown. The deepest known species is *Situla pelliculosa* Vinogradova, 1969 (Octacnemidae), found in the Kuril–Kamchatka Trench between 5000 and 8400 m.

The Brachiopoda (lamp shells) are generally limited to the abyssal depths at ~5500 m and no live brachiopods have ever been found in the hadal trenches. However, empty brachiopod shells were discovered at 6160 m in the northwest trough and at ~7500 m at the Romanche Trench fault (Belyaev, 1989). However, Lemche *et al.* (1976) noted small, flattened specimens of *Articulata brachiopoda* at 6758–6776 m in the New Hebrides Trench. These anecdotal observations are yet insufficient to prove or refute the presence of hadal brachiopods.

Belyaev (1989) notes that although Bryozoa had been collected by the *Galathea* and *Vitjaz* expeditions from the Kermadec (8210–8300 m), Java (6487 m), Kuril–Kamchatka (6090–8400 m), Izu-Bonin (8800–8830 m) and Peru–Chile Trenches

(7000 m) and from the Romanche Trench fault (7340 m), most of the specimens were not processed. He noted that some specimens are probably of the genus *Kinetoskias* or *Bugula*, from the Bicellariidae (now Bugilidae) family and that all the Bryozoa found at <3000 m belong to the order Cheilostomatida. Therefore, it is likely that the Bryozoa found in the trenches belong exclusively to Cheilostomatida.

9 Crustacea

In total, 11 orders of Crustacea have been found at hadal depths (Table 9.1). The Crustacea are an extremely important facet of the hadal community, particularly the orders Isopoda and Amphipoda. In fact, they are found in almost every sample and every trench. In terms of numerical abundance and diversity, isopods are one of the most important macrofaunal, benthic taxa in deep-sea communities (Hessler and Sanders, 1967; Hessler and Strömberg, 1989) and the number of species found at hadal depths exceeds that of all other Crustacea and any other class of multicellular organism (Belyaev, 1989).

The importance of the Amphipoda is evident from different sampling methods than the other invertebrates. The use of baited camera and traps have shown that amphipods completely dominate the mobile scavenging fauna at depth greater than 8000 m and are often recovered from traps in numbers of tens of thousands (Blankenship *et al.*, 2006); they are prolific scavengers that thrive at full ocean depth (Hessler *et al.*, 1978; Jamieson *et al.*, 2009a; Eustace *et al.*, 2013). Their importance is also two-fold. In addition to their overwhelming presence at >8000 m, they have been consistently shown to be a major prey item to larger predators in the upper trench at <8000 m (Jamieson *et al.*, 2009a, b; 2012a).

There are, of course, other less conspicuous and less diverse Crustacea found at hadal depths, including the Cirripedia, Ostracoda, Mysidacea, Pantopoda, the exceptionally rare Acariformes and the Leptostraca. The Acariformes are seldom found and are mostly restricted to the abyssal–hadal boundary. Similarly only a single, individual leptrostracan has ever been recovered from hadal depths; from 7100 m in the Japan Trench (Jamieson *et al.*, 2010). Furthermore, specimens from the cosmopolitan subclass Copepoda have been recovered from the hado-pelagic zone, but due to the fact that it is notoriously difficult to obtain quantitative samples from hado-pelagic waters, extraordinarily few samples and thus extraordinarily few specimens have ever been collected (Vinogradov, 1962). The fact that the orders of Crustacea described above appear to be low in abundance and diversity in the hadal zone, or seem to be restricted to shallower depths, may be a result of sampling bias. Hopefully, their role and the true extent of their existence at hadal depths will one day be realised, as more comprehensive data is obtained.

The final crustaceans worthy of a mention here are the Decapoda and the 'supergiant' amphipod *Alicella gigantea*. Up until 2009, one of the inexplicable findings to emerge from the extensive sampling in the 1950s was the complete absence of

Table 9.1 Each group of Crustacea currently known to inhabit the hadal zone with the total number of known species (+ number of undescribed species) with the corresponding maximum known depth for the order. Given the large number of undescribed species, these species number should be interpreted as indicative only.

Group	Number of known species (+ number of undescribed species) at hadal depths	Maximum known depth (m)
Copepoda	27 (+5)	10 000
Cirripedia	6 (+3)	7 880
Ostracoda	9 (+5)	9 500
Mysidacea	2 (+10)	8 720
Cumacea	6 (+10)	8 042
Tanaidacea	43 (+10)	9 174
Isopoda	94 (+39)	10 730
Amphipoda	63 (+14)	10 994
Decapoda	2*	7 703
Acariformes	1	6 850
Pantopoda	8 (+1)	7 370

decapods at hadal depths. In hindsight, employment of the wrong sampling gear for the task was probably to blame for this erroneous finding. Since then, baited cameras have consistently observed large (over 20 cm long), benthesicymid prawns in multiple trenches, as deep as 7703 m (Jamieson *et al.*, 2009b). Also, the supergiant amphipod has always been something of an enigma, with few specimens taken over decades from disparate, geographical locations. Recently, however, they were found in relatively high numbers using cameras and traps at 7000 m in the Kermadec Trench and some measured up to nearly 30 cm (Jamieson *et al.*, 2013). The discoveries of both the decapods and the supergiants are testament to how much more there is still to learn about this environment through hadal sampling. It is suprising that an entire order of such large Crustacea and an animal dubbed 'supergiant' has gone unnoticed in the trenches until very recently. In a sense, this provides some confidence that the other smaller crustaceans, or perhaps even the larger ones, are more extensive and diverse in the hadal zone than is currently known today.

The following provides an overview of the orders of Crustacea currently known to exist in the hadal zone.

One of the most intriguing characteristics of Crustacea at hadal depths is their increased body size relative to their shallow-water counterparts. Both Belyaev (1989) and Wolff (1960) refer to this phenomenon as 'gigantism', although Wolff (1962) stipulated that this actually meant a 'tendency towards larger dimensions'. True gigantism at hadal depths is perhaps only really evident in the 'supergiant' amphipod, *Alicella gigantea* (Jamieson *et al.*, 2013). The most striking example of increased body size with depth is the Asellota suborder of the Isopoda (Wolff, 1960), while others such as some Cumacea, Tanaidacea and Mysidacea genera also show this increase.

The possible reasons for the increased body size with depth and the greatest dimensions of hadal Crustacea remain unresolved. However, the increase in body size in some cases occurs in closely related species within the same genus at comparatively shallow depths in the polar regions and thus is likely to be related to low temperature, but in other cases, it is related solely to increased depths not temperature, suggesting hydrostatic pressure as a potential explanation in some cases. For the Isopoda at least, Wolff (1962) also attributed the larger body dimensions to a longer life span under trench conditions but also suggested that 'gigantism may be due to the pressure effects on metabolism (Wolff, 1960, also suggested by Zenkevitch and Birstein, 1956). This tendency for a larger body size has not been reported for any groups other than Crustacea.

9.1   Copepoda

Copepoda of the hadal zone are represented by the pelagic order Calanoida and the benthic order Harpacticoida. Information on the hadal Calanoida is almost exclusively derived from the RV *Vitjaz* expeditions. The first samples were obtained in 1953 by closing plankton nets, from 8500 to 6000 m in the Kuril–Kamchatka Trench (Vinogradov, 1962). These hauls produced 20 species belonging to 17 genera and 10 families, including 2 new genera; *Zenkevitchiella* and *Parascaphocalanus*. These plankton hauls revealed that, despite a relatively diverse catch, the composition was often dominated by just a few species. For example: 64 *Spinocalanus similis profundalis* Brodsky, 1950 (Spinocalanidae), 97 specimens of *Parascaphocalanus zenkevitchi* Brodsky, 1955 (Scolecitrichidae) and 37 *Metridia similis abyssalis* Brodsky, 1955 (Metridinidae), and the rest comprised single specimens of the other species.

Of the 32 species collected from the Kuril–Kamchatka Trench, 15 (47%) are endemic to hadal depths, whilst 10 are known to extend to abyssal depths and are generally only found in the northwest Pacific Ocean (with the exception of *Lucicutia curvifurcata* Heptner, 1971 (Lucicutiidae), also found in the Bougainville Trench). The other species are eurybathic and geographically widespread.

Deep-sea, benthic representatives of Harpacticoida have been described from as deep as 10 000 m in the Kermadec Trench (Belyaev, 1989) and they are, generally, the second most abundant meiobenthic taxon in marine samples, after nematodes (Giere, 2009). Of the Harpacticoida obtained from the *Vitjaz* expeditions, only one has been formally described as a new genus and species from the family Cerviniidae, *Herdmaniopsis abyssicola* Brotskaya, 1963 (Aegisthidae), from 6071 m (Belyaev, 1989). The same study described three new species of the genus *Cervinia*, from the same family (*C. brevipes*, *C. tenuicauda* and *C. tenuiseta* Brotskaya, 1963) from 5700 m, close to the Izu-Bonin Trench. Belyaev (1989) suggested that these species would also probably be found at hadal depths. In 2009, baited traps deployed in the Kermadec Trench collected 40 copepods from 5173 m, one from 6000 m and 19 individuals from 7561 m, but unfortunately, these specimens are still unidentified (Jamieson *et al.*, 2011a).

The most detailed study to date regarding harpacticoids is Kitahashi *et al.* (2012), who examined the spatial changes in the assemblages of harpacticoids at the family level, around the Ryukyu region (encompassing the trench and surrounding abyssal plain) and Kuril region (encompassing the Kuril–Kamchatka Trench and adjacent abyssal plains). In the Ryukyu region, they found high average dissimilarities in the assemblages between the trench, trench slope and abyssal plain, indicating that the assemblage structures differ substantially between these topographic settings at the family level. The dominant families from the 18 found in this region were Ectinosomatidae (15.8%), Psuedotachidiidae (15.1%), Zosimeidae (14.2%), Ameiridae (12.5%), Argestidae (12.1%) and Neobradyidae (9.3%).

In the Kuril region, the average assemblage dissimilarity between topographical locations suggested that the hadal assemblage is a transition zone between the slope and the abyssal plain. Sixteen families of harpacticoid were found in this region and were dominated by Ectinosomatidae (23.9%), Ameiridae (17.3%), Psuedotachidiidae (14.3%), Idyanthidae (13.3%), Argestidae (9.8%) and Cletodidae (6.1%).

The data from each region suggested that the composition of harpacticoid assemblages was influenced by the quantity of organic matter in the Ryukyu region, while in the Kuril region, sediment properties played a key role. Comparisons of the two assemblages showed that the average dissimilarities between the trenches and abyssal plains were higher than those between the adjacent slopes, suggesting that migration between regions is difficult for deep-sea, benthic harpacticoid copepods due to topographical barriers (Kitahashi *et al.*, 2012).

Unfortunately, as Kitahashi *et al.* (2012) point out, species-level analyses present a major challenge because more than 95% of deep-sea harpacticoid specimens are new to science. In addition, morphological identifications should be tentative, as molecular studies on shallower harpacticoids have revealed that some cosmopolitan harpacticoid species are actually 'species complexes' (e.g. Schizas *et al.*, 1999; Rocha-Olivares *et al.*, 2001). Therefore, much more research, in particular molecular analyses of deep-sea harpacticoids, is required to provide a clear overview of some taxonomic groups, in this case, the harpacticoids.

9.2   Cirripedia

Cirripedia are not a particularly characteristic hadal fauna, as most are found only around the 6000 m contour and in very low numbers. The majority have been found between 6000 and 7000 m in the Kuril–Kamchatka, Japan, Izu-Bonin, Ryukyu, Kermadec, Peru–Chile and Philippine Trenches but in most instances, they were only found in quantities of one or two individuals. All of the nine species of Cirripedia found at hadal depths belong to the family Scalpellidae; three have only been classified to genus or subfamily level. All of them have been described from a single trench only, with the exception of *Annandaleum japonicum* (Hoek, 1883), which has also been found in the neighbouring Kuril–Kamchatka, Japan and Ryukyu Trenches (6156–6810 m; Belyaev, 1989).

9.3   Ostracoda

There are currently 14 species of ostracod known from the hadal zone. These comprise mostly undescribed specimens belonging to five families of two orders (Podocopida and Halocyprida). The most common species is *Juryoecia* (*Metaconchoecilla*) *abyssalis*

(Rudjakov, 1962) of the Halocyprididae family. These have been found in the Kuril–Kamchatka, Mariana, Bougainville and Kermadec Trenches, from 4200 to 8500 m deep (Belyaev, 1989). The maximum known depth of a benthic ostracod is 8100 m in the Puerto-Rico Trench (*Retibythere scaberrima* Brady, 1886; Bythocytheridae) and the deepest pelagic ostracods are *Archiconchoecilla maculata* Chavtur, 1977, and *Paraconchoecia vitjazi* (Rudjalov, 1962), both Halocyprididae, from 9500 m in the Kuril–Kamchatka Trench.

Three individual ostracods were also recovered using baited traps in the Kermadec Trench in 2009, albeit from only 5173 m (Jamieson *et al.*, 2011a). These were identified as two *Bathyconchoecia* sp. n. and one *Metavargula cf. adinothrix* Kornicker, 1975.

9.4 Mysidacea

The Mysidacea are another group of crustaceans found in the hadal zone, about which little is known. There are only two that have been described to species level; *Amblyops magna* from the Kuril–Kamchatka Trench (6435–7230 m) and *Mysimenzies hadalis*, from the Peru–Chile Trench (6146–6352 m) (Bacescu, 1971). There are, however, records of various unidentified species, all from the family Mysidae from 14 trenches, comprising five genera (*Amblyops*, *Birsteiniamysis*, *Mysimenzies*, *Paramblyops* and *Mysidacea*) of which all are known from shallower zones. Of the 12 known species from hadal depths, only 3 species are known from the lower abyssal zone (below 4500 m), and the other 9 (75%) are endemic to hadal depths. The deepest Mysidacea found to date was from 8560–8720 m, in the Yap Trench (*Paramblyops* sp. n.).

There are also various records of hado-pelagic catches of mysids from the Kuril–Kamchatka, Japan, Izu-Bonin and Ryukyu Trenches (*Boreomysis incise*; 6000–7000 m) and *Dactylamblyops tenella* from 6600 m in the Ryukyu Trench (Birstein and Tchindonova, 1958).

While descending into the Puerto-Rico Trench in the *Archimède* bathyscaphe, Pérès (1965) reported observing euphausids between 6100 and 6450 m and another individual below 6600 m. However, Belyaev (1989) notes, based on the opinions of others at the time, that these euphausids were more likely to be mysids (although there are currently no records of hadal mysids from the Puerto-Rico Trench). Mysids were also tentatively identified from images taken at four stations in the Palau, New Britain, Bougainville and New Hebrides Trenches at depths between 6758 and 8662 m (Lemche *et al.*, 1976). They documented 37 individuals at 8021–8042 m in the Palau Trench that were, on average, 1 cm long and they noted other possible sightings at 8258–8260 m in the New Britain Trench (one specimen), at 7847–8662 m in the North Solomon Trench (four specimens) and at 6758–6776 m in the New Hebrides Trench (two specimens which were 3–4 cm long).

Perhaps the most striking and indicative examples of trench mysids are those observed using the Hadal-Lander B which was deployed at 4602, 5329, 6173 and 7050 m in the Peru–Chile Trench, in 2010 (Fig. 9.1). The lander took a photograph every 1 min of a 0.35 m^{-2} area of seafloor that was baited with tuna. At the abyssal

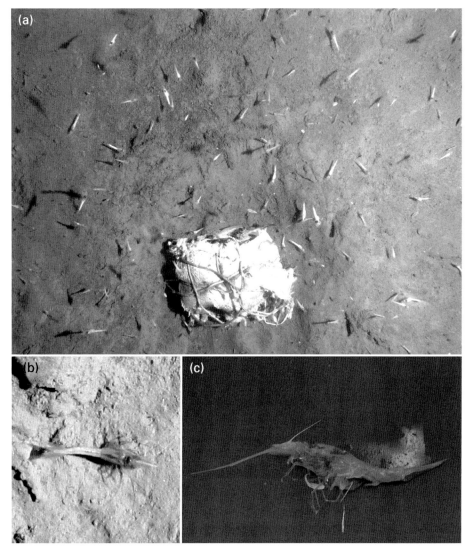

Figure 9.1 Examples of trench-dwelling mysids, where (a) is an *in situ* image of mysids aggregating at bait at 4602 m in the Peru–Chile Trench; (b) is a magnified view of an individual at 7050 m in the Peru–Chile Trench; and (c) is a specimen of *Amblyops* sp. recovered from 6709 m in the Kermadec Trench. Images courtesy of HADEEP.

stations, the number of mysids outnumbered the usually dominant amphipods and at the 6173 and 7050 m sites, they were still present but in lower numbers. At each of these four stations, the first arrival times of the mysids were 7, 67, 27 and 28 minutes, respectively. Their maximum numbers (in each image) reached 76, 5, 27 and 15, respectively, and they were present in 99.4, 59.7, 56.2 and 27.7% of the images taken (Fig. 9.2). Their mean body length was 2.21 cm $\pm$ 0.5 ($n = 40$), 1.69 cm $\pm$ 0.8 ($n = 10$),

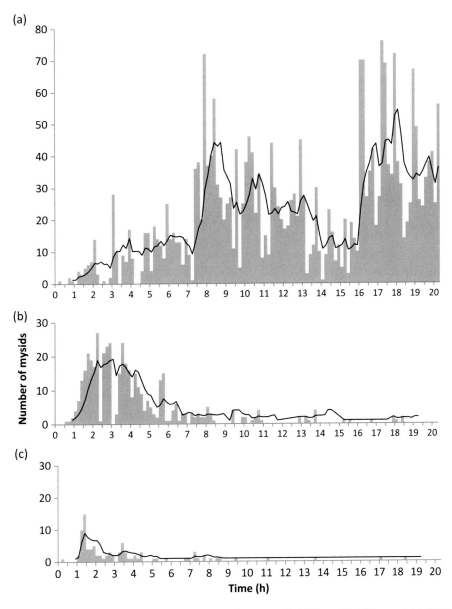

Figure 9.2 Total number of mysids observed per image at (a) 4602 m, (b) 6173 m and (c) 7050 m in the Peru–Chile Trench. The black line depicts a 1 h moving average. Data obtained using Hadal-Lander B, HADEEP.

1.88 cm ± 0.4 S.D. ($n = 24$) and 1.77 cm ± 0.4 S.D. ($n = 10$), respectively. Unfortunately, despite recovering hundreds of amphipod specimens per deployment, not a single mysid was caught and, therefore, the species is still unidentified. This observed hadal mysid species did, however, closely match the location and depth (6146–6354 m) of *Mysimenzies hadalis* (Bacescu, 1971).

Use of the same lander in other trenches, notably the Kermadec and Tonga Trenches, has produced many observations of smaller mysids, but normally in numbers of fewer than five individuals. The baited trap, *Latis*, did, however, recover two specimens from 6265 and 6709 m in the Kermadec Trench and these were identified to genus level as *Amblyops* sp. (unpublished data, HADEEP; Fig. 9.1).

9.5   Cumacea

The Cumacea are not a common or well-known group at hadal depths. There are records from 14 trenches comprising two identified species (*Makrokylindrus hadalis* Jones, 1969 and *Makrokylindrus hystrix* Gamo, 1985; Diastylidae) from the Java and Japan Trenches, respectively, three identifications to genus level (*Bathycuma* sp. from the Bougainville Trench, *Lamprops* sp. from the Java Trench and *Leucon* sp. from the Aleutian Trench) and one presumed species (*Vaunthompsonia* aff. *cristata* Bate, 1858, Bodotriidae) from the Kuril–Kamchatka Trench. The other 10 records have not been identified. All of the records are reported from depths between 5650 and 8042 m (Belyaev, 1989). Other reported sightings include those of Lemche *et al.* (1976) who saw an elongated cumacean in the Palau Trench (8021–8042 m), three distinctively different cumaceans at 7875–7921 m in the New Britain Trench (one of which was '*Leucon*-like') and a very elongate individual at 6758–6776 m in the New Hebrides Trench. There are currently no indications that any of the species or genera previously reported are endemic to hadal depths.

9.6   Tanaidacea

The Tanaidacea order are a relatively well-known group of Crustacea, having been found in most trenches and studied to a maximum known depth of 9174 m; *Akanthophoreus* (*Leptognathia*) *longiremis* (Lilljeborg, 1864) (Akanthophoreidae); Kermadec Trench (Fig. 9.3). The taxonomic composition of hadal Tanaidacea are quite diverse (Table 9.2), comprising 63 species of 26 genera and 13 families defined to date (derived from Belyaev, 1989, updated from Larsen and Shimomura, 2007a). The most diverse family is the Leptognathiidae that accounts for 52% of the total known hadal species. Leptognathiidae also contains 20 species of the genus *Leptognathia* and thus this genus accounts for one-third of all the Tanaidacea species found at hadal depths. Larsen and Shimomura (2007b) removed several species from the *Leptognathia* but did not give replacement family designation. Trench endemism of the Tanaidacea is currently around 40%, but increases with depth, from 16% between 6000 and 6500 m up to 75% between 8000 and 8500 m, and there are no endemic genera.

The distribution of tanaids prompted some discussion by Belyaev (1989). At hadal depths tanaids are thought to be distributed unevenly since the encounter frequency in trawl catches was less than 30% and in the bottom grab samples it was ~40% (based on *Vitjaz* and *Galathea* expedition collections). Furthermore, he doubted their apparent

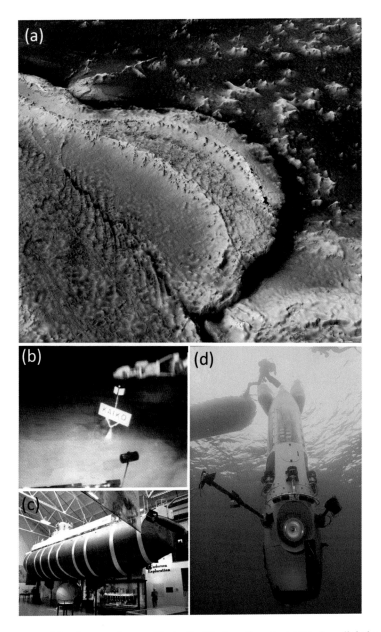

Plate 1 The deepest place on Earth: the Mariana Trench. (a) Modern digital swathe bathymetry showing the three-dimensional topography of the Mariana Trench. Image courtesy of P. Sloss, NOAA/NGDC (retired). (b) Video frame grab of the Japanese full ocean depth rated ROV *Kaikō* placing a flag to mark the deepest place on Earth: Challenger Deep at 10 911 m. Image © JAMSTEC, Japan. (c) The bathyscaphe *Trieste* residing in the National Museum of the US Navy, Washington, DC. Image courtesy of P.H. Yancey, Whitman Collage, USA. (d) The *Deepsea Challenger* manned submersible on its 2012 dive to the Challenger Deep. Image courtesy of Charlie Arneson.

Plate 2 Hadal-Landers. (a) Large haul of hadal amphipods (*Hirondellea gigas*) being emptied from a small baited funnel trap after 12 h at 9316 m deep in the Izu-Bonin Trench. Image courtesy of HADEEP. (b) University of Aberdeen's Hadal-Lander A being deployed to 10 000 m in the Tonga Trench in 2007. Image courtesy of J.C. Partridge, University of Bristol, UK. (c) DOV *Mike*, or 'Alpha Lander' being deployed to over 10 500 m deep in the Challenger Deep. Image courtesy of K. Hardy Scripps Institution of Oceanography, USA. (d) The microelectrode lander being deployed to Challenger Deep. Image courtesy of R.N. Glud, University of Southern Denmark.

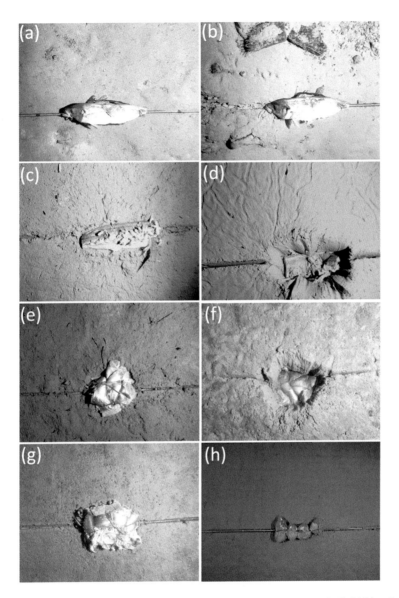

Plate 3 Trench sediments. (a) 6000 m, (b) 7561 m, (c) 8215 m and (d) 9281 m in the Kermadec Trench, (e) 6173 m, (f) 7050 and (g) 8074 m in the Peru–Chile Trench, and (h) 9729 m in the Tonga Trench. All images courtesy of HADEEP.

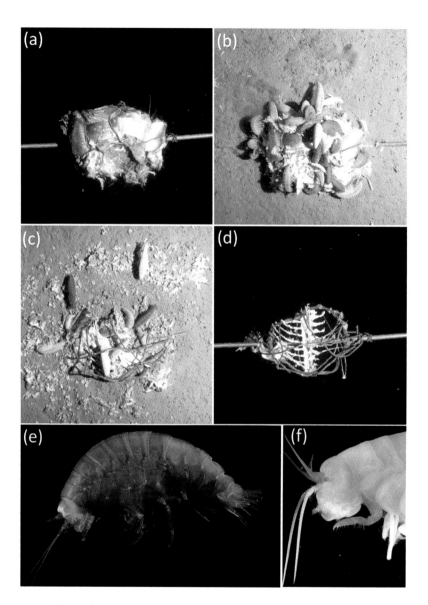

Plate 4 Scavenging amphipods. Rapid consumption of simulated carrion-fall at 8074 m in the Peru–Chile Trench, where (a) shows ~1 kg of tuna just before reaching the seafloor; (b) shows an aggregation of scavenging amphipods (*Eurythenes gryllus*) 2 h later; (c) shows the remnants of the bait scattered around the seafloor after 18 h; and (d) shows the skeletal remains after leaving the seafloor. (e) *E. gryllus* collected by baited trap with a close-up shown in (f). Images (a) – (d) courtesy of HADEEP, images (e) and (f) courtesy of J.C. Partridge and C. Sharkey, University of Bristol, UK.

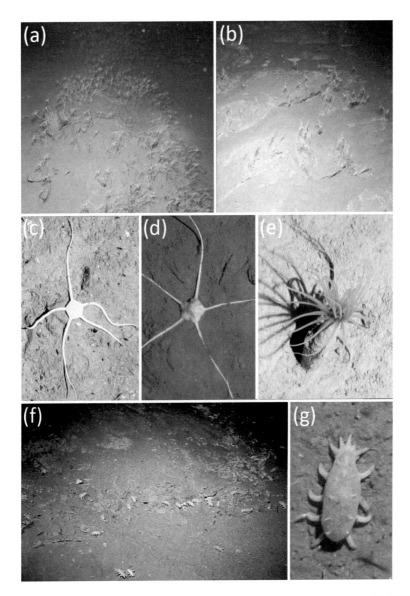

Plate 5 Hadal epifauna. Examples of 'crinoid meadows' at (a) 9092 m and (b) 9095 m in the Izu-Bonin Trench. (c) *Ophiura* aff. *loveni* from 7199 m in the Kermadec Trench. (d) Unidentified ophiuroid from the Peru–Chile Trench. (e) Tube dwelling anemone (Ceriantharia, Anthozoa) from 5173 m in the Kermadec Trench. (f) Mass aggregation of holothurians (Elpididae) at 7323 m in the Japan Trench. (g) *In situ* image of *Elpidia atakama* from 8074 m in the Peru–Chile Trench. Images (a), (b) and (f) © JAMSTEC, Japan; (c), (d), (e) and (g) courtesy of HADEEP.

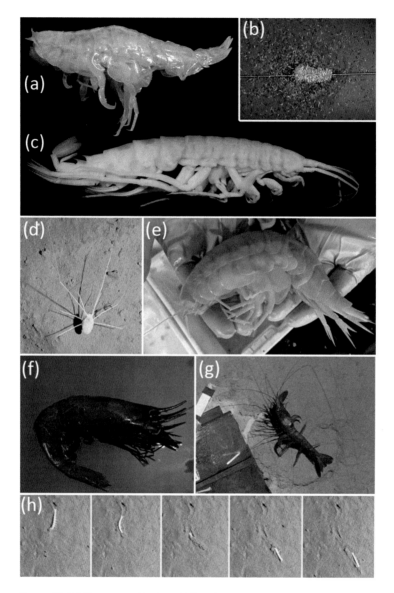

Plate 6 Hadal Crustacea. (a) *Hirondellea dubia* recovered from 9104 m in the Kermadec Trench. (b) Scavenging amphipods at nearly 10 000 m in the Tonga Trench showing nearly a 100% coverage of bait after 2 h. (c) The predatory amphipod *Princaxelia jamiesoni* (Pardaliscidae) from 7703 m in the Japan Trench. (d) Isopod from 5469 m on the edge of the Mariana Trench. (e) The 'supergiant' amphipod *Alicella gigantea* (27.8 cm in length) captured from 7000 m in the Kermadec Trench. (f) The decapod *Heterogenys microphthalma* from 6709 m in the Kermadec Trench. (g) The decapod *Benthesicymus crenatus* from 6474 m in the Kermadec Trench. (h) A burrowing tanaid at 7501 m in the Kermadec Trench: each image is 1 minute apart showing a tanaid emerging from a burrow and traversing under the sediment surface and reappearing. All images courtesy of HADEEP except (c) taken by T. Karanovic, Hanyang University, S. Korea.

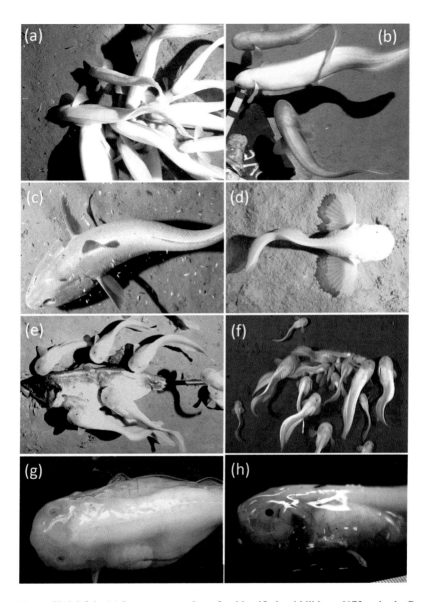

Plate 7 Hadal fish. (a) Large aggregation of unidentified ophidiids at 6173 m in the Peru–Chile Trench. (b) A group of ophdiids, *Bassozetus* sp. (possibly *B. robustus*) at 6474 m in the Kermadec Trench. (c) The macrourid *Coryphaenoides yaquinae* at 5469 m on the edge of the Mariana Trench. (d) An unidentified liparid from 7050 m in the Peru–Chile Trench. (e) Large aggregations of *Notoliparis kermadecensis* (liparid) at 7561 m in the Kermadec Trench and (f) *Pseudoliparis amblystomopsis* (liparid) from 7703 m in the Japan Trench and close-ups of samples of *N. kermadecensis* (7000 m; (g)) and *P. amblystomopsis* (7703 m; (h)).

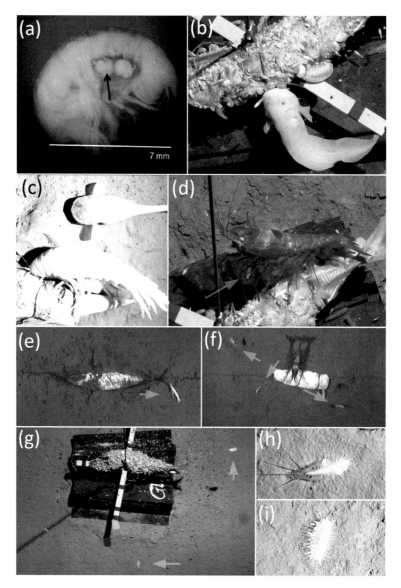

Plate 8 Behaviour and interactions. (a) A brooding female lysianassoid (subfamily Tryphosinae) from 7349–9273 m in the Tonga Trench. This is the only hadal amphipod found to attend bait whilst brooding (eggs arrowed). (b) The liparid *Notoliparis kermadecensis* feeding on amphipods at 6474 m in the Kermadec Trench. (c) The 'supergiant' amphipod *Alicella gigantea* feeding at bait at 6979 m in the Kermadec Trench, undeterred by the presence of *N. kermadecensis* that would otherwise prey upon amphipods. (d) The natantian decapod *Benthesicymus crenatus* removing amphipod from bait (arrowed) from 6474 m (Kermadec Trench). (e) The pardaliscid amphipod *Princaxelia jamiesoni* (arrowed) searching for prey at 7703 m in the Japan Trench. (f) Three isopods, *Rectisura* cf. *herculea*, in mid-escape bursts (arrowed, in direction of travel) following perturbation from *B. crenatus* at 7115 m in the Japan Trench. (g) Swimming polynoid polychaetes circling a swarm of amphipods at 8631 m in the Kermadec Trench, with close-ups of these Polynoidae from 9300 m (h) and 6979 m (i) in the Kermadec Trench. All images courtesy of HADEEP, except (a) courtesy of L. Levin, Scripps Institution of Oceanography, USA.

Table 9.2 Tanaid families found at hadal depths with numbers of genera, species and bathymetric range.

Family	Genera	Species	Depth range (m)
Apseudidae	2	3	6065–7657
Collettidae	2	2	3146–7433
Gigantapseudidae	1	1	6920–7880
Neotanaidae	2	12	5986–8330
Incertae sedis	3	2	5733–8006
Paratanaidae	1	1	7370–7370
Pseudotanaidae	2	5	6675–6890
Tanaidae	1	1	6090–6135
Agathotanaidae	1	1	6770–6890
Anarthruridae	2	2	3146–8015
Leptognathiidae	5	29	3853–9174
Paratanaoidea incertae sedis	1	1	2600–6850
Typhlotanaidae	3	3	3610–7370

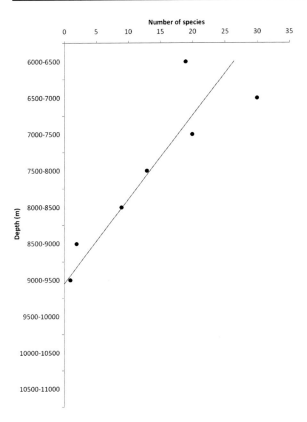

Figure 9.3 Number of species per 500 m depth stratum for the Tanaidacea.

vertical distribution, since many eurybathic species found in the hadal Pacific Ocean are also known from shallower depths in the Atlantic Ocean (but not shallower in the Pacific). He commented that it was more likely that these were morphologically similar or identical species, but living as reproductively isolated populations. Until these issues are resolved, the number of seemingly eurybathic and endemic species remains speculative. In support of this, others have stated that the depth range of most deep-sea tanaids does not exceed 2000 m and their geographic distribution is limited to far narrower regions (Wolff, 1956).

Tanaids are typically benthic, and usually inhabit the surface layer of the bottom sediments. However, some Tanaidacea are capable of floating, rising considerable distances above the bottom. For example, *Leptognathia* sp. (Sars, 1882; Leptognathiidae) was once caught by plankton net in the Kuril–Kamchatka Trench, between 50 and 100 m above bottom at a depth of 8700–7000 m (Belyaev, 1989).

Tanaids are seldom reported from *in situ* observations and this may partly be due to (a) their small body size, (b) their apparent patchy distribution as suggested by Belyaev (1989) and (c) when using baited cameras, they are difficult to distinguish among swarms of amphipods (Fig. 9.4). However, Lemche *et al.* (1976) reported observing a single tanaid, which he thought was *Neotanais*, enter a burrow in the New Britain Trench at 7057–7075 m. He also observed another solitary, unidentifiable tanaid (0.5–1 cm) at 6758–6776 m in the New Hebrides Trench. Despite multiple deployments of the Hadal-Lander B in the Kermadec Trench, tanaids were only observed during one deployment at 7501 m (unpublished data, HADEEP). During this deployment, tanaids of ~1–2 cm long were frequently seen, burrowing in and out of the sediment, often leaving long, visible tails on the sediment surface as they traversed just under the surface (Fig. 9.4). They were observed in 99 frames out of 599 (17%) and no more than three individuals were observed in one frame. Their absence from all other deployments in this trench, even at similar depths, supports Belyaev's idea of a sparse or patchy distribution.

Based on the specimens from the *Galathea* expeditions, Wolff (1956) noted that the mean body size of tanaids is correlated to depth, whereby the mean length in all environments <200 m was <10 mm, while the mean length across the abyssal–hadal boundary was greater than 20 mm (Fig. 9.5). Since then, there have been other notable findings regarding body size; in the Philippine Trench at 6290 to 7880 m, 77 specimens of *Gigantapseudes adactylus* Kudinova-Pasternak, 1978 (Gigantapseudidae) were found that measured up to 37 mm in length, 1.5 times longer than the next largest crustacean of this order (Kudinova-Pasternak, 1978). However, even larger representatives of this genus, *G. maximus* (length up to 75 mm), were described from 5460 to 5567 m deep, to the east of the southern Philippine Trench, close to the location of *G. adactylus* (Gamô, 1984). These additional data would, in fact, further increase the average body size of the abyssal and hadal tanaids shown in Figure 9.5 (Wolff, 1956).

One other peculiarity regarding hadal tanaids was reported by Wolff (1956) who observed that none of the 30 females recovered from hadal depths had eggs in the marsupium. This suggested that their egg-bearing period was either very short, seasonal or, perhaps more likely, the egg-bearing females exhibit a fossorial (buried) period which could not be detected by trawling. However, he did discuss the idea that the absence of egg-bearing females may be due to the fact that deep-sea crustaceans live to a relatively old age. Based on observation of *Apseudes galatheae* (Apseudidae) and

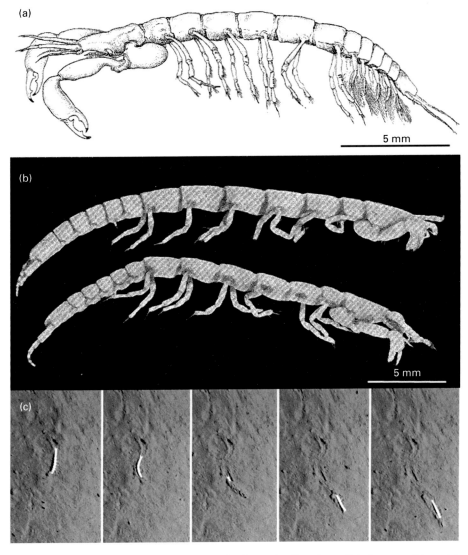

Figure 9.4 Examples of tanaids collected from the Kermadec Trench on *Galathea* expeditions. (a) *Neotanais serratispinosus hadalis* from 8210 m and (b) *Herpotanais kirkegaardi* from 7150 m (top specimen is a female, the bottom is a male). (c) A burrowing tanaid at 7501 m in the Kermadec Trench: each image is 1 min apart showing a tanaid emerging from a burrow and traversing under the sediment surface and reappearing. Images (a) and (b) are from Wolff (1956) reproduced with the permission of *Galathea Report* and (c) courtesy of HADEEP.

Herpotanais kirkgaardi (Neotanaidae), he concluded that each female must complete several egg-bearing periods, each potentially lasting at least 3 months. These periods may occur every 2–3 years, assuming that they live to be 15–20 years old and this would minimise the chances of collecting an egg-bearing female.

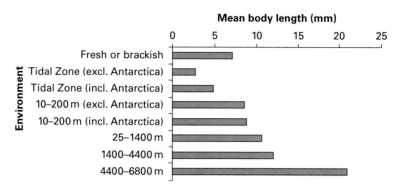

Figure 9.5 Mean body size of tanaids from environments increasing in depth. Modified from Wolff (1956).

9.7   Isopoda

The Isopoda are a very successful order of Crustacea, occurring in terrestrial, freshwater and marine environments (Schotte *et al.*, 1995). In terms of numerical abundance and species diversity, isopods are one of the most important macrofaunal benthic taxa in deep-sea communities (Hessler and Sanders, 1967; Hessler and Strömberg, 1989). In the hadal zone, isopods are plentiful and more isopod species are found at hadal depths than any other order of Crustacea and, indeed, any other class of multicellular animal (Belyaev, 1989). There are currently 15 families, encompassing 34 genera containing 135 species residing at depths greater than 6000 m (Table 9.3). The overwhelming majority of the species found at hadal depths belong to the suborder Asellotta (Fig. 9.6), as is also the case in the wider deep sea (Hessler *et al.*, 1979).

The vertical distribution of hadal Isopoda shows a typical decrease in diversity with increasing depth, but even at the greatest depths five different species still prevail (Fig. 9.7). Belyaev (1989) stated that the species he found to be endemic to the hadal zone accounted for 63% of those captured, while species that cross the abyssal–hadal boundary accounted for over 35% and only about 1.5% (two species) were eurybathic (2400–6200 m). However, of the hadal endemic species, approximately 75% were described from a single finding. Therefore, as is the case with many of the hadal fauna, caution must be taken when interpreting vertical distribution until further sampling is undertaken.

Of the endemic species discussed by Belyaev (1989), 50% had a depth range of less than 1000 m, 40% of 1000–2000 m and only 10% had a depth range of 2000–3000 m, suggesting that the hadal isopods are largely stenobathic. Of the non-endemic species, there was still a relatively high presence of stenobathic species, where 54% of them had depth ranges of <1000 or 2000 m and 29% had ranges of 2000–3000 m and only 17% had larger depth ranges though none of them exceeded a range of 4000 m.

Based on the *Galathea* and *Vitjaz* samples, isopods were found in approximately 70% of trawls (in some instances accounting for 10–40% of the catch) and 36% of the bottom grabs. They have also been found in almost all of the trenches sampled to date and at 10 700 m deep in the Mariana Trench.

Table 9.3 Isopod families found at hadal depths with numbers of genera, species and bathymetric range.

Family	Genera	Species	Depth range (m)
Acanthaspidiidae	1	3	5650–7216
Antarcturidae	2	3	6090–7370
Arcturidae	1	2	7200–7370
Cirolanidae	1	1	5986–6134
Desmosomatidae	1	2	5986–6710
Echinothambematidae	1	1	5800–6850
Haploniscidae	3	19	5986–10 415
Ischnomedidae	3	21	6050–8830
Janirellidae	1	9	6150–8430
Laptanthuridae	1	1	6580
Macrostylidae	1	15	5986–10 730
Mesosignidae	1	6	5986–7880
Munnidae	3	3	5986–6450
Munnopsidae	10	41	5345–10 687
Nannoniscidae	4	8	5986–9043

The most diverse families are the Munnopsidae (10 genera, 41 species), Ischnomedidae (3 genera, 21 species), Haploniscidae (3 genera, 19 species) and Macrostylidae (1 genus, 15 species). In instances where mass findings of isopods were collected, they typically consisted of two genera of the Munnopsidae family; *Eurycope* and *Storthyngura* (although it is worth noting that several of the species within *Storthyngura* were later reclassified into other genera, but still within the same family; Malyutina, 2003).

Examples of mass findings reported by Belyaev (1989) include one from 6200 m in the Japan Trench, where 159 specimens of 7 species, including 150 specimens of *Vanhoeffenura* (*Storthyngura*) *bicornis* (Birstein, 1957) were caught and the isopods accounted for 27% of the entire catch. Furthermore, in the same trench at 7200 m, 41% of the catch were of the Isopoda, comprising two species including eight *Storthyngura herculea* (Birstein, 1957; later reclassified as *Rectisura herculea* Malyutina, 2003). At 8000 m in the Kuril–Kamchatka Trench, 13% of the catch comprised two isopod species (40 *Rectisura* (*Storthyngura*) *vitjazi* Birstein, 1957 and 4 *Eurycope magna* Birstein, 1963).

Based on the *Galathea* findings, Wolff (1956, 1970) noted some other aspects of the hadal Isopoda, mostly concerning their size. He observed a tendency towards a larger body size in species that occurred at the greatest depths (Fig. 9.8). Of a total of 47 species and subspecies, only 10 were found to be smaller than the average size for the genus, whereas in half the genera, the mean size of the hadal species was much larger than that of the abyssal species. With the exception of the genus *Haploniscus*, only 7 species (4%) out of the 186 non-hadal species of the 11 genera were larger than the largest hadal species (Wolff, 1970). This 'gigantism', or rather increase in body size, in hadal Isopoda was attributed to the effect of hydrostatic pressure on metabolism (Wolff, 1960), as suggested earlier by Zenkevitch and Birstein (1956) and Birstein (1957).

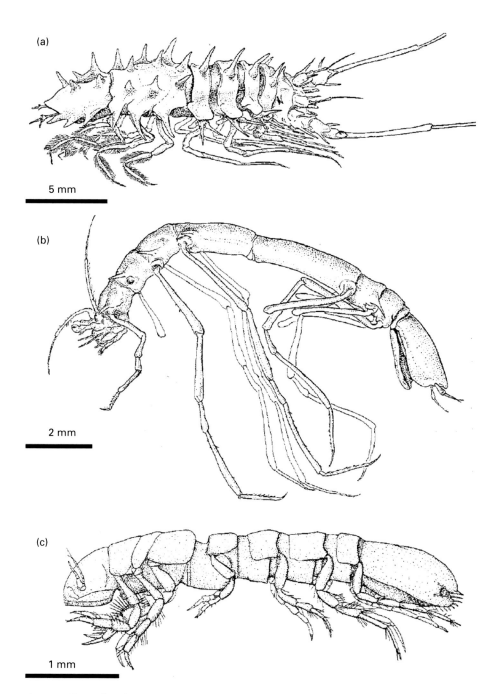

Figure 9.6 Examples of hadal isopods collected from the *Galathea* expeditions. (a) *Storthyngurella*
(*Storthyngura*) *benti* from 7000 m in the Kermadec Trench; (b) *Ischnomesus bruuni* from
7000 m in the Kermadec Trench; and (c) *Macrostylis hadalis* from 7280 m in the Banda Trench.
Images from Wolff (1956). Images reproduced with the permission of *Galathea Report*.

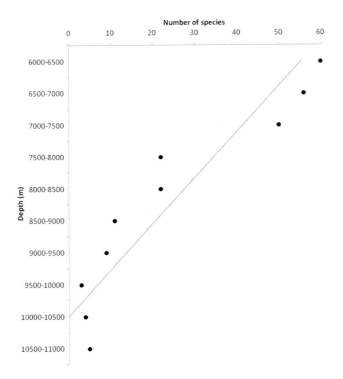

Figure 9.7 Number of species per 500 m depth stratum for the Isopoda.

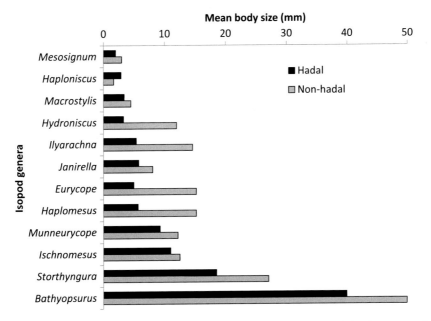

Figure 9.8 Mean body size for each of the isopod genera found at hadal depths compared to the non-hadal representatives. Data derived from Wolff (1970).

Upon inspection, Wolff also noted a complete lack of eyes in the hadal isopods, although this is normal for the Asellota. He also reported that the bodies of some species were extremely brittle, more so than their bathyal or abyssal counterparts. Conversely, he noted that the samples of *Storthyngura* were very robust and were no less calcified than similar abyssal species. Generally, he concluded that the hadal Isopoda were remarkably similar to their abyssal, bathyal and sublittoral relatives, and no endemic genera were found. They were, however, characterised as being larger, less spiny and more strikingly white in colour than bathyal and abyssal specimens of the same genera (Wolff, 1956). None of female isopods collected by the *Galathea* were found to be egg bearing, thus information regarding their fecundity or size of eggs within the marsupiums was lacking.

Due to their relatively small body size and diversity, isopods are notoriously difficult to identify from bottom photographs or video footage and, therefore, little is known about the behaviour of deep-sea isopods in general (Jamieson *et al.*, 2012b). Despite their position as a very characteristic component of the hadal community, even less is known about isopods at hadal depths, other than information derived from physical samples. On the PROA expeditions, Isopoda were observed several times on bottom images (Lemche *et al.*, 1976). They described seeing about seven isopods at 7847–8662 m in the North Solomon Trench (1.5–2 cm long), mostly resembling janirids, while one may have been a eurycopid. Another possible eurycopid was observed at 6758–6776 m in the New Hebrides Trench, alongside a 1.5 cm long individual thought to be of the genus *Ilyarachna*. At 8021–8042 m in the Palau Trench, Lemche *et al.* (1976) saw elongated and slightly cruciform isopods of 1–1.5 cm in length and in numbers of up to seven in a single image, apparently all facing in the same direction. In addition to these, another species (0.7–0.8 cm) was also observed but not identified.

In 2009, during the HADEEP project, two species of isopods were frequently seen in the Kermadec Trench (Jamieson *et al.*, 2011a and also Fig. 9.9). Two species of munnopsid isopods were observed by Hadal-Lander B but were too small to be identified to species level. The first (munnopsid) was seen once at 7199 m but was more conspicuous at 7561 m where a maximum number of three individuals were observed. These isopods (lengths = ~10–20 mm) were observed approaching the bait and climbing on it. The isopods were noted to leave the vicinity once the snailfish *Notoliparis kermadecensis* (Liparidae) arrived or were, perhaps, even consumed by the fish. Munnopsid B, possibly of the genus *Storthyngurinae*, was photographed at 7199 m. They were less conspicuous throughout the images than munnopsid A and tended to sit motionless, some distance from the bait with no apparent interaction with other species or the bait.

When the same lander was deployed in the Peru–Chile Trench in 2010, a much greater diversity of Isopoda was observed. Five putative species were observed between 4602 and 8074 m, two of which belonged to munnopsid. The three other species remain unidentified, however, one shared an uncanny resemblance to munnopsid A from the Kermadec Trench and one other was a long-legged species. The long-legged isopod was seen at 4602 m, whereas the other munnopsid was seen frequently at 5329, 7050 and

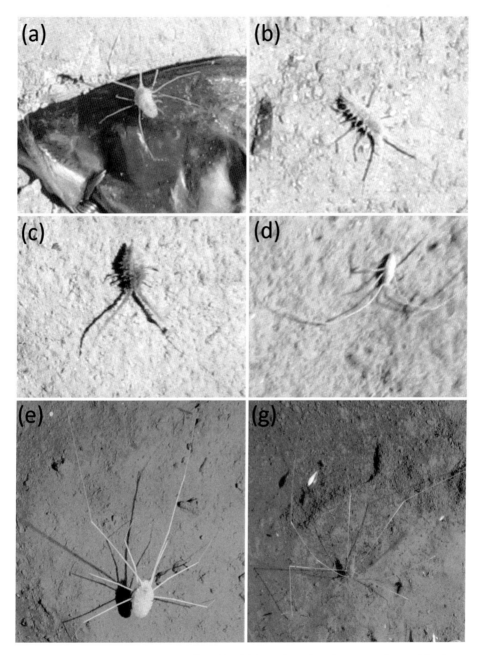

Figure 9.9 Examples of *in situ* observation of hadal isopods, although none can be confidently identified from photographs, all are thought to be from the Munnopsidae family. (a) and (b) are from 7561 and 7199 m in the Kermadec Trench, respectively. (c) and (d) are from 7050 m in the Peru–Chile Trench; (e) is from 5469 m on the edge of the Mariana Trench; and (g) is from 5329 m in the Peru–Chile Trench. All photos were taken using Hadal-Lander B, images courtesy of HADEEP.

8074 m (similar depths to the Kermadec Trench isopod). The other three species (sp. 1, 2 and 3) were seen at 4602–5329 m, 5329–6173 m and 7050 m, respectively. None of the isopod species observed in the Peru–Chile Trench were seen in numbers greater than one.

Identifying isopods to species level or detailing their behaviour is extremely difficult from *in situ* imaging due to their relatively small body size. However, one species of isopod was filmed at 6945 and 7703 m during HADEEP cruises to the Japan Trench in 2007 and 2008, respectively. This species, *Rectisura* (*Storthyngura*) *herculea*, was of a much larger body size than many others and this enabled a reasonably confident *in situ* identification. Second, the resolution of the camera on Hadal-Lander A was of a sufficient quality to track the locomotive speeds and other behaviours with confidence.

The video data showed *R*. cf. *herculea* aggregating towards the bait, particularly at 6945 m, where they reached a maximum number of 12 after 6 h (Fig. 9.10). Their behaviour towards the bait indicated that these isopods directly exploited the bait, as the largest percentage of observations (55%) showed individuals situated directly on the bait itself. This was interesting because Wolff (1962) had previously studied the gut contents of 36 specimens of 19 asellote isopod species from the *Galathea* expedition and concluded that deep-sea benthic asellotes were primarily detritus feeders, although some species are known to directly consume plant debris (Wolff, 1976). Previously, Wolff (1962) had reported Foraminifera in the guts of some asellote species and it was assumed that these had been accidentally swallowed along with detritus. However, Svavarsson *et al.* (1993) reported that *Ilyarachna hirticeps* Sars, 1870 and *Eurycope inermis* Hansen, 1916 (family Munnopsidae), from 1200 to 2000 m were seen to prey selectively on benthic forams, rather than consuming them indirectly with detritus, a strategy reiterated in mouthpart morphology (Wilson and Thistle, 1985). Scavenging is a feeding strategy used by some shallower isopod genera, notably *Natatolana* (e.g. Wong and Moore, 1995) that also occur, and scavenge, as deep as 2500 m (Albertelli *et al.*, 1992). Similarly, the giant isopod *Bathynomus giganteus* Milne-Edwards, 1879 (Soong and Mok, 1994) is known to occasionally consume fish and squid remains (Barradas-Ortiz *et al.*, 2003). However, these species belong to another suborder, Flabellifera, and the family Cirolanidae, a specialised group of sighted and active carnivorous scavengers that are not represented in the hadal zone.

The combined results of Wolff (1962), Svavarsson *et al.* (1993) and Jamieson *et al.* (2012*b*) may indicate that deposit-feeding and/or predation on forams may be routine, but the isopods are capable of exploiting the temporarily nutrient-rich presence of a carrion-fall (trophic plasticity); a strategy also adopted by lysianassoid amphipods (Blankenship and Levin, 2007). The observations from the Kermadec Trench also supported this indication; one specimen was recovered in a baited trap that did not contain sediment (Jamieson *et al.*, 2011a). Furthermore, images obtained from 8075 m in the Peru–Chile Trench not only show the presence of munnopsid isopods attending the bait, but also captured an individual isopod dragging a small piece of fragmented bait out of the field of view (Jamieson *et al.*, 2012b). These observations are further

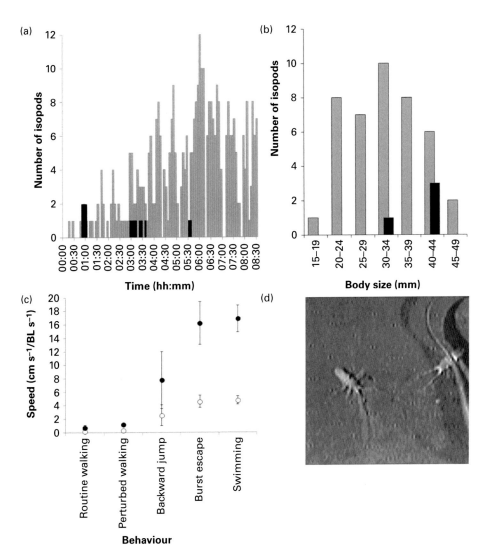

Figure 9.10 Behaviour and locomotion of *Rectisura* cf. *herculea*, a hadal isopod from 6945 and 7703 m in the Japan Trench. (a) Numbers aggregating over time; (b) size frequency (grey = 6945 m and black lines = 7703 m); (c) is their absolute speed (black dots) and size-specific speed (white dots) for various locomotion types; and (d) is a frame grab showing two individuals at 6945 m near the tail of a macrourid. Modified from Jamieson *et al.* (2012b).

supported by anecdotal evidence of munnopid isopods of the subfamily Bathyopsurinae approaching and feeding on bait at 4185 m in the Peru Basin (Brandt *et al.*, 2004). Therefore, it seems appropriate to conclude that *Rectisura* cf. *herculea* is a bait-attending species, most likely of the facultative scavenging guild, however, the extent of this feeding behaviour among the hadal Isopoda requires more *in situ* observations using baited systems.

In addition to noting these feeding behaviours, the Jamieson *et al.* (2012b) data also provided the first measurement of locomotion of a hadal isopod, albeit there are no other such measurements for other deep-sea isopods for comparison. For the *Rectisura* cf. *herculea* isopod, its routine mode of locomotion is walking using pereopods I–IV at mean of $0.19\,\mathrm{BL\,s^{-1}} \pm 0.04$ S.D. If perturbed by the presence of predators, most notably the decapod *Benthesicymus crenatus* Bate, 1881, the isopods accelerated to $0.33\,\mathrm{BL\,s^{-1}}$ ± 0.04 S.D. (an increase of 74% from routine). When the threat became more urgent ($<80\,\mathrm{mm}$ away), the isopods jumped backwards, propelling themselves with pereopods V–VII at $2.6\,\mathrm{BL\,s^{-1}} \pm 1.5$ S.D. to a distance of 1.9 (± 0.8 S.D.) body lengths away. When the threat is imminent ($<50\,\mathrm{mm}$), the isopods initiated a burst swimming escape response, whereby they retracted pereopods I–IV and swam backwards in a spiralling motion using pereopods V–VII, either vertically or horizontally across the seafloor at $4.63\,\mathrm{BL\,s^{-1}} \pm 0.9$ S.D. After a $<0.6\,\mathrm{s}$ burst of acceleration they continued swimming at a mean velocity of $4.8\,\mathrm{BL\,s^{-1}} \pm 0.6$ S.D. for at least $2\,\mathrm{s}$ until reaching an estimated distance of 30–40 cm away from the threat. During vertical burst escape, the descent back to the seafloor was dampened by a full extension of pereopods I–IV and an increase in projected body area (and thus drag) by descending (gliding) ventral side down to the seafloor; a 'burst and coast' tactic. These data suggested no significant reduction in locomotory capability despite the extreme depths in which they inhabit and findings were comparable to the swimming speeds of shallow-water isopods (Alexander, 1988).

9.8   Amphipoda

The Amphipoda are a very characteristic element of the hadal fauna. Their ability to detect, intercept and consume baits placed on the seafloor, particularly at depths exceeding 8000 m, is exceptional. They have been found in all the trenches sampled to date, at all depths including over 10 500 m (e.g. Blankenship *et al.*, 2006; Fig. 9.11). The vast majority of hadal amphipods are prolific scavengers, albeit facultative, and given their propensity to swarm at bait they have become a well-known fauna through the use of baited traps (e.g. Hessler *et al.*, 1978; Blankenship *et al.*, 2006; Jamieson *et al.*, 2011a). They also normally occur in high numbers and are one of the few highly mobile species to be frequently photographed using drop-cameras in trenches (Lemche *et al.*, 1976). The baited trap catch frequency within the HADEEP projects was 100% based on 46 deployments in six trench locations and the trawl catch frequency of Amphipoda was 70% (based on 124 successful trawls; Belyaev, 1989). Across the board, the vast majority of deep-sea amphipods that have been collected using baited systems have belonged to the superfamily Lysianassoidae (Dahl, 1979). At hadal depths, there are currently 77 known species belonging to 42 genera comprising 23 families, all of which are from the suborder Gammaridea (Table 9.4). Of these, 61% are considered benthic or bentho-pelagic, while the remainder are pelagic.

The most diverse amphipod family is the Lysianassidae with 5 genera and 10 known species. Based on the findings of HADEEP and other studies, there are six main benthic

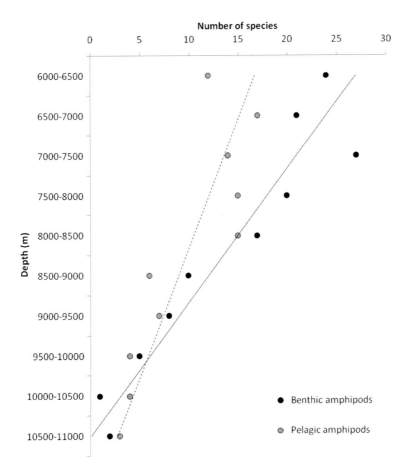

Figure 9.11 Number of species per 500 m depth stratum for the benthic and pelagic Amphipoda.

families found at hadal depths; Eurytheneidae, Hirondellidae, Allicellidae, Uristidae, Scopelocheiridae and Pardaliscidae (*Princaxelia*). Based on the Soviet and Danish expeditions there are three main hado-pelagic families; Pardaliscidae (*Halice*), Hyperiopsidae and Lanceolidae. It is also worth noting that many of the species therein are bentho-pelagic and can be found both on the seafloor and at considerable distances above (e.g. *Eurythenes gryllus* Lichtenstein, 1822).

Eurythenes gryllus is one of the most bathymetrically and geographically widespread marine species (Thurston, 1990; Fig. 9.12). It is known throughout all oceans and is thought to be a cold water stenotherm; inhabiting shallower depths at the poles but restricted to deeper colder waters at low latitudes (Thurston *et al.*, 2002). *E. gryllus* forms an important part of the deep-sea benthic community because of its ability to rapidly intercept and consume carrion-falls (Ingram and Hessler, 1983; Hargrave *et al.*, 1995) and also as a predator (Sainte-Marie, 1992). The geographic and bathymetric coverage of *E. gryllus* is also reflected in its occurrence at hadal depths. Earlier literature reported *E. gryllus* from 6770–6850 m in the Izu-Bonin Trench (northwest Pacific;

Table 9.4. Amphipod families found at hadal depths with numbers of genera, species and bathymetric range.

Family	Genera	Species	Depth range (m)
Stilipedidae	1	1	7210–7230
Maeridae	2	2	6600–8900
Lysianassidae	5	10	6007–10 500
Ischyroceridae	1	1	6324–6328
Ampeliscidae	1	1	6475–6571
Epimeriidae	1	1	6156–7230
Eurytheneidae	1	1	4329–8074
Eusiridae	3	6	6090–9120
Pardaliscidae	3	9	4000–10500
Phoxocephalidae	3	3	6324–7550
Hirondellidae	1	5	6000–10787
Hyperiopsidae	3	6	4200–8500
Lanceolidae	2	3	4000–10500
Atylidae	1	3	6475–8015
Liljeborgiidae	1	1	6156–6207
Alicellidae	2	4	4329–8480
Cyclocaridae	1	2	6007
Scinidae	1	2	6000–9400
Scopelocheiridae	2	2	6000–8723
Stegocephalidae	3	9	6000–8500
Uristidae	2	3	5173–6173
Valettiopsidae	1	1	6007
Vitjazianidae	1	1	4200–8480

Kamenskaya, 1981) and at 7230 m in the Peru–Chile Trench (southeast Pacific; Ingram and Hessler, 1987). It was later found in large numbers at 7800 m in the Peru–Chile Trench (Thurston *et al.*, 2002). In the southwest Pacific, it was found in the Tonga Trench, albeit in low numbers between 5155 and 6252 m (Blankenship *et al.*, 2006) and at 4329–6007 m in the neighbouring Kermadec Trench (Jamieson *et al.*, 2011a). More recent and unpublished sampling from the Kermadec Trench found more specimens, also restricted to ~6000 m (unpublished data, HADEEP). Several large specimens of *E. gryllus* were observed using baited cameras by Fujii *et al. (*2010), at 7703 m in the Japan Trench. The unidentified specimens observed by Hessler *et al.* (1978), also in the Peru–Chile Trench, were later suspected to be *E. gryllus* (Thurston *et al.*, 2002).

The occurrence of E. *gryllus* in the Peru–Chile and Japan Trenches appears to slightly contradict the cold water stenotherm hypothesis discussed by Thurston *et al.* (2002), since although *E. gryllus* appears to dominate the hadal depths of the warmer Peru–Chile Trench, it is apparently unable to penetrate far into the colder Kermadec Trench (below 6000 m), albeit this trench is only ~0.75°C colder than the other two trenches. The reasons for this anomaly remain unresolved but it appears likely that *E. gryllus* distribution is driven by food supply or a combination of this and temperature, rather than temperature alone (Fujii *et al.*, 2013).

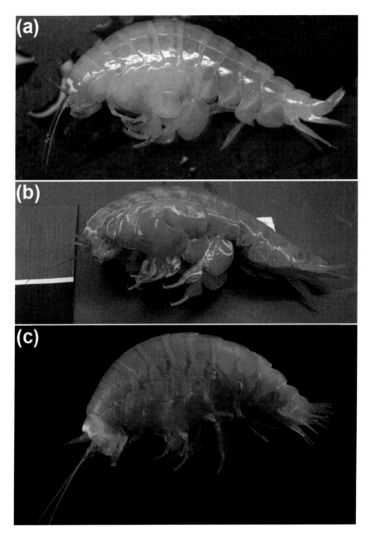

Figure 9.12 Examples of *Eurythenes gryllus* caught from: (a) 7703 m in the Japan Trench; (b) 6079 m in the Kermadec Trench; and (c) 6173 m in the Peru–Chile Trench. Images (a) and (b) courtesy of HADEEP, image (c) taken by Camilla Sharkey, University of Bristol, UK.

The dominance of *E. gryllus* at upper hadal depths is clear in video footage from the Japan Trench (Fujii *et al.*, 2010). However, the significance of this species is, perhaps, most striking in the time-lapse images from the Hadal-Lander, taken across the trench depths of the Peru–Chile Trench, at 8074 m in particular, at the deepest point (Richards Deep; Fujii *et al.*, 2013; Fig. 9.13). At this station, the deepest at which they have ever been found, *E. gyllus* arrived within an hour of the bait reaching the seafloor and consumed over 1 kg of tuna in less than 20 h.

    E. gryllus is also known to occur tens of metres above the seafloor (up to 500 m above bottom in the abyssal North Pacific; Ingram and Hessler, 1983). During sampling for the

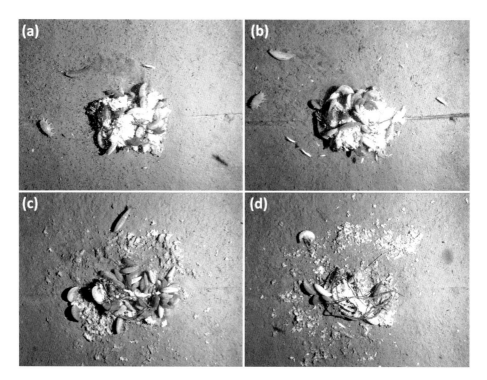

Figure 9.13 Example of the response to bait places at 8074 m in the Peru–Chile Trench by *Eurythenes gryllus*; (a–d) are images taken 5 h apart, the field of view is $0.29\,\mathrm{m}^{-2}$ (62×46.5 cm). Images were taken using Hadal-Lander B, images courtesy of HADEEP.

HADEEP project in the Peru–Chile Trench, traps were set at 1, 2, 20, 30, 40, 50, 60 and 90 m above bottom. The only species caught in any of the traps set at 20 m or greater was *E. gryllus*, albeit in numbers far lower than those obtained in the abyssal North Pacific (Smith *et al.*, 1979; Smith and Baldwin, 1984) or North Atlantic (Charmasson and Calmet, 1987). Although somewhat anecdotal, it appears that *E. gryllus* play a lesser role in the pelagic environments at hadal depths than at abyssal depths.

Eurythenes gryllus also exhibits bathymetrically stratified populations. Abyssal populations are known to be genetically diverse across immense geographic locations, yet a distinctive, divergent population was identified between those on the abyssal plains and those inhabiting guyots within the same area (Bucklin *et al.*, 1987), suggesting the populations were bathymetrically stratified. Thurston *et al.* (2002) also commented that specimens found at 7800 m in the Peru–Chile Trench appeared to be outside of the known morphological variability for *E. gryllus*, and that they may be undergoing incipient speciation. Genetic analysis currently arising from the HADEEP project is revealing that the Peru–Chile Trench population is highly divergent from the abyssal populations, even more so than the guyot population described by Bucklin *et al.* (1987). Furthermore, these two depth-stratified populations are found in the same area and appear to be separated at the 6000 m mark.

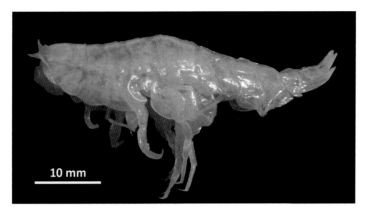

Figure 9.14 A specimen of *Hirondellea dubia* recovered from 9104 m in the Kermadec Trench. Image courtesy of HADEEP.

Hirondellea of the family Hirondelleidae (Fig. 9.14) are named after the yacht *Hirondelle* that first used baited traps at abyssal depths in 1888 and are, perhaps, the most conspicuous and ubiquitous hadal amphipod genus with distinctive phylogeographic structure. *Hirondellea gigas* (Birstein and Vinogradov, 1955) are prolific scavengers, found in the trenches of the northwest Pacific; Kuril–Kamchatka, Philippine, Mariana, Japan, Izu-Bonin, Volcano, Yap and Palau Trenches (Birstein and Vinogradov, 1955; Dahl, 1959; Hessler *et al.*, 1978; Kamenskaya, 1981; France, 1993; unpublished data, HADEEP). This species is known to attend baited systems at depths greater than 10 000 m, in extraordinarily large numbers, in fact, these swarms increase in size with increasing depth. Very few have been found at abyssal depths and the shallowest record to date is 6770 m (Belyaev, 1989). The genetic homogeneity of *H. gigas* was examined by France (1993) who studied abyssal-partitioned populations from the Mariana, Philippine and Palau Trenches. He concluded that these geographically isolated populations may have reduced levels of gene flow causing them to diverge morphologically. Nonetheless, *H. gigas* is currently known only in the northwest Pacific trenches.

In the southwest Pacific trenches (Kermadec and Tonga) the most common scavenging amphipod is *Hirondellea dubia* (Dahl, 1959). *H. dubia* exhibits many of the same characteristics as *H. gigas*, in that it appears to be mostly restricted to the southwest quadrant of the Pacific. It is also found in ever-increasing numbers with increasing depth (found at depths exceeding 10 500 m) and is often the sole amphipod species caught in baited traps >9200 m (Blankenship *et al.*, 2006). While *H. dubia* is primarily a species restricted to the southwest Pacific, with a minimum depth recorded in the Kermadec Trench of 6000 m, albeit only four specimens (Jamieson *et al.*, 2011a), the DNA sequencing of the HADEEP samples found several specimens of *H. dubia* on the abyssal plains, east of the Mariana Trench at 5469 m, some 6000 km north of the Kermadec and Tonga Trenches (unpublished data, HADEEP). This study suggests that *H. dubia* are not entirely restricted to the southwest Pacific trenches, nor are they, therefore, an exclusively hadal species, but are, in fact, found in the open abyssal plains, albeit in very low numbers. Their status as the dominant, scavenging amphipod of the southwest Pacific trenches is not refuted, since catches in baited traps at >9500 m have been known to exceed 12 000 individuals (Blankenship *et al.*, 2006). It is interesting to note that no *H. gigas* were found in the same abyssal location, despite being known as the dominant scavenging amphipod of the Mariana Trench. The gene flow, albeit limited between the trenches of the northwest Pacific, also suggests an abyssal fraction of *H. gigas* that may become more apparent as more sampling is undertaken between trenches.

Across in the southeast Pacific, Perrone *et al.* (2002) documented a new, but unnamed, species of *Hirondellea* from 7800 m, in the Peru–Chile Trench, based on conventional morphological taxonomy. This finding added another piece to the puzzle concerning the *Hirondellea* distribution pattern in the hadal Pacific Ocean; it appears that at hadal depths, each quadrant of the Pacific is inhabited by a different, single species of *Hirondellea* where *H. gigas* and *H. dubia* inhabit the northwest and southwest quadrants, respectively. The new Peru–Chile specimens accounted for 64.7% of the 945 amphipods collected by Perrone *et al.* (2002). The results were, however, taken from a single location. In 2010, the HADEEP project sampled five stations at 4602, 5329, 6173, 7050 and 8075 m in the same trench and found that, unlike the *Hirondellea* species of the western trenches, three morphologically distinct species were identified: species 1 was found in low numbers at 6173 m ($n = 4$), species 2 was found at the deepest three sites and increasing in numbers with depth ($n = 2$, 15 and 104, respectively) and species 3 was found only at 7050 m ($n = 33$) (Kilgallen, in press). Which of the three species found in the HADEEP study corresponds to that reported in Perrone *et al.* (2002) is currently unknown. However, based on the closest geographical and bathymetric deployment, *Hirondellea* species 2 is likely to be the unnamed species that Perrone *et al.* also found because it accounted for 62.3% of the catch at 8074 m, whereas *E. gryllus* comprised only 32.2%. Perrone *et al.* (2002) and Thurston *et al.* (2002) reported a catch of *Hirondellea* sp. nov. and *E. gryllus* from 7800 m as approximately 50:50.

The amphipod family Alicellidae was named after the vessel *Princess-Alice*, one of the first vessels to trawl at hadal depths, in 1901. The baited trap campaigns of the HADEEP project have often recovered species of *Paralicella* from the Alicellidae family, notably *P. tenuipes* Chevreaux, 1908 and *P. caparesca* Shulenberger and Barnard, 1976. These have been recovered from 4329–7000 m and 4329–6007 m, respectively, in the Kermadec Trench (Jamieson *et al.*, 2011a). They have also been found in the Peru–Chile Trench at 5329–7050 m and 4602–6173 m, respectively (Fujii *et al.*, 2013) and at similar depths in the Japan Trench and on the abyssal plains surrounding the Mariana Trench (unpublished data, HADEEP). While these species account for a large fraction of the abyssal and upper hadal depths, they are a cosmopolitan, abyssal species and no specimens have been found deeper than ~7000 m. However, *Paralicella microps* (Birstein and Vinogradov, 1958) exhibits a more 'hadal' presence and has been discovered in the Kuril–Kamchatka, Japan and Izu-Bonin Trenches down to depths of 8000 m (Belyaev, 1989).

Perhaps the most striking species of Alicellidae is *Alicella gigantea* (Chevreux, 1899), also known as the 'supergiant' amphipod (*sensu* Barnard and Ingram, 1986). It is the largest known amphipod, originally identified in the northern hemisphere (Chevreux, 1899; Hessler *et al.*, 1972; Barnard and Ingram, 1986; DeBroyer and Thurston, 1987). Despite its conspicuous and inexplicably large body size (up to 340 mm), the 'supergiant' has remained somewhat enigmatic as a result of sparse findings spanning an enormous bathymetric and geographic range (both North Atlantic and North Pacific, 1720–6000 m).

The largest known amphipod, measuring up to 340 mm total body length was found among the regurgitated stomach contents of an albatross in Hawai'i (Harrison *et al.*, 1983). The species is known to inhabit the deep abyssal plains of the northern hemisphere, in the North Atlantic Ocean (off the Canaries, Cape Verde and in the Demerara Basin) and has been found in the vicinity of the Hawai'i Islands in the North Pacific Ocean (Barnard and Ingram, 1986; De Broyer and Thurston, 1987). These two localities are approximately 12 800 km (6900 nautical miles) apart and separated by the American continental land mass. Furthermore, in both these known areas of occurrence, the specimens have been captured (albeit in low numbers) multiple times, but have never been found in the more frequently studied areas around the associated ocean rims. There is also a report of 61 individuals recovered by baited traps at 6200 m off the coast of Japan which weighed a combined 1.1 kg, but no other information is available (Hasegawa *et al.*, 1986). In 2011 and 2012, a total of nine specimens of *A. gigantea* were recovered using the *Latis* fish trap, from depths of 6265 and 7000 m and upwards, to where a further nine individuals were observed by the Hadal-Lander at 6890 m in the Kermadec Trench (Jamieson *et al.*, 2013; Fig. 9.15). These were the first and only findings of the supergiant in the southern hemisphere and at hadal depths. The specimens ranged from 102 to 278 mm long. The largest specimens were mature males with an overall length to weight relationship of length $= 0.968$ weight $+ 113.87$ ($n = 8$; $R^2 = 0.9166$). The individuals caught on camera were estimated to be between 175 and 349 mm long. Shortly after the Kermadec Trench findings, *A. gigantea* was photographed by a baited camera at 5160 m in the Sargasso Sea, in the Atlantic (Fleury and Drazen, 2013).

The samples from the Kermadec Trench were identified as *A. gigantea* using three methods: (1) comparison with the re-description of type specimens of Chevreux (1899) and description of specimens from both the North Atlantic and Pacific by De Broyer and Thurston (1987); (2) direct comparison with a single male specimen (TL $= 240$ mm), 5851 m, central North Pacific Ocean (30° 18.0' N, 157° 50.9' W, ID C10951; University of California, San Diego Benthic Invertebrate Collection); and (3) DNA sequence comparisons between the Kermadec Trench samples and the confirmed central North Pacific individual. The results showed no significant morphological differences between the specimens from the Kermadec Trench and previous descriptions (Chevreux, 1899; De Broyer and Thurston, 1987) or any genetic differences from the male specimen from the central North Pacific (Jamieson *et al.*, 2013). These results support the observation that there is no significant morphological variation between the geographically widespread populations (De Broyer and Thurston, 1987).

Figure 9.15 The supergiant amphipod *Alicella gigantea*. The top image was taken *in situ* at 6979 m in the Kermadec Trench, showing a large individual next to the snailfish *Notoliparis kermadecensis*. The bottom image was captured from 7000 m in the same trench and measured 27.8 cm in length. Images courtesy of HADEEP.

The majority of samples containing *A. gigantea* have been recovered from the lower abyssal to hadal zone (4850–7000 m), however, a single juvenile female was captured at 1720 m in the central North Pacific Ocean (Bernard and Ingram, 1986). This record provides the bathymetric range for *A. gigantea* of 5280 m. The limited observations of *A. gigantea* are, therefore, surprising given its apparent vast bathymetric and geographic range. The question then arises as to why a relatively large deep-sea animal with such a large bathymetric and geographic range is so infrequently found, while other smaller amphipods, with similarly wide ranges are so frequently caught in high abundance (e.g. *Eurythenes gryllus*; 184–7800 m; Barnard, 1961; Thurston *et al.*, 2002; De Broyer *et al.*, 2004; Stoddart and Lowry, 2004). The scarcity of findings for this large crustacean can probably be attributed to the low number of samples taken from depths of >5000 m, particularly when sampling is reliant only upon traps with a small opening

that could prevent the capture of the large-bodied A. *gigantea*. However, in the case of
the Kermadec Trench, the same vehicles that captured and imaged A. *gigantea* during
previous sampling campaigns have been used frequently (eight and nine times, respectively), within the same depth range, without detecting the presence of A. *gigantea*
(Jamieson *et al.*, 2009a, b, 2011a). What is, perhaps, even more perplexing is that
despite the fact that these amphipods were readily captured and observed 2 km apart on
the same day, further attempts to observe or capture these animals on the same voyage
at the same locations failed, despite having detected the predictable occurrences of fish,
decapods and smaller lysianassoid amphipods. The low frequency of capture could have
occurred because either: (1) A. *gigantea* has a very patchy distribution or (2) it is very
sparsely distributed. It would take a relatively long time for a very sparsely distributed,
scavenging animal to arrive at a bait, perhaps even longer than the typical baited camera
deployment time of <12 h and, therefore, it is rarely trapped or imaged. In the Kermadec
Trench study (Jamieson *et al.*, 2013), individuals of A. *gigantea* were among the last
species to arrive at the bait beneath the camera (>5 h 33 min), and their maximum
number was not achieved until towards the end of a typically long lander deployment
(after 16 h of a 25 h deployment). Thus, it is probable that A. *gigantea* is very sparsely
distributed in its habitat.

Genetic homogeneity among the known locations of A. *gigantea* span thousands of
kilometres, albeit in relatively slowly evolving DNA regions. This implies that genetic
connectivity between the Kermadec Trench and Hawai'i populations is at a level that is
difficult to reconcile with a seemingly disjunct distribution. Similar patterns of large-
scale distribution within the same depth zone have been shown for the cosmopolitan
amphipod E. *gryllus* (France and Kocher, 1996). These findings indicate that the
disjunct distribution of A. *gigantea* may be an artefact of both small-scale patchiness
in population density and a lack of sampling across the adjoining abyssal plains at
depths of greater than 5000 m. For example, the central Pacific basin, particularly in the
equatorial waters, is an area where very little deep-sea biological sampling has been
undertaken. It can therefore be construed that if further sampling campaigns are
undertaken using appropriate methods on the deep abyssal plains and abyssal–hadal
transition zones, it is likely that findings of the supergiant A. *gigantea* may become
more frequent and geographically widespread. Issues such as patterns of phylogeographic structure and gene flow across oceans could then be assessed more readily and
with greater reliability.

Perhaps the most conspicuous feature of A. *gigantea* is their extraordinary large body
size relative to all other deep-sea amphipods; an explanation for this is yet to be
unequivocally established. It is known, however, that the *Alicellidae* are a primitive
family of gammaridean Amphipoda, with broad unpleated gills on coxa 2–7 and
accessory lobes on gills 5–6. Gammarideans are believed to have undergone a reduction
in body size accompanied by a reduction in respiratory surface (Steele and Steele,
1991). The presence of a gill on coxa 7 and tubiliform accessory lobes in A. *gigantea*
may suggest a need for additional respiratory surfaces, retained from when amphipods
first evolved the larger body size (Steele and Steele, 1991). Pressure, temperature and
oxygen have all been hypothesised to be related to gigantism. The most recent

discussions have focused on oxygen availability as a driver for maximum potential size, as a product of both concentration of oxygen and partial pressure (Chapelle and Peck, 1999, 2004; Peck and Chapelle, 1999; Spicer and Gaston, 1999). However, this does not adequately explain deep-sea gigantism where little to no variation has been recorded in temperature vertically through the water column. Further discussion has considered the ecology of *A. gigantea* and *E. gryllus* as bentho-pelagic scavengers, since they are more closely related to pelagic specialists than the benthic scavengers with whom they compete for carrion, suggesting that they may experience a higher oxygen threshold on body size by swimming above more oxygen-impoverished bottom waters (Chapelle and Peck, 2004).

The Uristidae family, particularly the genus *Abyssorchomene*, share a similar distribution to the Alicellidae, in that they include cosmopolitan abyssal species that often dominate the abyssal and sometimes upper hadal depths of a trench, in particular *A. chevreuxi* (Stebbing, 1906) and *A. distinctus* (Birstein and Vinogradov, 1960) in the Peru–Chile and Kermadec Trenches, and *A. musculosus* (Stebbing, 1888) also in the latter (Jamieson *et al.*, 2011a). In the Kermadec Trench, *Orchomenella gerulicorbis* Shulenberger and Barnard, 1976 (Lysianassidae) is also present at similar depths. Phylogentic analysis of the HADEEP amphipod samples did, however, refute the monopoly of the two genera, suggesting that they are one of the same (a sentiment reiterated in Antarctic studies; Havermans *et al.*, 2010). Species belonging to each of these groups are identified based on acute differences in their feeding appendages, which have been shown to exhibit high degrees of evolutionary plasticity during periods of range expansion (MacDonald *et al.*, 2005). Therefore, there is clearly a need to revise the morphological characters that are used for the classification of *Abyssorchomene* and *Orchomenella* species, ensuring traits that exhibit developmental or evolutionary plasticity are avoided. Blankenship *et al.* (2006) also documented a new species of *Uristes* from the Tonga and Kermadec Trenches, between 7349 and 9273 m, which was previously unreported from hadal depths. This species was later described as *Uristes chastaini* and some of the specimens were found to have eggs nestled in the brood pouch. This was the first recording of hadal lysianassoid females attending bait while brooding (Blankenship and Levin, 2009; Fig. 9.16). However, the name *Uristes chastaini* is *nomen nudum* nor does this species belong to the genus *Uristes* (Lysianassoidea: Uristidae), but rather it appears to belong to a new genus within the subfamily Tryphosinae (Lysianassoidea: Lysianassidae) (M.H. Thurston and T. Horton, pers. comm.).

The Scopelocheiridae are not a diverse family at hadal depths, comprising only two 'hadal' species of which one, *Scopelocheirus (Bathycallisoma) pacifica* Dahl, 1959, is known only from 6960–7000 m in the Kermadec Trench. However, the second species *Scopelocheirus schellenbergi* Birstein and Vinogradov, 1958 (Fig. 9.16) exhibits a seemingly enormous geographical range. It has been recorded in the North Pacific trenches, the Indo-Pacific trenches, the southwest Pacific trenches, the Puerto-Rico Trench in the North Atlantic and in the Java Trench in the Indian Ocean (Lacey *et al.*, 2013). It appears to inhabit depths of 6000 m to 9104 m and is listed by Belyaev (1989) as being a pelagic species, although it has been recovered in fairly large numbers from benthic traps (Blankenship *et al.*, 2006).

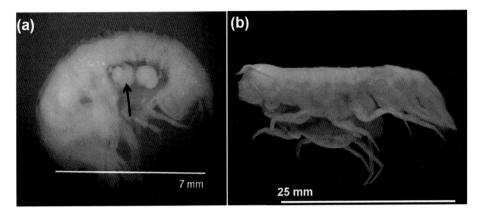

Figure 9.16 (a) The new species of the Lysianassidae subfamily Tryohosinae (formally *Uristes* sp. nov; Blankenship *et al.*, 2006) from 7349–9273 m in the Tonga Trench with arrow indicating the developing embryo. (b) *Scopelocheirus schellenbergi* from 6252–9104 m in the Tonga Trench. Images courtesy of L. Levin, Scripps Institution of Oceanography, USA.

The family Pardaliscidae is largely represented in the trenches by the benthic genus *Princaxelia* (named after Prince Axel of Denmark, 1888–1964, following the Danish *Galathea* expedition), and the pelagic genus *Halice*. The first known species of *Princaxelia* was *P. abyssalis*, sampled from the Kermadec Trench (Dahl, 1959) and later on *P. stephenseni* Dahl, 1959 and *P. magna* Kamenskaya, 1977 were also discovered, the latter of which was claimed to occur in the Aleutian, Kuril–Kamchatka, Izu-Bonin, Yap, Japan, Philippine, Bougainville and Kermadec Trenches (Kamenskaya, 1981). With the exception of *P. stephenseni* (known to occur in the shallower North Atlantic), all other species of *Princaxelia* are, so far, considered to be endemic to hadal trenches. The HADEEP projects discovered a new species, *Princaxelia jamiesoni* Lörz, 2010, from 7703 m in the Japan Trench (Fig. 9.17) and 9316 m in the neighbouring Izu-Bonin Trench. Amphipods of this genus appear to be represented in most trenches and are clearly visible in lander video footage from the Japan (*P. jamiesoni*, 7703 m), Izu-Bonin (*P. jamiesoni*, 9316 m), Kermadec (*P. abyssalis*, 7966 m) and Tonga Trenches (*P. abyssalis*, 8798 m; Jamieson *et al.*, 2012a). A further, unidentified species was also found in the Peru–Chile Trench at 5329 and 7800 m (unpublished data, HADEEP and Perrone *et al.*, 2002, respectively). The princaxelids differ somewhat from the other benthic amphipods encountered in the trenches because they do not consume the bait but rather, they prey upon smaller necrophagous species of amphipods.

The *Princaxelia* spp. appear large and distinct enough in *in situ* video recordings to permit analyses of behaviour and locomotion (Jamieson *et al.*, 2012a). The Hadal-Lander A made observations of *P. jamiesoni* in the Japan Trench (7703 m) and Izu–Ogasawara Trench (9316 m) and of *P.* aff. *abyssalis* in the Kermadec Trench (7966 m) and the Tonga Trench (8798 m). The body lengths of four *P. jamiesoni* from 7703 m and four from 9316 m were 57–71 mm (mean = 65 mm ± 6 S.D.) and 25–32 mm (mean = 29 mm ± 3 S.D.). In the Kermadec Trench, the body lengths for

Figure 9.17 The predatory amphipod *Princaxelia jamiesoni* (Pardaliscidae) recovered from 7703 m in the Japan Trench. Modified from Jamieson *et al.* (2011a), original image taken by Tomislav Karanovic, Hanyang University, S. Korea.

P. aff. *abyssalis* were 19–51 mm (mean = 29 mm ± 8 S.D., *n* = 14) and in the Tonga Trench specimens ranged from 18 to 37 mm (mean = 28 mm ± 5 S.D., *n* = 14) in length. The *P. jamiesoni* specimens caught in the traps at 7703 m in the Japan Trench were 56.2 mm (female; holotype), 57.5 mm (male; paratype) and 61.0 mm (female; paratype) long. At 9316 m in the Izu–Ogasawara Trench, the catch included one female at 36 mm and one 24 mm long juvenile.

These *in situ* observations and the specimens caught in the traps confirm the assumptions made by Kamenskaya (1981) and Lörz (2010) that princaxelid amphipods have well-developed olfactory senses and are carnivores and efficient swimmers. Their bait interception times of between 15 and 41 min were relatively fast compared to other taxa at similar depths (e.g. decapods, fish; Jamieson *et al.*, 2009a, b; Fujii *et al.*, 2010). It is assumed that olfaction is their primary detection sense, since princaxelid amphipods, particularly in the Japan and Izu–Ogasawara Trenches, arrived at the bait long before the other amphipod species (mainly lysianassoids) had aggregated in large numbers. Their status as carnivorous can be expanded further to confirm that they are not scavengers but rather predatory, preying on small scavenging amphipods; a common strategy at these depths (e.g. liparid fish and natantian decapods; Jamieson *et al.*, 2009a, b).

Princaxelia exhibits impressive mobility and flexibility in swimming, not only in a forward direction in the typical horizontal posture but also backwards with the head up, swimming vertically. As suggested by Lörz (2010), *Princaxelia* have an optimal body shape for efficient swimming, as the telson and uropods can be hydrodynamically

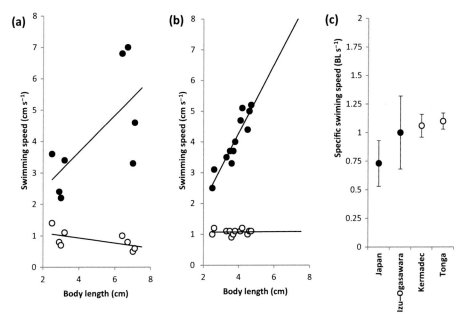

Figure 9.18 Swimming speeds of *Princaxelia* amphipods. (a) Absolute overground swimming speed (closed dots) and size-specific swimming speeds (open dots) of the amphipod *Princaxelia jamiesoni* from the Japan and Izu–Ogasawara Trenches (7703 m and 9316 m, respectively). (b) Absolute overground swimming speed (closed dots) and size-specific swimming speeds (open dots) of the amphipod *Princaxelia* aff. *abyssalis* from the Kermadec and Tonga Trenches (7966 m and 8798 m, respectively). (c) A summary of mean specific swimming speeds (BL s^{-1}) for *P. jamiesoni* (closed dots) and *P.* aff. *abyssalis* (open dots). Modified from Jamieson *et al.* (2011a).

streamlined to reduce drag and therefore increase locomotive efficiency. In the case of *P. jamiesoni* (Fig. 9.17), the wide rami of uropod III can be used for routine locomotion and burst acceleration, which is particularly useful during fast predatory attacks; a strategy all the more useful given their strong mouthparts which enable rapid immobilisation of prey, in conjunction with large maxilliped and strong gnathopods. Efficient swimming ability is a trait common to other deep-sea pardaliscid amphipods. Individuals have been observed holding station in mid-water, where current speeds reached up to $10 \, \text{cm} \, \text{s}^{-1}$ ($>10 \, \text{BL} \, \text{s}^{-1}$) (Kaartvedt *et al.*, 1994). Mean absolute swimming speeds were calculated for *P. jamiesoni* and *P.* aff. *abyssalis* as $4.16 \, \text{cm} \, \text{s}^{-1} \pm 1.8$ S.D. and $4.02 \, \text{cm} \, \text{s}^{-1} \pm 0.87$ S.D., respectively (Jamieson *et al.*, 2012a; Fig. 9.18). The observations showed that these amphipods have the capacity for long-range swimming, high manoeuvrability in close range and efficient predatory behaviour. Also, burst swimming speeds for *P.* aff. *abyssalis* were 9 and $10 \, \text{cm} \, \text{s}^{-1}$ with accelerations up to 22–$25 \, \text{cm} \, \text{s}^{-2}$.

In addition to the benthic *Princaxelia* species, Dahl (1959) also recorded two occurrences of *Pardaliscoides longicaudatus*, from 10 000 m in the Philippine Trench and 6180 m in the Kermadec Trench.

Moving further away from the seafloor, the Pardaliscidae genus *Halice* comprises five species that are known to occupy the hado-pelagic zones of multiple trenches. For example, *H. aculeate* has been described from four trenches in the western Pacific Ocean, from the Tonga in the south to the Kuril–Kamchatka Trench in the north (4000–10 500 m). *H. quarta* Birstein and Vinogradov, 1955 has a similarly wide geographic range, spanning the western Pacific trenches (6000–10 000 m). *H. secunda* (Stebbing, 1888), *H. rotundata* Birstein and Vinogradov, 1960 and *H. subquarta* Birstein and Vinogradov, 1960 are all known from a range of western Pacific trenches, from depths of 6960–10 190, 4050–9120 and 7190–10 500 m, respectively.

The Hyperiopsidae are a pelagic family comprising three genera (*Hyperiopsis*, *Paragissa* and *Protohyeriopsis*) and six species, and the Lanceolidae family include two genera (*Lanceola* and *Metalanceola*) and three species. Both of these families are known only from the western Pacific trenches. Most of these species were identified from single trawl samples, while others show considerable bathymetric range. For example, *Paragissa arquarta* is found between 4200 and 8500 m in the Kuril–Kamchatka Trench (range = 4300 m) and *Lanceola clausi gracilis* is found in the same trench between 4200 and 8000 m (range = 3800 m). As a result of the limited number of pelagic trawls at hadal depths, information regarding the pelagic Amphipoda is limited relative to that of the benthic species.

Extraordinarily large numbers of amphipods are often recovered from the hadal zone using baited traps. Similarly large numbers are also observed in still photographs and video from baited systems (Fig. 9.19). Mobile scavenging amphipods, particularly at depths exceeding 8000 m account for the entire suite of natatory fauna (Hessler *et al.*, 1978). However, the relatively large numbers of amphipods relative to, for example, fish at abyssal depths makes accurate population density and biomass estimates difficult at best. However, the absence of abundance and biomass estimates does not detract from the overwhelming dominance of these organisms that is witnessed when using baited sampling techniques. Their ability to detect, intercept and consume bait is extraordinary. Moreover, the deeper the experiment, the more individuals are present and the faster they arrive. Given that food-falls should, theoretically, occur in the deep sea irrespective of depth, the role of scavenging amphipods in the dispersal of organic matter is significant, even more so at depths exceeding 8000 m where there is little evidence of any other taxa of the scavenging guild. Also, at these depths, the redistribution of organic matter by hadal amphipods in the deep trenches may also provide nutrients to the wider hadal fauna; in the absence of predators, amphipods will eventually die and will presumably be distributed uniformly across the seafloor, thus playing their part in another mechanism for the dispersal of food within the hadal zone.

The reasons why amphipods play such a dominant role within the scavenging guild at the greatest depths is not clear. In most of the data from depths exceeding 10 000 m, only *Hirondellea* are found (Hessler *et al.*, 1978; Blankenship *et al.*, 2006). Whether this trend is a direct testament to the success of the Amphipoda or simply due to the absence of larger predators due to physiological limitations is unclear. However, their dominance at full ocean depth cannot be understated.

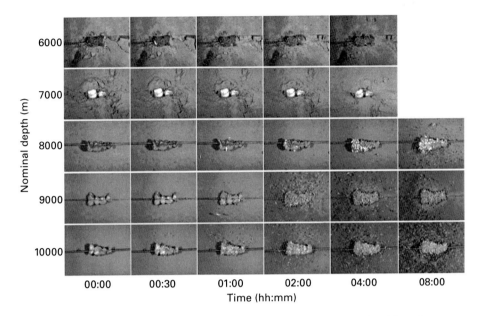

Figure 9.19 Succession of scavenging amphipods at 1000 m intervals over time. These images are frame grabs from the Hadal-Lander A (each image $= 0.35\,\mathrm{m}^{-2}$ with fish bait in the centre) showing relatively low amphipod activity at the shallower depths, with an increase with both depth and time, culminating in nearly a 100% coverage of bait at 9000 and 10 000 m after 2 h and the greatest activity after 8 h at 10 000 m. Images are taken from the Kermadec and Tonga Trenches, courtesy of HADEEP.

The success of the Amphipoda at hadal depths may lie in the array of adaptations that they possess in order to cope with a low food environment. Deep-sea scavengers occupy a very specific ecological niche, whereby they must meet the requirements necessary for survival by consuming large meals that may arrive at long time intervals and that are randomly dispersed over large areas of seafloor (Dahl, 1979). In order to do survive in this environment, a scavenger must (1) have the ability to localise and recognise potential food sources, (2) have the ability to feed on large muscular food sources, (3) be capable of consuming large quantities of food in relatively short periods of time, (4) store the energy obtained for gradual utilisation over extended periods (survive for long periods of starvation) and (5) supplement their diet with alternative food sources that become available between large carrion-falls.

All of these requirements are met by the hadal Amphipoda. Amphipods use chemo-sensory stimulation to detect a food-fall via the emanating odour plume (e.g. Tamburri and Barry, 1999), as is the case for most deep-sea scavengers (Wilson and Smith, 1984; Sainte-Marie and Hargrave, 1987; Hargrave *et al.*, 1995). While the foraging strategy for amphipods is as yet unclear, it is likely to encompass either the 'sit and wait' method or the cross-current drifting method (Bailey and Priede, 2002). Furthermore, the strategy for detecting odour plumes is potentially species-specific. Dahl (1979) discussed the possibility of mechanoreception as an indicator of the arrival of food, however, the idea was mostly discounted based on the fact that amphipods are known to continue arriving at bait after 24 h.

Chemosensory adaptations are evident in the lysianassid amphipods, for example, some species sweep water over the proximial part of the antennulae, mouthparts and into the branchial region when beating their pleopods (Dahl, 1977), presumably to increase the chance of detecting chemical stimuli. Lysianassids also have a dense array of chemosensor-type setae on the ventral side of the first flagellar article of the antennae (Dahl, 1979). They also have short and stout antennae which are kept depressed to increase exposure to chemosensory stimuli, and when swimming, these sensors are prominently exposed to the body of water through which the amphipod swims.

Smith and Baldwin (1984) provided estimates of sound intensity and spherical spreading of the noises created by *Eurythenes gryllus* whilst feeding. They calculated that the resulting noise intensity (75 dB re 1 µPa) could potentially produce 15 dB (assuming spherical spreading) at a distance of 1 km. This hydroacoustic stimuli, although unproven, may attract further visitors to the carrion-fall (whether other amphipods or other taxa), but it is generally agreed that the detection of the odour plume is the primary strategy.

The amphipods' ability to consume the flesh of carrion is self-evident when images and video from the hadal zone are viewed. Often, carcasses are stripped of every visible shred of flesh within 24 h and often, when no other taxa are present (e.g. Fig. 9.13). This efficient removal of flesh is a result of highly adapted mouth morphology (Fig. 9.20). The basic gammaridean amphipod has a strongly and irregularly serrated incisor part with well-developed lacina mobilis on both mandibles (Dahl, 1979). When biting, the left incisor passes in front of the right incisor which, in turn, moves between the left lacinia mobilis, which then sits in front of the right one. However, for the three most dominant amphipod genera in the hadal zone (*Eurythenes*, *Hirondellea* and *Paralicella*), this feeding motion is slightly different; the right incisor slides in behind the left and the 'shape' of the bite is bowl-shaped allowing these genera to remove larger pieces of food than those with a flattened mandible (Dahl, 1979). Furthermore, *E. gryllus* and *H. gigas* have distinctively shaped molars, such that when they are closed, they form an almost complete funnel from the mouth to the stomach, thought to aid in guiding larger food particles to the digestive tract.

The ability of amphipods to store large food items is clearly evident when examining specimens that have just been captured. Often, the guts of the amphipods look like they are 'ready to burst'. The alimentary tracts of the above-mentioned genera are adapted for the accumulation and storage of large volumes of food relative to their size and thus are capable of storing more food than their shallower-water counterparts. This ability is particularly evident in *Paralicella* where the body wall can extend ventrally to two or three times its body size (Shulenberger and Hessler, 1974; Thurston, 1979). In *Eurythenes*, *Hirondellea* and *Paralicella*, food can be stored in the midgut which can expand to fill the entire body cavity.

The feeding strategy exhibited by amphipods is species-specific. Based on the morphology of the *Eurythenes gryllus* mandibles, their capacious guts and high assimilation rates, Sainte-Marie (1992) characterised them (and *Paralicella* spp.) as

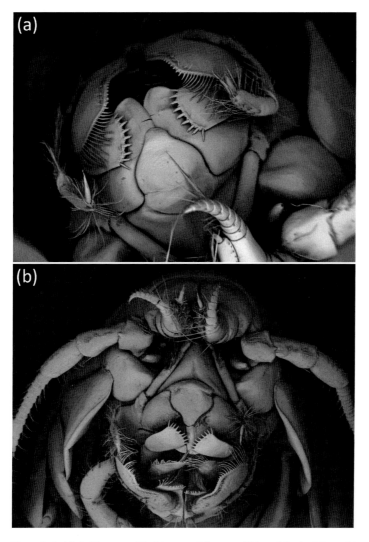

Figure 9.20 (a) A 40× magnified image of the mandibles of the hadal amphipod *Hirondellea dubia* from 9908 m and (b) *Scopelocheirus schellenbergi* from 8487 m in the Kermadec Trench. Image (a) taken by Nichola Lacey and (b) by Kevin MacKenzie, University of Aberdeen, UK.

'batch-reactor-type' feeders that are able to survive prolonged starvation periods. In contrast, other deep-sea species such as *Orchomene* sp. were described as a 'plug-flow reactor-type', whereby, they process food continuously and, thus, are more dependent on a consistent food supply.

Whilst in contact with a carrion-fall, amphipods are capable of characterising the food source, presumably using gustatory seta on the gnathopods and periods (Kaufmann, 1994). The majority of gustatory sensors (a combination of chemical and physical) are found on the ventral margins of appendages and are thought to be used for 'tasting' the food items upon which the amphipod is crawling, or items held by the gnathopods. Thus, it appears that amphipods are capable of discriminating between food items that differ either chemically or texturally. This is further supported by observations of selective feeding on liver and gonads during necrophagy (Scarratt, 1965), presumably in order to maximise their energetic intake per unit feeding by consuming the most energy-rich tissues first (Kaufmann, 1994). This behaviour was also observed during a 2011 HADEEP cruise to the Kermadec Trench, whereby a single juvenile snailfish was recovered from 7012 m in a baited trap and found to be largely intact except for its liver which had been removed by amphipods (pers. obs.).

After gorging themselves on a carrion-fall, it is likely that the amphipods must wait for an extended period of time before the next carrion-fall arrives. There are two strategies to cope in this scenario. The first is simply energy management: making the most of the last meal. Smith and Baldwin (1984) showed that necrophagous, deep-sea amphipods (*Paralicella caperesca* and *Orchomene* sp.) may drastically reduce their metabolic activity during starvation periods. Tamburri and Barry (1999) demonstrated that an amphipod, *Orchomene obtuse* Sars, 1895 could survive without any food whatsoever for 4–6 weeks in laboratory conditions, although repeating similar experiments with hadal amphipods is not yet possible. Yayanos and Nevenzel (1978) reported that *Hirondellea gigas* specimens from the Philippine Trench had appreciable stores of lipids in their bodies (26.1% of the total dry weight), presumably an energy cache to cope with long periods of starvation.

The storage methods of the three deep-sea, scavenging amphipod species were investigated by Bühring and Christiansen (2001). The *Paralicella* spp. and *Orchomene* sp. were found to store triacylglycerols whereas *Eurythenes gryllus* stored wax esters. All three amphipod species were identified as necrophagous and the lower total lipid content of *Orchomene* sp. compared to the other two amphipod species supported the idea that, unlike the other species, *Orchomene* sp. process food continuously, rather than gorging after a period of starvation.

The success of *Hirondellea* in the trenches may be attributed to several such energy-conserving strategies that maximise reproductive success. Hessler *et al.* (1978) found that as *H. gigas* specimens increased in size, the number of individuals that had bacteria and sediment in their guts decreased. This was thought to be a potential strategy for protecting their young, i.e. smaller individuals and juveniles may have more difficulty in feeding at food-falls, whereas for larger individuals bottom feeding may require too much energy to be profitable. Likewise, the large amount of stored lipids, a slowing of growth in later instars, the disproportionate number of females and the absence of brooding females attending food-falls all appear to be adaptations in hadal amphipods in order to optimise reproductive success in an energy-poor environment, like the trenches.

A second strategy that hadal amphipods use to survive is to supplement a scavenging diet with other smaller food items, beyond the limits of necrophagus scavenging on carrion. The existence of obligate scavengers has been a matter of debate for some time (Britton and Morton, 1994; Kaiser and Moore, 1999; Tamburri and Barry, 1999) and it seems likely that even amphipods, the prolific scavengers of the deep sea, are also facultative in this nature. Blankenship and Levin (2007) examined the nutritional strategies of four lysianassoid amphipod species from the Tonga Trench: *E. gryllus, S. schellenbergi, H. dubia* and former *Uristes* sp. nov. Their results revealed a remarkable trophic plasticity in that, supplementary to necrophagy, these amphipods exhibited detrivory and predation. Furthermore, the nutritional strategies of some species appeared to change with age and depth. *E. gryllus,* and *S. schellenbergi* were shown to employ predation and possibly detrivory in the absence of carrion. These species were found to have digested tunicates, ascidians, pelagic salps or larvaceans and other amphipods. A more recent study by Kobayashi *et al.* (2012) showed that *H. gigas* from 10 897 m in the Mariana Trench possessed a unique digestive enzyme capable of digesting wood debris. Given the morphology of the mouthparts, it is unlikely that *H. gigas* can remove pieces of wood from a larger object, but rather, it may consume small detrital pieces of wood that are delivered to the trench from terrestrial origin amongst plant debris.

Although it is unclear whether these food/prey items were dead or alive at the time of consumption, or what the exact mechanisms involved in the consumption of wood debris are, the results align well with the idea that amphipods are one of the most trophically diverse taxa in the marine environment (Nyssen *et al.*, 2002).

Gammaridean amphipods develop primary and secondary sexual characteristics that indicate a particular stage of maturity, resulting in distinct morphological characteristics for a particular developmental stage or 'instar' (Sexton, 1924; Steele and Steele, 1970; Hessler *et al.*, 1978). *Hirondellea gigas,* collected from 9600–9800 m in the Philippine Trench, were classified into developmental stage and sex (Hessler *et al.*, 1978). With the exception of female and male stages 1 and 2, respectively, the stages are the equivalent of instars that exhibit a relatively constant growth ratio. The females had seven to eight instars and the males had four. An exception occurred in stage 6 females, where growth decreased, coinciding with the development of reproduction products. This decrease is probably an effect of nutritional intake being converted into reproduction. As no brooding females or individuals with spent gonads were caught in this study, it is therefore assumed that stage 6 is followed by a brooding instar with spent gonads and morphologically mature oostegites. This study suggested that females breed only once and, if so, then the total fecundity averages ~97 oocytes per female. The study also found that the females accounted for 63% of the population, which, given the distinct absence of brooding females, was likely to be higher.

The absence of ovigerous females suggested that during this period, females do not partake in feeding bouts, presumably because the expansion of the stomach may expel or crush the eggs (Blankenship *et al.*, 2006), although it was speculated that if post-brooding females were larger than 12.5 mm in height, then they may not have readily entered the traps used by Hessler *et al.* (1978). However, larger traps deployed in the Mariana Trench still failed to recover any brooding females. Alternatively, the absence of brooding instars may be because if sufficient nutrients have already been accumulated to ensure a successful brooding, partaking in feeding bouts may pose unnecessary danger to the female (as reported in other amphipods; Fulton, 1973), favouring a period of fasting. Furthermore, large stores of lipid are present in *H. gigas*

(Yayanos and Nevenzel, 1978) but it is not known if these are sufficient for more than one brooding cycle.

The afore-mentioned studies highlighted the dominance of *Hirondellea* at the greatest depths of the ocean, but as the deployments were all undertaken at the deeper end of trenches, larger scale vertical ontogenetic patterns could not be identified. Even so, Hessler *et al.* (1978) did record individuals of all stages (except brooding instars) within his samples from the deeper sites, suggesting a lack of ontogenetic vertical structure. A slightly larger dataset focusing on *H. gigas* was obtained in the nearby Izu-Bonin Trench from 8172 and 9316 m (Eustace *et al.*, 2013). Like the Hessler *et al.* (1978) study, data from this trench also showed that a similar pattern of all instars was present at the corresponding depth of ~9300 m, however, the shallower depth samples were comprised mostly of juveniles. These data suggest that although all instars can be found at a single depth, vertical stratification in the form of shifting male:female:juvenile (M:F:J) ratios (and size frequency) still occurs (Fig. 9.21). Similar trends in M:F:J ratio shifting with depth, in favour of mature individuals at the deeper depths, is also becoming apparent in preliminary HADEEP data for other hadal amphipods (N.C. Lacey, unpublished). For example, *H. dubia* and *S. schellenbergi* both show a bias towards juveniles at the shallower end of their depth range, although this trend is perhaps most pronounced in *H. dubia*.

Blankenship *et al.* (2006) carried out 11 baited trap deployments in the Tonga Trench from 5155 to 10 787 m. They found four main scavenging amphipod species with distinct vertical zonation; *E. gryllus*, 5155–6252 m; *S. schellenbergi*, 6252–8723 m; an unidentified Tryphosinae (listed as *Uristes*), 7349–9273 m; and *H. dubia*, 7349–10 787 m. The most dominant amphipod, *H. dubia*, ranged from 2.8 to 20.9 mm with females reaching 20.9 mm and males reaching 18.6 mm. Although, again, no ovigerous females were found and the vast majority of juveniles were found in the shallower stations (7329–8723 m), corresponding to a larger size-frequency distribution at the greater depths, thus indicating an ontogenetic vertical structure. Similarly, although based only on one sampling depth at 7800 m in the Peru–Chile Trench, Perrone *et al.* (2002) reported that the majority of the *Hirondellea* sp. nov. were close to the maximum possible size. Furthermore, the size frequency distribution of these *Hirondellea* sp. nov. resembled that of *H. dubia* in the Tonga and Kermadec Trenches (Blankenship *et al.*, 2006), suggesting a similar pattern of ontogenetic structuring in *Hirondellea*.

Eustace *et al.* (2013) stated that, despite the presence of both mature and juvenile *H. gigas* towards the deepest parts of the Izu-Bonin Trench, ontogenetic vertical stratification was evident. This statement is contrary to the findings of Hessler *et al.* (1978), who based their conclusions on an extremely narrow depth range (9600–9800 m). The presence of both mature and juvenile individuals at all stations reported by Eustace *et al.* (2013) and Hessler *et al.* (1978) suggests that the vertical stratification was not driven by depth (or pressure) per se, but rather by another environmental driver in combination. It is also unlikely that this trend was influenced by temperature, as despite a slight warming of bottom waters with depth, the temperature range was very low; ~1°C increase across the depth range of the trench. Salinity

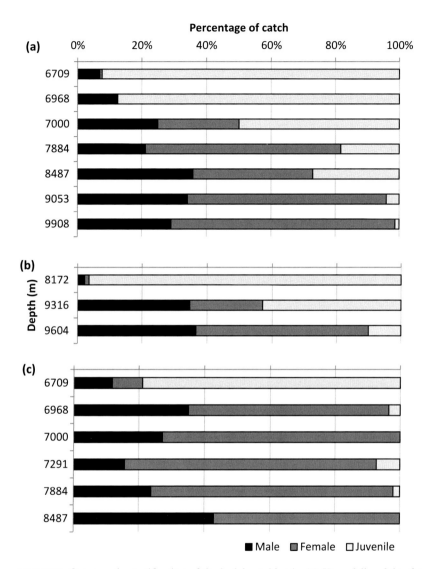

Percentage of catch

Figure 9.21 Ontogenetic stratification of the hadal amphipods. (a) *Hirondellea dubia* from
the Kermadec Trench (Lacey, unpublished), (b) *Hirondellea gigas* from 8172 and 9316 m in the
Izu-Bonin Trench (Eustace *et al.*, 2013) and 9604 m from the Philippine Trench (Hessler et al.,
1978) and (c) *Scopelocheirus schellenbergi* from the Kermadec Trench (Lacey, unpublished).

was also found to be constant (34.69 ppt) throughout the hado-pelagic water column.
It is, therefore, more likely that food supply and other associated ecological interactions
drive ontogenetic stratification.

Eustace *et al.* (2013) found that juveniles were the most dominant life stage of
H. gigas at 8172 m and that their percentage of the total population diminished with
depth. This raises the question as to why juveniles are most abundant at the shallower
depths. Likewise, if amphipods descend with age, culminating in the onset of brooding

females at the deepest point, what mechanisms drive juveniles to culminate at shallower depths? It is not known if females migrate up the trench slopes to release their brood, or if juveniles independently move up the trench. If juveniles are to move such relatively large distances, then they would need to have high fat reserves in order to fuel such a large migration. This is partially supported by the known high lipid content of *H. gigas* (Yayanos and Nevenzel, 1978). However, it seems impractical for a female to supply such a high quantity of energy to a large brood in an energy-poor environment and since large food-falls are scarce and ephemeral, there is little opportunity for juveniles to build up their own lipid stores in order to make such a migration (Blankenship *et al.*, 2007). The most logical explanation, therefore, is that the ovigerous females migrate to a shallower depth and then release their brood (Blankenship *et al.*, 2006), however, direct evidence for this theory is currently lacking.

The associated advantages of *H. gigas* spending at least part of its life cycle at shallow depths are numerous and of key consequence. Decreasing pressure is known to increase the rate at which metabolic reactions proceed (Blankenship *et al.*, 2006), therefore, juveniles who mature at shallow depths are able to assimilate food more rapidly and thus increase the rate at which they grow and become reproductive. Food-falls are also of higher nutritional quality and quantity at shallower sites than at the deeper trench depths. This further provides juvenile individuals with the opportunity to rapidly increase both their weight and size in order to better cope with the competition and predation pressures at greater depths, where *H. gigas* abundance is far higher, i.e. the shallower depth hosts a greater number of food-fall items and therefore less competition. Furthermore, at shallower depths, competition with conspecifics for resources is vastly reduced as well as the risk of cannibalism, due to the reduced abundance of larger adult *H. gigas* at shallow depths. One disadvantage to such a strategy is that in the shallower depths of the trenches (6000–8000 m), amphipod predators such as fish and natant decapods are frequently encountered (Jamieson *et al.*, 2009a, b, 2011a). However, this may be negated by their relatively small size and thus potentially low energy transfer, since predators may selectively choose larger prey items than amphipods if available. Upon reaching notable size and sexual maturity, the amphipods descend to depths where predator abundance is low (7700 m) (Fujii *et al.*, 2010).

Female *H. gigas* show the opposite trend in vertical structure to juveniles; their presence is minimal at shallow sites and increases with depth towards the trench axis. This trend may also be related to the correlated increase in pressure. Neither this study nor any of the previous studies performed on *H. gigas* have found egg-laden females (Hessler *et al.*, 1978; France, 1993). One proposed theory for this is that ovigerous females do not attend baited traps in order to help reduce the risk of mortality through cannibalism. Alternatively, once females become ovigerous they may stop feeding to prevent the expulsion of eggs via the swelling of their midgut (Hessler *et al.*, 1978; Blankenship *et al.*, 2006). If females do not attend baited traps then it may be likely that they undergo a period of starvation, during which they are maintained by their lipid reserves (Hessler *et al.*, 1978). Therefore, the vertical patterning shown by female *H. gigas* may be an attempt to survive the starvation period by decreasing their metabolism through exposure to increasing pressure. However, this would also suggest that the developing embryos would suffer a decreased metabolism and the rate at which they grow would decrease, further increasing the brooding period and thus the starvation period, negating the advantages of descending to greater depths.

It has also been postulated that *H. gigas* females are semelparous (Hessler *et al.*, 1978). The data reported in Eustace *et al.* (2013) support this hypothesis, where no females with spent oostegites were captured. Furthermore, no stage 6 females were captured, a rarity noted in previous studies (Hessler *et al.*, 1978; France, 1993). However, the collection of amphipods from 9316 m by Eustace *et al.* (2013), was found to contain a copious amount of loose oocytes. It is unknown how the oocytes came to be extruded from any of the specimens, but may have been a result of cannibalism or other damage to the ventral surface of females within the trap. None of these loose oocytes were held by females in any great number and were simply loose in the containers or caught in between the pereopods and ventral surface. The oocytes varied in size from 0.27 to 1.1 mm but all were of the same yellow-orange colour and morphology. The origin of the loose oocytes remains unresolved. They were not genetically identified, although given the overwhelming dominance of *H. gigas* (99.6%) and the large number captured ($n = 3968$) it was postulated with some certainty that the oocytes did indeed, belong to *H. gigas*.

The same ontogenetic structuring pattern as seen in *H. gigas* has also been documented in *H. dubia* and *S. schellenbergi* from the Kermadec and Tonga Trenches (Blankenship *et al.*, 2006). In these trenches, juveniles were found to reside at shallower depths and upon maturation, they descended to greater depths. This appears to be a mechanism for survival exploited by bentho-pelagic amphipods that, as adults, reside in the deepest depths of hadal trenches. It can, therefore, be construed that pressure per se does not drive the observed trends but rather an interaction between depth (pressure) and topography-influenced distribution of resources, in terms of both quality and quantity. Juveniles are found in the shallower depths where a greater net quantity and higher quality of POM is present and the net total of food-falls is the highest (and competition is lowest) and pressure-induced effects on metabolism are beneficial for growth. Mature individuals are found deeper where predator abundance is low, where their large body size means that they are no longer a target for predators, and where their metabolic rate can be reduced due to the reassurance of high lipid stores. For both males and females the ability to maintain high lipid stores is advantageous during periods of starvation such as occur in between food-falls, but even more so for females during brooding, who undergo prolonged periods of starvation.

9.9   Decapoda

Since the first major hadal sampling efforts in the 1950s, crustaceans of the order Decapoda have been thought absent from the hadal zone, with no representatives documented below 5700 m (Wolff, 1960). There were brief mentions of decapods at hadal depths in George and Higgins (1979), from 7600 m in the Puerto Rico Trench (eight *Plesiopeneus* and *Nematocarcinus*) and by Hessler *et al.* (1978) in the Peru–Chile Trench, who reported 'occasional natantian decapods' from between 6767 and 7196 m. However, despite these reports, it was assumed until recently that decapods had no hadal representatives (Herring, 2002; Blankenship *et al.*, 2006). This conclusion was largely based on the results of a series of trawls during the *Galathea* and *Vitjaz* expeditions, in the 1950s (Wolff, 1960, 1970). Although >700 species of invertebrates and fish were described from the ~33 000 individuals that were recovered from 80 hadal trawls on the *Galathea*, not a single decapod was found (Wolff, 1970). The apparent absence of hadal decapoda was attributed to the physiological limitations of hydrostatic pressure and the deepest decapod (*Parapagurus* sp.) was recorded at 5160 m. Following the expeditions of the 1950s, the deepest findings of decapods were 4785, 4986, 5060, 5413, 5440 and 5700 m (Tiefenbacher, 2001; Haedrich *et al.*, 1980; Gore, 1985b; Bouvier, 1908; Domanski, 1986; Kikuchi and Nemoto, 1991, respectively).

One of the most surprising results to emerge since the very beginning of the HADEEP project was the presence of decapods at almost every site, at depths of <7700 m in all the trenches studied (Jamieson *et al.*, 2009b; Fig. 9.22).

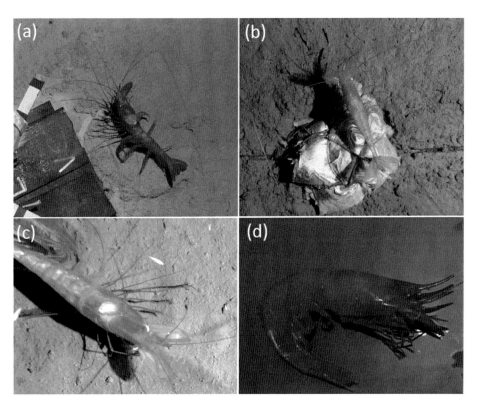

Figure 9.22 Examples of decapods from the trenches. (a, b and c) *Benthesicymus crenatus* from 6474 m, 6173 m and 5545 m in the Kermadec, Peru–Chile and Mariana Trenches, respectively. (d) *Heterogenys microphthalma* from 6709 m in the Kermadec Trench. Images courtesy of HADEEP.

The Hadal-Lander A obtained video footage of the natantian decapod *Benthesicymus crenatus* Bate, 1881 (Benthesicmidae), at 6007 and 6890 m in the Kermadec Trench, 6945 and 7703 m in the Japan Trench and at 5469 m on the edge of the Mariana Trench (Jamieson *et al.*, 2009b). Furthermore, Hadal-Lander B later photographed the same species at 5172 and 6000 m in the Kermadec Trench (Jamieson *et al.*, 2011a), and at 5329 and 6173 m in the Peru–Chile Trench (unpublished data, HADEEP). In addition to *B. crenatus*, *Hymenopenaeus nereus* (Faxon, 1893) of the family Solenoceridae was photographed at 4602 m in the Peru–Chile Trench (but not at hadal depths), and the smaller Caridean decapod *Heterogenys microphthalma* (Smith, 1885; originally described in Jamieson *et al.*, 2009b as *Acanthephyra* sp.) was photographed at 6007 and 6890 m in the Kermadec Trench. In 2012, a single specimen of *H. microphthalma* was recovered from 6709 m in the same trench with the *Latis* fish trap but, to date, no physical specimen of *B. crenatus* has been recovered from hadal depths.

The video footage of *Benthesicymus crenatus* showed that they were readily and consistently attracted to bait. The appearance of this decapod is similar to that of the common deep-sea Aristaeid prawn, *Plesiopenaeus armatus* Spence, Bate, 1881 (Aristeidae), perhaps more commonly seen at baits (Thurston *et al.*, 1995; Janßen *et al.*, 2000). However, the resolution of the camera was sufficiently high to allow *B. crenatus* to be distinguished from *P. armatus* (shorter rostrum).

The mean body length of *B. crenatus* in the Kermadec Trench at 5172 m was 18.7 cm $\pm$ 0.8 S.D. ($n = 3$), at 6000 m it was 17.9 cm $\pm$ 3.2 S.D. ($n = 3$), at 6007 m it was 22.0 cm $\pm$ 3.9 ($n = 10$) and at 6890 m it was 22.4 cm $\pm$ 2.6 S.D. ($n = 4$) (see Fig. 9.23). The estimated total number of individuals ranged from 10 at 6007 m, to 3 at 5172 m and at 6000 m. At 6945 m in the Japan Trench, 29 sightings of 20 individual *B. crenatus* were made (mean body length 15.3 cm $\pm$ 2.9 S.D.). At the abyssal Mariana location, two sightings were made of one individual decapod (body length = 23.5 cm) among a succession of the scavenging demersal fish *Coryphaenoides yaquinae* Iwamoto and Stein, 1974 (Macrouridae).

B. crenatus was observed as it preyed upon small scavenging amphipods already present on the bait, rather than feeding on the bait itself. The confirmation of predation of amphipods by *B. crenatus* was made difficult because its mouth is ventrally located and seldom in view. In one instance, however, an individual was observed removing and handling a large lysianassoid amphipod (~2 cm body length). The decapod, 20 cm in length, approached the bait up-current with its pereopods trailing below and outwards. It rapidly decelerated once contact was made with the bait. It then reached down, clasped and shuffled its pereopods while drifting slightly down-current away from the bait. During this time, the distinctive orange body of the amphipod clearly contrasted against the red-coloured underside of the decapod. At the same time, the site on the bait that was previously occupied by the large amphipod became visible and vacant, indicating that the decapods were preying upon small amphipods.

The behaviour of *B. crenatus* was similar to other baited camera observations from the shallower abyssal plains. One study, where the numbers of decapods (*Plesiopenaeus armatus*) were high, was undertaken in the Arabian Sea at 4000–4500 m (Janßen *et al.*, 2000). Although in this study, the decapods were the first to arrive at bait (within 1 h),

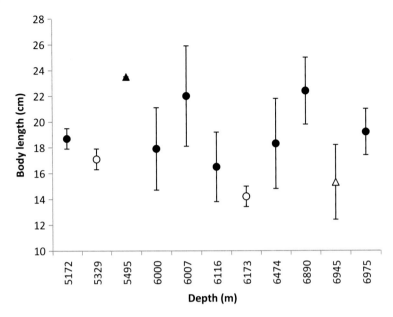

Figure 9.23 Mean body length (S.D.) for the decapod *Benthesicymus crenatus* from 5172–6975 m in the Kermadec Trench (black dots), Peru–Chile Trench (white dots), Japan Trench (white triangle) and Maraiana Trench region (black triangle).

they did not appear in the same place in consecutive images and no loss of bait was visible when only decapods were present. Similarly, only 40% of individuals were seen to be in direct contact with the bait. This also suggests that the decapods may have been exploiting the temporarily high density of amphipods, rather than feeding at the bait itself. In the Atlantic (4000–5000 m) small clusters of *P. armatus* have been viewed directly on the bait, however, in both instances, still photography could not confirm predatory behaviour (Thurston *et al.*, 1995). The lack of bait consumption by decapods in this study and others (Janβen *et al.*, 2000) suggests that exclusive dependency on carrion-falls is unlikely. Stomach contents from abyssal specimens have comprised phytodetritus, small bivalves and ground-up crustacean parts (Domanski, 1986; Thurston *et al.*, 1995); further evidence of facultative necrophagy and possibly active epibenthic predation is suggested by these observations and others (Gore, 1985a, b).

The swimming speeds of *B. crenatus* in the Kermadec Trench were recorded by Jamieson *et al.* (2009b) and did not show any obvious signs of hydrostatic pressure-induced limitations. Overground swimming speeds from 6007 and 6890 m in the Kermadec Trench were measured as 7.4 cm s^{-1} ± 1.8 S.D. and 6.9 cm s^{-1} ± 1.6 S.D., respectively (translating as 0.34 BL s^{-1} ± 0.08 S.D and 0.35 BL s^{-1} ± 0.11 S.D) and at 6890 m in the Japan Trench they were 6.9 ± 2.0 cm s^{-1} (0.49 BL s^{-1} ± 0.17 S.D.).

The other decapod, *H. microphthalma*, has, to date, only been observed swimming off the bottom, with no apparent interest in the bait and it is, therefore, likely that these observations were simply chance encounters. Combined, these observations do provide unequivocal proof that decapod crustaceans are active at hadal depths, inhabiting the upper slopes of the Pacific Ocean trenches in both the northern and southern hemispheres. They are capable of preying upon the abundant scavenging amphipod community that thrives at these depths and form a major component of the hadal food-web (Blankenship et al., 2006; Blankenship and Levin, 2007), thus adding a new element to the hadal food-web as a top predator in the upper trenches.

B. crenatus and H. microphthalma (Jamieson et al., 2009b) are, however, not endemic to the hadal zone but simply transcend the abyssal–hadal transition zone. The absence of decapods at depths greater than 7700 m bears a strong similarity to the depth ranges of teleost (bony) fish. It is thought that the depth limit for fishes is ~8000–8500 m, where the deepest fish ever seen alive was observed during the same deployment as the deepest decapod (Jamieson et al., 2009a; Fujii et al., 2010). The maximum depth for fish is thought to be limited by the content of the osmolyte trimethylamine oxide (TMAO) within their cells, which is predicted to reach isosmosis at 8000–8500 m (Jamieson and Yancey, 2012), i.e. the depth where TMAO can no longer counteract the perturbing effects of high hydrostatic pressure (Yancey et al., 2001, 2004). It is also worth noting that the TMAO content for fish is very close to those already found in decapods (Kelley and Yancey, 1999), suggesting that decapods, like fish, are perhaps also limited to depths of 8000–8500 m.

The reason decapods have, until recently, remained elusive in hadal sampling campaigns is that they appear to be extremely difficult to physically catch. The decapods have evolved highly efficient threat detection (disproportionately large sensory antennae) and evasion techniques (burst fast-start escape responses) that make them highly adapted to evade a trawl, particularly a slowly moving trawl at hadal depths. Furthermore, since the discovery of the decapods using baited cameras, baited traps and tangle nets have been deployed multiple times in their known vicinity but to no avail. The trade-off is that the baited camera method is effective in finding decapods at hadal depths but does not provide the necessary physical specimen to prove or refute the TMAO hypothesis. These observations do, however, highlight the point that for an entire and relatively conspicuous order of Crustacea to remain undiscovered from such a large depth zone, the need for further exploration of the hadal trenches, with appropriate techniques, is paramount in order to reveal the true structure of the hadal community.

9.10   Acariformes

Acariformes are the most diverse of two superorders of mites. Acarina (family Halacaridae) were first found in the abyssal zone in the Pacific Ocean, at approximately 4000 m (Newell, 1967). They do not represent a characteristic hadal group as they are currently known from only a few trench samples. A new genus and species of Halacaridae, *Bathyhalacarus quadricornis* Sokolov and Yankovskaya, 1968, was described from two findings in the vicinity of the Kuril–Kamchatka Trench at 5100–5200 m and later from the Izu-Bonin Trench at 6770–6850 m (Belyaev, 1989).

9.11   Pantapoda

The Pantopoda, of the class Pycnogonida, are rarely found at hadal depths and have not yet been found greater than 7370 m. There are currently nine known species previously found in hadal trenches, belonging to five genera and three families, each known from just a single specimen. They were found at 13 stations in five Pacific Ocean trenches and in the South Sandwich Trench. Two other undescribed specimens of Pantopoda were caught in the Peru–Chile Trench (Menzies, 1964, cited in Belyaev, 1989). To date, all the Pantopoda collected from hadal depths are typical deep-sea species. Of the nine known species, only two have been found in more than one trench; *Heteronymphon profundum* Turpaeva, 1956 (Nymphonidae) is known from the Kuril–Kamchatka Trench and neighbouring Japan Trench, at depths of 6860 m and 6156–6380 m, respectively, and the other, *Pantopipetta longituberculata* (Turpaeva, 1955; Austrodecidae) is, rather curiously, known only from the Kuril–Kamchatka Trench (6090–6710 m) in the North Pacific and the South Sandwich Trench in the Southern Ocean (6052–6150 m).

10 Cnidaria and fish

10.1 Cnidaria

Representatives of the Cnidaria phylum have been retrieved from hadal depths (Hydrozoa, Scyphozoa and particularly Anthozoa; Table 10.1, Fig. 10.1). Hydroid polyps have been identified in most trenches, the deepest of which were discovered at 8210–8300 m in the Kermadec Trench (*Halisiphonia galatheae*; Kramp, 1956), at 8185–8400 m in the Kuril–Kamchatka Trench and at 8950–9020 m in the Tonga Trench (the latter two are not listed or named in Belyaev, 1989). There are 12 species listed by Belyaev (1989), where only one species has been found in more than one trench (*Crossota* sp.; Palau and New Hebrides Trenches). These 12 species belong to 7 families (two of them putative). Hydroids of the genus *Branchiocerianthus* have been collected from the Kermadec and New Hebrides Trenches and photographed *in situ* in the New Hebrides and Peru–Chile Trenches, where they exceeded 25 cm in length (Lemche *et al.*, 1976). Hydroids are, however, somewhat rare in numbers and species at depths exceeding 6500 m and are, therefore, not considered characteristic of the hadal zone.

The Medusae are a rare group but are known to occur in the trenches. There are five known species belonging to three putative families (Mitroconidae, Anthomedusidae and Phopalonematidae) of three orders (Anthomedusae, Leptomeusae and Trachymedusae). The first hadal Hydromedusae were sampled by plankton net by the *Vitjaz*, between 6800–8700 m (*Voragonema profundicola* Naumov, 1971; Rhopalomatidae) in the Kuril–Kamchatka Trench. Other Hydromedusae have been sampled from the Palau, New Britain and New Hebrides Trenches; these organisms were present in 17 images obtained by Lemche *et al.* (1976), from between 6758 and 8260 m who calculated the density of the Trachymedusae to be approximately one individual per 100 m^2. The occasional Hydromedusae have also been seen in baited camera data from the HADEEP projects, as deep as 6945 m in the Japan Trench. Unfortunately, identification of such small and gelatinous organisms is, at best, difficult (unpublished data, HADEEP).

There are three known species of hadal Scyphozoa, all belonging to the genus *Stephanoscyphus* (Atorellidae), with the exception of the *Ulmaridae* sp., identified in Bougainville Trench samples from 7847–8662 m. *Stephanoscyphus simplex* Kirkpatrick, 1890 has been recorded from the Banda and Kermadec Trenches from

Table 10.1 Maximum depth of each order and family of Cnidaria and the number of genera and species therein.

Class	Order	Family	Genera	Species	Depth range (m)
Hydrozoa	Anthoathecata	Corymorphidae	1	1	6260–6776
Hydrozoa	Leptothecata	Lafoeidae	1	1	6860
Hydrozoa	Leptothecata	Hebellidae	1	1	8210–8300
Hydrozoa	Leptothecata	Aglaopheniidae	1	3	6300–7000
Hydrozoa	Leptomedusae	Mitroconidae?	1	1	8258–8260
Hydrozoa	Anthomedusae	Anthomedusidae?	1	1	8258–8260
Hydrozoa	Tracymedusae	Rhopalonematidae	3	3	6758–8700
Scyphozoa	Coronatae	Atorellidae	1	2	6000–10000
Scyphozoa	Semaeostomeae	Ulmaridae	1	1	8746–8662
Anthozoa	Alcyonacea	Alcyonaria	1?	1?	6758–8662
Anthozoa	Alcyonacea	Primnoidae	1	1	8021–8042
Anthozoa	Pennatularia	Kophobelemnonidae	1	2	5650–6150
Anthozoa	Pennatularia	Umbellulidae	2?	5	5650–6730
Anthozoa	Hexacorallia	Actinosolidae	2	2	6660–8230
Anthozoa	Hexacorallia	Bathyphelliidae	1	1	7250–7290
Anthozoa	Hexacorallia	Edwardsiidae	1	1	7160
Anthozoa	Actiniaria	Galatheanthemidae	2	4	10730
Anthozoa	Actiniaria	Actiniidae	1	1	5650–6780
Anthozoa	Antipatharia	Unknown	1	1	7200–8840
Anthozoa	Scleractinia	Fungiidae	1	1	6090–6328

6490–6650 m and 6180–7000 m, respectively (Kramp, 1959). By far the most common 'hadal' scyphozoan is an unidentified species of *Stephanoscyphus*, recorded in the hado-pelagic zone of 10 trenches. However, Belyaev (1989) notes that, based on size, colouration and morphology, there are likely to be at least different species among them, with the added possibility of more families. Furthermore, findings of scyphopolyps (*S. simplex*) at depths of greater than 6000 m indicate that Scyphomedusae must also exist in the trenches, although they have yet to be found. Images of Scyphomedusae, presumably Ulmaridae, ranging in size from 5 to 7 cm diameter, were obtained in the PROA expedition to the Bougainville Trench at depths between 7847 and 8662 m (Lemche *et al.*, 1976). The distinct lack of pelagic sampling at hadal depths means that the densities and ecological significance of Scyphomedusae is still unresolved and is another area requiring further examination.

Of the Anthozoa subclass of Octocorallia, representatives of the orders Alcyonacea and Pennatulacea corals have been recorded in images from several trenches, including the New Britain and New Hebrides Trenches (Lemche *et al.*, 1976). With the exception of 3 out of 21 species, all are restricted to depths of <7000 m (the remaining three are found between 7500 and 8000 m). Several *Pennatula* species of the genus *Khophbelemnon* and *Umbellula* have been found in multiple trenches but are still generally limited to <7000 m. In the northwest Pacific

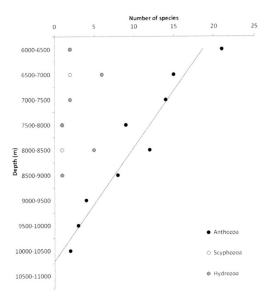

Figure 10.1 Number of species per 500 m depth stratum for the Cnidaria: Anthozoa, Scyphozoa and Hydrozoa. Note: Not shown are 10 findings of pelagic Scyphozea (*Stepanoscyphus* sp.) caught somewhere between 6000 and 10 000 m.

Trenches where the greatest numbers of samples have been taken, e.g. in the Kuril–Kamchatka Trench, octocorals were only found at three stations and only on the upper slopes. The most common octocorals were recorded from the Peru–Chile and South Sandwich Trenches. A single trawl in the Peru–Chile Trench by the *Vitjaz* recovered 26 specimens and another in a bottom grab, both from 6040 m (Belyaev, 1989). A single specimen of the order Ceriantharia was recorded in time-lapse in the Kermadec Trench, albeit at abyssal depths (Jamieson *et al.*, 2011a; Fig. 10.2).

Of the six known and one unknown families of Hexacorallia, the genus *Galatheanthemum* (family Galatheanthmidae, order Actinaria) is the most dominant, having been found in nearly every trench sampled (45 findings from 16 Pacific and Atlantic trenches >6000 m), including occasional mass findings. Two specimens of these tube-forming sea anemones have been found at full ocean depth; *Galatheanthemum hadale* Carlgren, 1956 (9820–10 210 m in the Philippine Trench) and *Galatheanthemum* sp. n. (10 170–10 730 m in the Mariana Trench). The bathymetric range of the other species covers all other depths and, therefore, these organisms are not restricted by hydrostatic pressure. Actinaria were also present in the images taken in the New Britain and New Hebrides Trenches (7057–7075 m and 6758–8930 m, respectively; Lemche *et al.*, 1976), as well as in the Puerto-Rico Trench (Heezen and Hollister, 1971). Most other known species are specimens from single findings (George and Higgins, 1979), including a photo from 7600 m in the Puerto-Rico Trench of 2 out of 64 specimens identified as 'Tube-dwelling

Figure 10.2 Although there are no high quality images of Cnidaria at hadal depths, this tube-dwelling anemone (Ceriantharia, Anthozoa) was imaged at 5173 m in the Kermadec Trench. Image courtesy of HADEEP.

actiniarian anthozoans (*Galatheanthemum*)', presumably *Galatheanthemum profundale* Carlgren, 1956 (Cairns *et al.*, 2007).

The *Vitjaz* and *Galathea* samples indicate that the family Galatheanthemidae is widespread in many trenches and is a family that is mainly restricted to the deep abyssal plains and hadal trenches, although shallower-water findings have been made in the Antarctic (3947–4063 m) (Dunn, 1983). Their distribution was later extended to the Cayman Trough (5800–6500 m; Keller *et al.*, 1975), the Puerto-Rico Trench (5749–8130 m) and the Virgin Islands Trough (4028–4408 m; Cairns *et al.*, 2007). These data suggest that Galatheanthemidae originated in deep Antarctica and spread into many deep-sea trenches. It is still not known why the Galatheanthemidae in all the regions, except for the Antarctic, are confined to depths greater than 5500 m.

10.2   Fish

The diversity and bathymetric range of fishes in the deep trenches has, until recently, been unresolved, partially due to a low number of records, some spurious records and some erroneous reports (Fujii *et al.*, 2010; Jamieson and Yancey, 2012). Fishes of the bathyal and abyssal zone are well documented across the world's oceans (Merrett and Haedrich, 1997) and are known to be a diverse and important component of the deep-sea community (Priede *et al.*, 2010). Records of hadal fish species are comparatively less numerous and several species have been recorded from solitary samples that were often in very poor condition (e.g. Stein, 2005). The lack of credible and high quality samples is a direct result of the technical challenges associated with capture. Many of the hadal fishes known to date were originally caught by trawl in the 1950s and 1960s.

Unfortunately, the complications associated with trawling at hadal depths meant that the number of high quality or high quantity samples obtained during this period were few and far between. Consequently, current understanding of the occurrence of hadal fishes has been primarily based upon a very limited number of fish specimens and has, until recently, been woefully inadequate (Fujii *et al.*, 2010). Furthermore, as a result of the dearth of fish samples, very little is known about their behaviour, ecology or abundance. For decades, it has been generally assumed that hadal fish diversity and population abundances are low and perhaps these fish are simply eking out an existence on the fringes of the abyssal plains (Wolff, 1961; Nielsen, 1964). Recently, this assumption has been entirely disproven (Jamieson *et al.*, 2009a, 2011b; Fujii *et al.*, 2010).

The first hadal fish was recovered from 6035 m in 1901, onboard the *Princess-Alice* in the east Atlantic Ocean. This ophidiid (cusk-eel), *Bassogigas profundissimus* (Roule, 1913) was considered the 'deepest fish' until the *Galathea* expedition trawled another specimen from 7160 m in the Java Trench (northeast Indian Ocean). The *Galathea* also recovered five snailfish liparids (Liparidae), described at the time as *Careproctus kermadecensis* (Nielson, 1964), from 6660–6770 m in the Kermadec Trench. The Soviet *Vitjaz* expeditions later captured another two liparid species in the northwest Pacific Ocean: a solitary *Careproctus amblystomopsis* (Andriashev, 1955) from the Kuril–Kamchatka Trench, at 7230 m and an individual, *Careproctus* sp., from the Japan Trench, at 7579 m. The two named species were later reclassified to *Notoliparis (C) kermadecensis* (Nielsen, 1964) and *Pseudoliparis (C) amblystomopsis*.

Bassogigas profundissimus was collected once again in 1970, from 8370 m in the Puerto-Rico Trench (Staiger, 1972) and was later reclassified as *Abyssobrotula galatheae* (Nielson, 1977). This specimen is still regarded as the deepest living fish on record. More recently, several other new fish species have been described from the hadal zone (e.g. Anderson *et al.*, 1985; Nielsen *et al.*, 1999; Chernova *et al.*, 2004; Stein, 2005), but they were typically singlular, often poor quality specimens that did not offer any new insights into the role of fishes in the hadal environment.

The confusion regarding the exact ocean depth at which fish can no longer survive can be directly traced back to the *Trieste* bathyscaphe dive in 1960. The publicity surrounding this dive catapulted the story of Jacques Piccard and his sighting of a flatfish at over 10 900 m in the Challenger Deep into the public domain. The story goes that when *Trieste* reached the bottom of Challenger Deep it crash-landed on the bottom and resuspended considerable seabed sediments. After 20 minutes on the seafloor it ascended back to the surface. Jacques Piccard was quoted as saying, 'Lying on the bottom just beneath us was some type of flatfish, resembling a sole, about 1 foot long and 6 inches across. Even as I saw him, his two round eyes on top of his head spied', he went on to add '. . .extremely slowly, this flatfish swam away. Moving along the bottom, partly in the ooze and partly in the water, he disappeared into his night' (Piccard and Dietz, 1961). However, the pilot of the *Trieste*, US Navy Lt. Don Walsh, recounted a less detailed version, 'As we landed, a cloud of sediment was stirred. This happened with all of our dives and usually after a few minutes it would drift away. Not this time

The cloud remained for the entire time on the bottom and showed no signs of moving away. It was like looking into a bowl of milk' (Walsh, 2009), and more recently he has added, 'In the half century since our dive, there has been some speculation that we did not see a flatfish. And this is entirely possible. Neither Jacques nor I were trained biologists and the critter could have been something else' (Burton, 2012).

The flatfish story was quickly refuted by the scientific experts. Wolff (1961) described the observation as 'somewhat dubious' since flatfish are rarely found beyond 1000 m, and none of the trench sampling campaigns up to that point had found any fish deeper than 7587 m, despite recovering representatives from most major taxa. He concluded that the flatfish was far more likely to be the bathypelagic holothurian *Galatheathuria aspera* (Thiel, 1886), which matches the description of the *Trieste* flatfish.

Other scientific literature, reporting on fishes from >6000 m have reiterated the point that the flatfish 'was, in reality, probably not a fish' (Nielson, 1964), and nearly 17 years later, the description of the *Abyssobrotula galatheae* (Ophidiidae) from 8370 m was published as the deepest living fish (Nielson, 1977). The *A. galatheae* from the Puerto-Rico Trench is still, to this day, considered to be the deepest living fish.

Around the same period as the *Trieste* dive, there were also other, credible reports of hadal fish. The pilot of the French *Archimède* bathyscaphe described seeing 200 small fish 'similar to liparids' and three individuals of two other fish species in the Puerto-Rico Trench, at 7300 m (Pérès, 1965). Despite the lack of images to corroborate these observations of diversity, density and behaviour, they are somewhat intriguing since the observations do match more recently gathered data, albeit from other trenches (Fujii *et al.*, 2010).

Despite overwhelming evidence that contradicts the likelihood of the *Trieste* flatfish actually being a fish (Wolff, 1961; Jamieson and Yancey, 2012), the story has continued to be perpetuated to this day in the popular media, leading to confusion and misinformation on the matter. Fortunately, a great deal of research has been undertaken in recent years that can, once and for all, dispel the myth of the *Trieste* flatfish and offer scientific insight into the occurrence of fish at hadal depths.

A comprehensive analysis of all fish species was reported in Priede *et al.* (2006b). The analysis was primarily conducted in order to show that Chondrichthyes were absent from abyssal depths. Linear regression of the log species numbers of all 9360 fish records in the database predicted a maximum depth for fish to be between 8000 and 8500 m (Priede *et al.*, 2006b, 2010). This coincides with the 8370 m record for *Abyssobrotula galatheae* (Nielson, 1977). This also placed the *Trieste* flatfish at nearly 3000 m deeper than any other fish and 7916 m deeper than the next deepest pleuronectiform (flatfish; Jamieson and Yancey, 2012). These depth-related trends are, of course, still based on preliminary data, owing to the sparse number of records at the deeper end of the scale.

The HADEEP project specifically aimed to investigate hadal fish populations using baited cameras and traps, thus providing unequivocal proof of the depths they inhabit and, at the same time, gathering behavioural information, population density estimates, high quality images and samples of fish (Jamieson *et al.*, 2009a, 2011c; Fujii *et al.*,

Table 10.2 Current list of fish species recorded at depths of >6000 m (updated and modified from Fujii *et al.*, 2010).

Species	Depth (m)	Trench	Record
Macrouridae **(Grenadiers)**			
Coryphaenoides yaquinae	6000	Kermadec	Jamieson *et al.*, 2011a
	6160	Japan	Horibe, 1982
	6945	Japan	Jamieson *et al.*, 2009a
	6380–6450	Japan	Endo and Okamura, 1992
Carapidae (Pearlfish)			
Echiodon neotes	8200–8300	Kermadec	Markle and Olney, 1990
Ophidiidae (Cusk-eels)			
Bassozetus zenkevitchi	0–6930	*Not specified*	Rass, 1955
Bassozetus cf. *robustus*	6116, 6474	Kermadec	Jamieson *et al.*, 2013
Leucicorus atlanticus	4580–6800	Cayman	Nielsen, 1975
Unidentified Ophidiid	6173*	Peru–Chile	Unpublished data, HADEEP
Barathrites sp.	6116*	Kermadec	Unpublished data, HADEEP
Abyssobrotula galatheae	3110–8370	Puerto-Rico	Nielsen, 1977
Holcomycteronus profundissimus	5600–7160	Sunda	Roule, 1913
Apagesoma edentatum	5082–8082	*Not specified*	Carter, 1983
Liparidae (Snailfish)			
Notoliparis antonbruuni	6150*	Peru–Chile	Stein, 2005
Notoliparis kermadecensis	6660–6770	Kermadec	Nielsen, 1964
	6890	Kermadec	Jamieson *et al.*, 2009a
	6474–7501	Kermadec	Jamieson *et al.*, 2013
	7199, 7561	Kermadec	Jamieson *et al.*, 2011a
Pseudoliparis amblystomopsis	7210–7230	Kuril–Kamchatka	Andriashev, 1955
	6945	Japan	Jamieson *et al.*, 2009a
	7420–7450	Japan	Horikoshi *et al.*, 1990
		Japan	Fujii *et al.*, 2010
Pseudoliparis belyaevi	7565–7587	Japan	Andriashev and Pitruk, 1993

* Indicates that *in situ* photography from HADEEP expeditions recorded this fish at either 4602 and 5139 m, or 7050 m in the Peru–Chile Trench.

2010). This project prompted a reappraisal of hadal fishes by Fujii *et al.* (2010), with an updated version shown in Table 10.2.

Data regarding the depth of occurrence and the diversity of hadal fish were extracted from global datasets, accessed through FishBase (available from www.fishbase.org; Froese and Pauly, 2009). The dataset for fishes found deeper than 6000 m documented 15 species belonging to six families (five Ophidiidae (cusk-eels), four Liparidae (snailfish), three Bathylagidae (deep-sea smelts), one Eurypharyngidae (gulpers), one Macrouridae (grenadiers) and one Carapidae (pearlfish)). On re-examination of the original references, it was found that many of these records were either erroneous or misleading.

The bathylagids, *Lipolagus ochotensis* Schmidt, 1938, *Bathylagus pacificus* Gilbert, 1890 and *Pseudobathylagus milleri* (Jordan and Gilbert, 1898) are well-known meso-pelagic fish that migrate vertically from deep (~1000 m) to shallow waters (~500 m) at night in order to feed (Radchenko, 2007). They happen to inhabit the pelagic waters overlying the Aleutian, Kuril–Kamchatka and Japan Trenches. These are relatively well-researched fish, known to rely heavily on the meso-pelagic and epi-pelagic zone for feeding and spawning. Therefore, it appeared more likely that these bathylagids were accidental catches from the shallower zones, captured as the trawl was hauled to the surface. Furthermore, the depth ranges were cited as 0–6000 m for the first two species of bathylagids and 230–7700 m for the latter, suggesting that the fish were captured by vertical trawling, a method that provides no means of determining the exact depth of capture. The same conclusion can be drawn on a record for the gulper eel *Eurypharynx pelecanoides* Vaillant, 1882. This gulper eel is another well-known bathypelagic fish (Gartner, 1983), generally found between 1200 and 1400 m, several thousand metres above the seafloor (Owre and Bayer, 1970; Masuda *et al.*, 1984).

Other records were found to be inconclusive, such as the record for the pearlfish *Echiodon neotes* Markle and Olney, 1990 (Carapidae), since all 12 species of *Echiodon* are otherwise known to inhabit depths of 18–2000 m (Markle and Olney, 1990; Williams and Machida, 1992). The hadal species recorded was a single specimen from 8200–8300 m in the Kermadec Trench. It was recorded as demersal but is debatably pelagic (Nielsen *et al.*, 1999). Furthermore, it seems unlikely that a species of this relatively shallow-water family would inhabit waters ~6300 m deeper than any other *Echiodon* species. Other spurious records include that of the Ophidiid *Apagesoma edentatum*, recorded as living between 5082–8082 m, although Anderson *et al.* (1985) and Nielsen *et al.* (1999) state its depth range at 2560–5082 m. The origins of the 8082 m entry are unclear, thus its status as a hadal species, living at hadal depths is unresolved.

It is also worth noting that there are some erroneous references to sharks at hadal depths. For example, the bigtooth cookie-cutter shark *Isistius plutodus* has been recorded from as deep as 6440 m (Kiraly *et al.*, 2003). In reality, the specimen in question was caught at a depth of 200 m mid-water over the Ryukyu Trench (6440 m deep).

The appraisal of Fujii *et al.* (2010) and the results of the HADEEP projects concluded that there are, in general, three families of fish that inhabit the hadal zone; Macrouridae (grenadiers or rat-tails), Ophidiidae (cusk-eels) and Liparidae (snailfish).

Macrourids are an abundant and diverse family of deep-sea gadiformes (Wilson and Waples, 1983; Fig. 10.3). There are two main macrourids observed within the vicinity of trenches around the Pacific Rim; *Coryphaenoides yaquinae* Iwamoto and Stein, 1974 and *Coryphaenoides armatus* (Hector, 1875). The deeper of the two macrourids, *C. yaquinae*, is restricted to the Pacific Ocean and is a well-documented species, observed down to 5900 m on the abyssal plains (Priede and Smith, 1986). Although primarily a deep abyssal scavenger, it has also been observed *in situ* at 6160 m (Horibe, 1982) and 6945 m (Jamieson *et al.*, 2009a) in the Japan Trench, at 6000 m in the Kermadec Trench and at abyssal depths in the northern sector of the Peru–Chile Trench (Jamieson *et al.*, 2012c). The deepest of these records, 6945 m, highlights its ability to transcend the abyssal–hadal boundary, although this has only been observed in the Japan Trench. The most ubiquitous abyssal macrourid is *Coryphaenoides armatus*

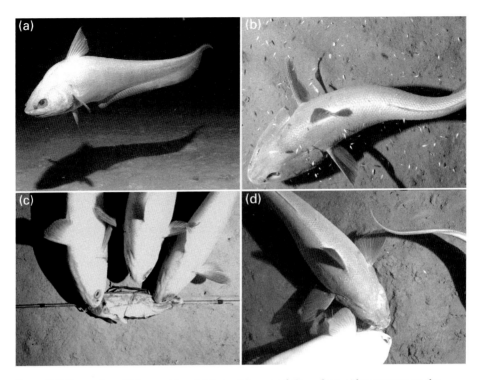

Figure 10.3 Examples of Macrouridae. (a) Lateral image of *Coryphaenoides armatus* on the abyssal plains of the northeast Atlantic Ocean; (b) *C. yaquinae* at 5469 m on the edge of the Mariana Trench; (c) *C. armatus* from 4329 m in the Kermadec Trench; (d) *C. yaquinae* (dark) and *C. armatus* (light) at 4602 m in the Peru–Chile Trench. Images taken by Hadal-Lander B, courtesy of HADEEP except (a) courtesy of the University of Aberdeen, UK.

but this fish is limited to depths of 5180 m (Cohen *et al.*, 1990). It is a wide-ranging, mostly eutrophic, deep-slope or upper-rise species that dominates the abyssal plains of the Atlantic and Indian Oceans, between 2000 and 4800 m (Wilson and Waples, 1983). In the Pacific Ocean, it is restricted to relatively food-rich zones between 2000 and 4600 m around the Pacific Rim, and it appears unable to penetrate into the large oligotrophic expanses of the abyssal plains, unlike *C. yaquinae* (Jamieson *et al.*, 2012c). Interestingly, Pérès (1965) reportedly saw a macrourid at ~7000 m in the Puerto-Rico Trench, in the Atlantic. If this identification was actually a macrourid, then perhaps *C. armatus* lives 2000 m deeper than previously thought or otherwise, there is a new and undescribed species of macrourid inhabiting hadal depths. While neither *C. yaquinae* nor *C. armatus* penetrate far into hadal depths, they do account for a significant presence in the abyssal–hadal transition zone.

There are descriptions of five ophidiids with a depth range exceeding 6000 m: *Bassozetus zenkevitchi* Rass, 1955, *Leucicorus atlanticus* Nielsen, 1975, *Holcomycteronus profundissimus* (Roule, 1913), *Apagesoma edentatum* Carter, 1983 and *Abyssobrotula galatheae* Nielsen, 1977; all of which are relatively rare. Various cusk-eels have also been observed by baited cameras in the hadal zone, however, they have yet to be identified to species level.

Bassozetus zenkevitchi is the only known pelagic *Bassozetus* species, although benthic captures have been acknowledged (Nielsen and Merrett, 2000). Its large depth range of 0–6930 m is again misleading because of the vertical trawling method used for sampling. More accurate trawl records suggest that it is an abyssal species (Machida and Tachibana, 1986; Nielsen and Merrett, 2000), although its presence in the upper depths of trenches is not unlikely. *B. zenkevitchi* is found mainly in the vicinity of the northwest Pacific trenches (Kuril–Kamchatka, Japan and Izu-Bonin) (Orr *et al.*, 2005).

Holcomycteronus profundissimus (Roule, 1913), formerly *Bassogigas* (Nielsen, 1964), is a geographically extensive species, found in the Atlantic, Pacific and Indian Oceans. It is a lower abyssal to upper hadal species (5600–7160 m; Nielsen *et al.*, 1999) and the deepest point at which it has been found was recorded in the Java Trench.

The deepest fish, and indeed the deepest vertebrate ever found, is the *Abyssobrotula galatheae* (8370 m in the Puerto-Rico Trench; Nielsen, 1977). Although this *A. galatheae* is widely accepted as the deepest fish, the accolade is still questionable. There are 17 records of this fish available on the fishbase.org database: 1 bathyal (2330 m; Shcherbachev and Tsinovsky, 1980), 14 abyssal and 2 hadal (3100–8370 m; Machida, 1989). Although the deeper records for *A. galatheae* are from or in the vicinity of trenches (Puerto-Rico, Japan and Izu-Bonin; Machida, 1989), the specimens were obtained using a non-closing trawl net. Therefore, there is the possibility of incidental capture in mid-water during hauling. This issue was raised by Nielson (1964) and Nielson and Munk (1964), as 15 genera of pelagic fish were reported from the same trawl that included *A. galatheae*. The stomach contents indicate that it is potentially bentho-pelagic. However, Shcherbachev and Tsinovsky (1980) reported on a total of 12 specimens of *A. galatheae*, all from bathyal or abyssal depths and noted that one was captured in a pelagic trawl 400–800 m above the seafloor. The true depth range of this species is, therefore, still debatable until further information becomes available, but for the time being it is assumed to be the deepest fish.

Leucicorus atlanticus was once considered a 'unique hadal fish species of the Caribbean region', having been described from a specimen recovered from the Cayman Trench (Rass *et al.*, 1975). It is now known to be a lower abyssal to hadal species (4590–8600 m; Anderson *et al.*, 1985). Aside from a few trawled specimens, very little is known about *L. atlanticus*. However, an ophidiid similar in appearance to *L. atlanticus*, but more likely of the genus *Barathrites*, was observed in high abundance at a baited lander in the Peru–Chile Trench (unpublished, HADEEP). The fish was infrequently observed at 5329 m (in only 2.3% of the 669 frames), however, its abundance at a deeper 6173 m site was extraordinarily high (Fig. 10.4). At this depth, it arrived at 2 h 41 min into the deployment and increased in numbers, reaching an estimated maximum of 20 individuals in the $0.29 \, \text{m}^{-2}$ field of view. The fish were clearly feeding at the bait, and did not show any signs of a decrease in numbers even 19 h later (the end of the deployment). In total, at least 7257 observations were recorded, accounting for a presence of 71.5% in 1120 images. This one deployment represents more observations of a single species of fish at these depths than found anywhere else, regardless of depth, using the same equipment.

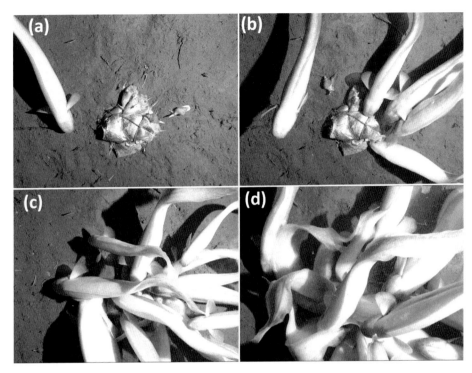

Figure 10.4 Large aggregation of an unidentified ophidiid at 6173 m in the Peru–Chile Trench, after (a) 3 h, (b), 6 h, (c) 9 h and (d) 14 h on the seafloor. Images taken by Hadal-Lander B, courtesy of HADEEP.

More recently, another two species of Ophidiidae have been photographed *in situ* at 6116 and 6474 m in the central Kermadec Trench (unpublished data, HADEEP; Fig. 10.5). At these two sites, no macrourids were observed, however, the ophidiid *Bassozetus* sp. was dominant. They reached a maximum number of three and six, respectively. These ophidiids were present in 80% and 73% of 840 frames (14 h; Fig. 10.6). However, the *Bassozetus* sp. is externally very similar to specimens photographed at 5469 m on the edge of the Mariana Trench (Jamieson *et al.*, 2009a), and at 5329 and 6173 m in the Peru–Chile Trench. It may be construed that the *Bassozetus* sp. is actually *B. robustus* Smith and Radcliffe, 1913 (pers. comm., A. Stewart, Te Papa Fish Collection, New Zealand), a globally ubiquitous abyssal species. Also observed at 6116 m was a solitary ophidiid of the *Barathrites* genus. Neither of the species observed entered the *Latis* fish trap, which was deployed at the same depths 2 km away. Therefore, identification of these ophidiids to species level was not possible.

There are at least six species of the Liparidae known to occur at hadal depths. Two of the species, *Notoliparis antonbruuni* Stein, 2005 and *Pseudoliparis belyaevi* Andriashez and Pitruk, 1993, were described from single specimens from 6150 m in the Peru–Chile Trench (Stein, 2005) and 6380–7587 m in the Japan Trench (Chernova *et al.*, 2004), respectively. The limited number of samples, and, in particular, the poor

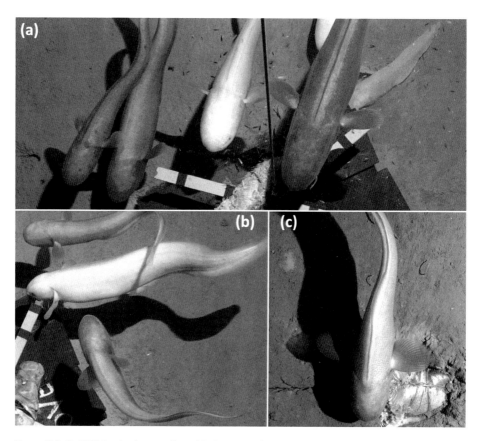

Figure 10.5 Ophidiidae in the trenches, (a) A group of *Bassozetus* sp. (possibly *B. robustus*) at 6474 m in the Kermadec Trench, (b) more *Bassozetus* sp. and one *Barathrites* sp. (white, centre) at 6116 m in the Kermadec Trench and (c) *Bassozetus* sp. at 6173 m in the Peru–Chile Trench. Images taken by Hadal-Lander B, courtesy of HADEEP.

quality of *N. antonbruuni*, unfortunately added little to our understanding of fish populations at these depths. Baited camera deployments in the Peru–Chile Trench at 4602 and 5329 m observed two small liparids preying upon amphipods that were attracted to the bait (Fig. 10.7). There was also another liparid observed relatively frequently at 7050 m. In the absence of physical samples, and given the poor condition of the deep liparid described from the region (Stein, 2005), it was not possible to determine if one of these two species was *N. antonbruuni*.

In addition to these rarer records, there is also the 1964 account of liparids at 7300 m from *Archimède* bathyscaphe in the Puerto-Rico Trench (Pérès, 1965). Two hundred individuals were observed over a short transect and were mostly 10–12 cm long, although some were as long as 25 cm. The report describes the fish as pale pink in colour, becoming darker with increasing body size and with distinct black eyes. Their swimming behaviour was described as 'spirally' and often discontinuous and some fish were observed to fall to the seafloor and remain stationary for short periods, with the

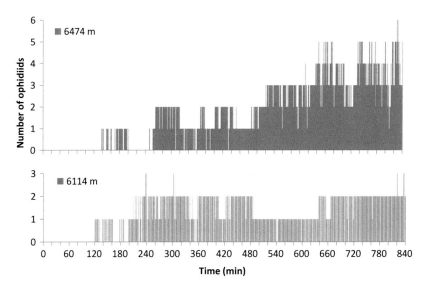

Figure 10.6 Time series of ophidiids (*Bassozetus* cf. *robustus*) attending bait in the central Kermadec Trench and 6116 and 6474 m.

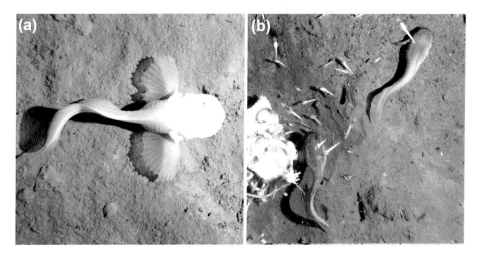

Figure 10.7 Liparids of the Peru–Chile Trench. (a) Unidentified liparid from 7050 m and (b) *Notoliparis* cf. *antonbruuni* found between 4602 and 5329 m. Images taken by Hadal-Lander B, courtesy of HADEEP.

body arched and lying on one side. There were no samples, video or still images taken to corroborate these observations at the time, however, the description and behavioural accounts reported by Pérès (1965) bear an uncanny resemblance to more recent observations of other liparids, taken *in situ* in other trenches (Jamieson *et al.*, 2009a, 2011a; Fujii *et al.*, 2010). Such observations suggested that the upper depths of the

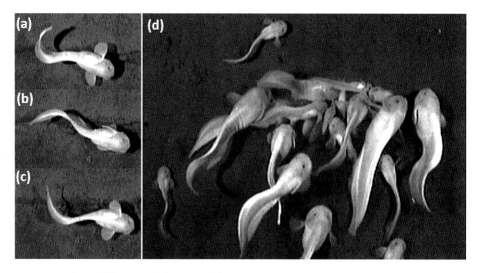

Figure 10.8 The snailfish *Pseudoliparis amblystomopsis* (Liparidae) in the Japan Trench (6945–7703 m). (a–c) show one tail-beat 0.3 s apart from 6945 m and (d) shows a mass aggregation at 7703 m, the deepest fish ever seen alive. Modified from Fujii *et al.* (2010).

Puerto-Rico Trench are, indeed, inhabited by liparids. Interestingly, a deep-living liparid, *Careproctus sandwichensis*, was also found in the vicinity of the South Sandwich Trench, in the Southern Ocean at 5453 m (Andriashev and Stein, 1998).

A number of comparative *in situ* observations of hadal fish were made during the HADEEP projects, in different trenches and they comprised significant observations of two species in particular; *Notoliparis kermadecensis* Nielsen, 1964 from the Kermadec Trench and *Pseudoliparis amblystomopsis* Andriashev, 1955 from the Japan Trench.

Pseudoliparis amblystomopsis is a relatively well-documented liparid, inhabiting some of the northwest Pacific trenches. It has been found between 7210 and 7230 m in the Kuril–Kamchatka Trench (Andriashev, 1955) and between 7420 and 7230 m in the Japan Trench (Horikoshi *et al.*, 1990). The first observation of this species alive occurred at 6945 m in the Japan Trench, where an individual (22.5 cm) was seen twice (Jamieson *et al.*, 2009a). In the Kermadec Trench at 6890 m, three individuals of *N. kermadecensis* (32.2, 33.3 and 28.7 cm long) were observed over a 6 h period. Both *P. amblystomopsis* and *N. kermadecensis* were observed suction feeding on small scavenging amphipods, at a rate of 2 and 9 min^{-1}, respectively (Fig. 10.8).

The initial observations undertaken during the first HADEEP project gave the impression that fish populations in the upper depths of the trenches were indeed low. However, the still photographs and video footage obtained using baited cameras provided additional information about the fish that the historical trawl samples could not. For example, Jamieson *et al.* (2009a) observed that *P. amblystomopsis* was capable of holding its position in near-bottom currents of ~3–7 cm s^{-1}, with a routine tail-beat frequency of 0.47 Hz ± 0.01 S.D. ($n = 2$). Both pectoral and caudal fins were used for propulsion with a tendency of 1:1 synchrony (mean = 0.76). *N. kermadecensis* was seen

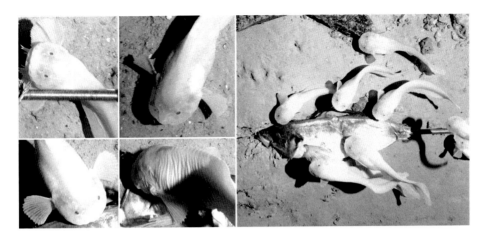

Figure 10.9 The snailfish *Notoliparis kermadecensis* (Liparidae) from the Kermadec Trench (6474–7561 m). (left) Close-ups of body and (right) mass aggregation at 7561 m. Images taken by Hadal-Lander B, courtesy of HADEEP.

to be capable of swimming against currents of ~10–14 cm s^{-1} with a mean tail-beat frequency of 1.04 Hz $\pm$ 0.11 S.D. ($n = 31$) and a caudal:pectoral frequency ratio of 2.08. Problems of size scaling, temperature differences and absence of data on shallow-water liparids makes it hard to assess whether these hadal fish show lower activity levels, as might be expected in fish living at extreme hydrostatic pressure. However, tentative comparison of the tail-beat frequency and theoretical swimming speeds of these liparids with those of shallower abyssal species of Moridae and Macrouridae (Collins *et al.*, 1999) showed no obvious signs of physiological limitations.

The idea that fish at hadal depths are rare and low in number was quickly challenged when further expeditions to the Japan and Kermadec Trenches revealed large aggregations of both species at slightly greater depths than had been previously observed (Fig. 10.8, Fig. 10.9). At 7703 m in the Japan Trench, *Pseudoliparis amblystomopsis* arrived at the bait within 1 h 15 min and subsequently represents the deepest fish seen alive to date (Fujii *et al.*, 2010). The number of *P. amblystomopsis* increased exponentially over the following 5 h, to a maximum number of 20 by the end of recording (6 h 35 min). Both adults and juveniles were present, with maximum and minimum body lengths of 30.0 cm and 7.4 cm, respectively (mean body length of 19.8 cm $\pm$ 5.2 S.D., $n = 10$). In the Kermadec Trench at 7199 and 7561 m, *N. kermadecensis* were present in 63% and 90% of the frames taken at these sites (779 and 813 frames, respectively) (Jamieson *et al.*, 2011a; Fig. 10.10). Their arrival times were 78 and 75 min, respectively. At 7199 m, their numbers increased to a maximum of 5, whereas at 7561 m, the maximum number was 13. Their body lengths ranged from 18.5 to 34.1 cm at 7199 m (mean = 25.4 cm $\pm$ 5.4 S.D., $n = 8$) and 13.5 to 22.7 cm (mean = 17.5 cm $\pm$ 2.3 S.D., $n = 11$) at 7561 m.

Since then, a further three deployments of the baited camera have detected *N. kermadecensis* in the central Kermadec Trench (~32° S), at 6474, 6979 and 7501 m,

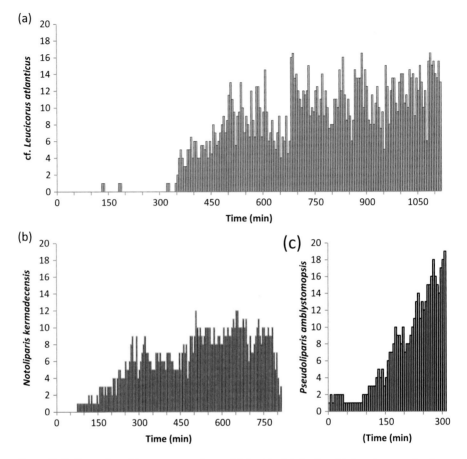

Figure 10.10 Examples of large aggregations of fish at hadal depths. (a) *Leucicorus atlanticus* (ophidiid) at 6173 m in the Peru–Chile Trench, (b) *Notoliparis kermadecensis* (liparid) at 7561 m in the Kermadec Trench and (c) *Pseudoliparis amblystomopsis* (liparid) from 7703 m in the Japan Trench. Data derived from (a) unpublished data, HADEEP, (b) Jamieson *et al.* (2011a) and (c) Fujii *et al.* (2010).

on the 2011 and 2012 RV *Kaharoa* expeditions (KAH1109 and KAH1202; HADEEP). The 6474 m record represents the shallowest record of occurrence for *N. kermadecensis*. The previous record was reported at 6660–6770 m when the *N. kermadecensis* holotypes were caught in 1951 on the *Galathea* expedition (Nielsen, 1977). Furthermore, the RV *Kaharoa* KAH1109 expedition was the first to use the full ocean depth rated fish trap, 'Latis', specifically designed for recovering hadal snailfish. *Latis* was deployed at 7000, 7012, 7291, 7844 and 9908 m and five specimens were recovered from the 7000 m site (mean length = 23.7 cm; Fig. 10.11) and small single specimens were recovered from 7291 and 7179 m, with body lengths = 19.0 and 12.8 cm, respectively. Subsamples were taken from each of these specimens for, among other things, visual function, DNA sequencing and cellular pressure adaption analysis.

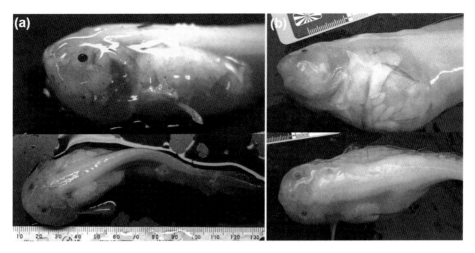

Figure 10.11 Specimens of Liparidae from the Pacific Ocean trenches. (a) *Pseudoliparis amblystomopsis* from 7703 m in the Japan Trench and (b) *Notoliparis kermadecensis* from 7000 m in the Kermadec Trench. Images courtesy of HADEEP.

None of the Japan or Kermadec liparids were observed to feed directly on the bait itself, but rather to suction feed on small amphipods ($< {\sim}2.0$ cm). On several occasions, they were seen sucking sediment from the seafloor and blowing it through the gills, presumably filtering out amphipods in the process. In the Japan Trench, large amphipods (*Eurythenes gyllus*) were observed interfering with the tails or body of the liparids. These perturbations were counteracted by sudden and repeated body-flicks, including flicking of the head and body against the seafloor, often resulting in turning upside-down and exhibiting a spiral-swimming gait. In contrast, some individuals temporarily became motionless and rolled onto one side as if forced by the current; a behaviour also observed among other individuals that were otherwise very active. On average, 36% of the fish exhibited these short periods of inactivity but there are no obvious explanations as to the initiation or cessation of this behaviour.

This spiral-swimming gait and motionless periods are intriguing, as they are rarely evident in deep-sea fish. Furthermore, it is unlikely that these behaviours could be deduced from captured specimens, thus, the reports of 'spirally' swimming and fish 'passing out' on the seafloor in an unbalanced arch position (as well as the descriptions of the size and appearance of the fish) suggest that the bathyscaphe findings of Pérès (1965) at *c.* 7000 m in the Puerto-Rico Trench are entirely credible. It can, therefore, be construed that a hadal species of Liparidae inhabits the same depths of the Puerto-Rico Trench as do the liparids observed in the Japan, Kermadec and Peru–Chile Trenches.

The repeated *in situ* observations of *P. amblystomposis* in the Japan Trench and *N. kermadecensis* in the Kermadec Trench have provided a sufficient spread of body length data to be included in the general depth trend for Actinopterygii maximum body size proposed by Priede *et al.* (2010). These data have been modified to account for the reappraisal of deepest fish records presented by Fujii *et al.* (2010) and to include the

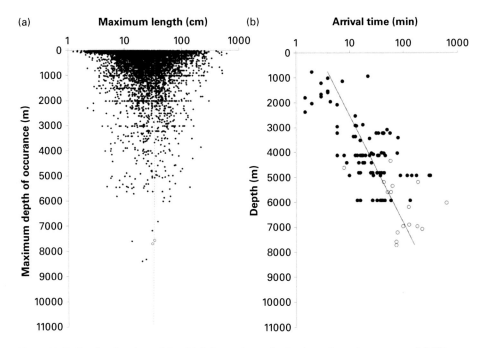

Figure 10.12 Depth-related trend for (a) fish maximum body size, where the average of 8686 records = 32.2 cm (dashed line). The hadal liparids (*P. amblystomopsis* and *N. kermadecensis*) are in open dots. (b) Fish arrival times (T_{arr}) at baited cameras (closed dots) with all HADEEP fish data from trenches or neighbouring abyssal plains in open dots (depth time relationship: $t_{arr} = 2.5793^{e0.0005D}$, n = 82, $R^2 = 0.42$). (a) modified from Priede *et al.* (2010) and (b) modified from Jamieson *et al.* (2009a).

latest maximum body sizes for the hadal liparids. These data show that the optimum body size for fish towards their deepest known depths, for all fishes approaches 30 cm (mean length = 32.2 cm, n = 8686; Fig. 10.12). Similarly, with increasing depth, the time it takes for fish to arrive at the bait also increases (Jamieson *et al.*, 2009a; Fig. 10.12b).

10.2.1   Distribution

The distribution of scavenging fish in the vicinity of the trenches is becoming more apparent as new data are collected from around the world. In the oligotrophic Sargasso Sea in the North Atlantic (annual chlorophyll biomass $<0.25\,\mathrm{mg\,m}^{-3}$; Longhurst, 2007), ophidiids, particularly *Bassozetus* sp., were the primary scavengers (Fluery and Drazen, 2013). Elsewhere in the Atlantic (Porcupine Abyssal Plain, Madeira Abyssal Plain and off Cape Verde), studies have shown an abundance of macrourids, mostly *Coryphaenoides armatus*, albeit at slightly shallower depths of 4000–4900 m (Nielsen, 1986; Armstrong *et al.*, 1992; Thurston *et al.*, 1995; Priede and Merrett, 1998; Henriques *et al.*, 2002). Around the continental slopes of the Pacific Rim the macrourids also dominate, where *C. armatus*, a wide-ranging, mostly eutrophic, deep-slope/upper-rise species dominates between 2000 and 4800 m (Wilson and Waples, 1983; Jamieson *et al.*, 2012c). As depth increases, *C. armatus* is replaced by another macrourid, *Coryphaenoides yaquinae*, which is confined to the Pacific and dominates under the vast expanse of the central gyres, typically at depths of 3400 to 5800 m (Wilson and Waples, 1983). Although these two species share a bathymetric overlap of 900 m, they are generally bathymetrically segregated (Endo and Okamura, 1992). *C. yaquinae* has also been reported on occasion, from the deeper depths of 6380–6450 m (Endo and Okamura, 1992), 6160 m (Horibe, 1982) and 6945 m (Jamieson *et al.*, 2009a) in the Japan Trench. It has also been seen at 5469 m on the edge of the Mariana Trench (Jamieson *et al.*, 2009a), at 5329 m in the Peru–Chile Trench (Jamieson *et al.*, 2012c) and at 6000 m in the Kermadec Trench (Jamieson *et al.*, 2011a). These data suggest that *C. yaquinae* is a deep-dwelling species that crosses the abyssal and hadal boundary, albeit by only a few hundred metres. However, the observations of macrourids in the Kermadec Trench by Jamieson *et al.* (2011a) were made at the southern tip of the trench in the vicinity of the continental slopes. Additional deployments made further offshore, in the central, more oligotrophic area of the trench, between 6000 and 7000 m, did not detect a single macrourid. Instead, relatively large numbers of ophidiids were found; mainly *Bassozetus* cf. *robustus* and a single *Barathrites* sp. at 6116 and 6474 m.

While data from all these locations are low relative to the abyssal depths, it appears that there may be a succession of scavenging fish with depth; *C. armatus* (2000–4800 m), *C. yaquinae* (3400–6945 m, although generally <6000 m) and ophidiids of the genus *Bassozetus* (5000–6500 m).

Fleury and Drazen (2013) suggest that, based on data obtained from the deep Atlantic, the distribution of macrourids and ophidiids in the deep Kermadec Trench is likely to be linked to the overlying surface productivity. Their findings from the deep Atlantic Ocean report that ophidiids dominated the more oligotrophic Madeira Abyssal Plain (annual chlorophyll biomass <0.50 mg m^{-3}; Longhurst, 2007) when compared to the more eutrophic Porcupine Abyssal Plain (annual chlorophyll biomass <1.5 mg m^{-3}; Longhurst, 2007), where macrourid abundance is often found to be high (Armstrong et al., 1992). The trend in abundances of ophidiids and macrourids has been explained by the different productivity regimes (Armstrong *et al.*, 1992; Thurston *et al.*, 1995) and seasonality of surface production (Merrett, 1987). This suggestion is supported by other observations which have indicated that macrourids are more common in eutrophic regions. For example, the California Current in the North Pacific is known to be dominated by macrourids (Priede *et al.*, 1994), while oligotrophic regions such as the North Pacific subtropical gyre are dominated by ophidiids (Yeh and Drazen, 2009).

These observations suggest that the scavenging fish communities in the vicinity of trenches are not determined by depth alone but probably by a combination of depth and trophic setting. Where trenches underlie eutrophic waters, such as the Japan Trench, macrourids are found to extend down the slopes into the upper trench (Horibe, 1982; Endo and Okamura, 1992; Jamieson *et al.*, 2009a, 2012c) and in the case of the Kermadec Trench, macrourids are found on or near the more productive continental slopes at abyssal depths, while ophidiids are found deeper and further offshore under more oligotrophic waters. Further sampling is still required to prove or refute this hypothesis and to disentangle the effects of depths and productivity as the driver for scavenging fish distribution.

The distribution of fish beyond the bait-attending species, such as the Macrouridae and Ophidiidae, comprises only that of the Liparidae. All the trench studies undertaken have shown that each trench (or adjoining trench) is inhabited by a single species of liparid. Also, there are relatively good data to show that the trench liparids inhabit depths of between approximately 6500 and 7500 m. While there are no comparative data from both flanks of any trench, it can be assumed that if liparids inhabit these depths on one slope, then presumably, they inhabit the adjacent slope also, which would suggest a peculiar distribution band that circumnavigates the trench. Moreover, the liparids are capable of swimming at high altitudes off the bottom, and, therefore, are theoretically incapable of moving directly from one slope to the adjacent one, across the trench axis. Thus, for the adjacent-slope populations to mix, they must have to circumnavigate the trench. This type of distribution is similar to that of organisms living on the slopes of a seamount or even a terrestrial mountain, where the population is locked into a circular band, unable to traverse the summit or extend beyond the base.

10.2.2   Depth limitations

All of the newly acquired data on fish occurrence with depth from the HADEEP project, combined with regression analysis from the databases, indicate that fish are limited to depths of less than approximately 8000–8500 m. For example, Figure 10.13 shows the maximum depths records of all known species of fish, as extracted from www.fishbase, org (comprised of over 9300 records, modified from Priede *et al.*, 2006b, 2010) and also shows all the Hadal-Lander deployments ($n = 29$) where fish were, and were not, found. Both sets of data show the same limit of around 8000 m.

The reason for the apparent 8000–8500 m bathymetric limits to fish is a direct effect of hydrostatic pressure. One of the major prerequisites to survival in the hadal zone is the ability to cope with extreme high hydrostatic pressure, which can have large perturbing effects on biological molecules. Membranes and proteins are known to have structural adaptations that provide pressure resistance (Hochachka and Somero, 1984). In recent years, a different adaptation to pressure has been hypothesised involving 'piezolytes' (*sensu* Martin *et al.*, 2002); small organic solutes first discovered as organic osmolytes. Solutes accumulated by most marine organisms prevent osmotic shrinkage of their cells by osmoconforming to the ambient environmental conditions (osmotic pressure of about 1000 mOsm). Among others present in marine organisms, one of the main osmolytes is trimethylamine oxide (TMAO), which is found in deep-sea fish. Most marine Osteichthyes (bony fish) are osmoregulators, which means that they maintain a relatively high internal osmotic pressure of about 300–400 mOsm, compared to that of shallower bony fish (~40–50 mOsm).

In bony fish, internal osmotic pressure increases with depth as a result of increasing TMAO content. Whereas in osmoconformers, other osmolytes decrease in osmotic

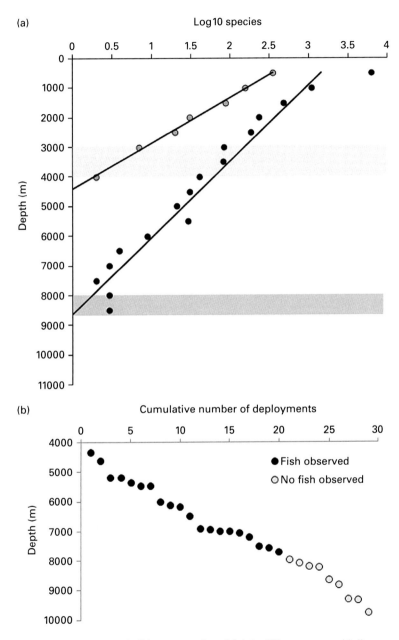

Figure 10.13 (a) Log$_{10}$ of all known species of fish in 500 m stratum with linear regression (black line), modified from Priede *et al.* (2006b, 2010). The light and dark shaded areas represent the depths where TMAO concentrations are predicted to reach isosmosis in both groups, respectively. Modified from Jamieson and Yancey (2012). (b). Baited camera deployments in and around hadal trenches from the HADEEP projects where solid dots represent where fish were observed and grey dots where no fish were observed.

compensation as TMAO increases with hydrostatic pressure, e.g. urea in elasmobranchs and glycine in decapods (Kelly and Yancey, 1999).

TMAO concentration analyses of deep-sea fishes revealed a linear relationship with depth, down to 4900 m (Gillett *et al.*, 1997; Kelly and Yancey, 1999; Yancey *et al.*, 2004; Samerotte *et al.*, 2007). Extrapolation of these data suggested that the depth at which the fishes' cells would become isosmotic with seawater is about 8000–8500 m, roughly the depth of the deepest fishes ever observed or captured (Nielsen, 1977; Jamieson *et al.*, 2009a; Fujii *et al.*, 2010). This was recently tested by Yancey *et al.* (in press) using samples from several *Notoliparis kermadecensis* specimens recovered by the *Latis* fish trap from 7000 m in the Kermadec Trench, who did indeed prove that fish are restricted to a maximum depth of ~8500 m.

Furthermore, it has been shown that Chondrichthyes (sharks, rays and chimaeras) are limited to bathyal depths, as predicted by linear regression of global database records for fish (Priede *et al.*, 2006b). They proposed that chondrichthyans are excluded from abyssal and hadal depths due to their high-energy demands, which are required to maintain, among other things, an oil-rich liver for buoyancy, which cannot be sustained in extreme oligotrophic conditions. However, Laxson *et al.* (2011) analysed the major organic osmolytes of 13 chondrichthyan species, caught between 50 and 2850 m. While the urea concentration of these chondrichthyans declined with depth, TMAO content increased from 85–168 mmol kg in the shallowest group to 250–289 mmol kg in the deeper group and a plateau was predicted at the greatest depths, suggesting that the deepest chondrichthyans may be unable to accumulate sufficient TMAO content to counteract the perturbing effects of hydrostatic pressure at upper abyssal depths (3000–4000 m; Laxson *et al.*, 2011). Although Chondrichthyes are not present at hadal depths, these findings provide further support for the TMAO hypothesis that fish cannot exceed 8500 m.

Part IV

Patterns and current perspectives

Introduction

Trying to put the trenches and the hadal community into some form of coherent ecological context is very difficult for many reasons. Beyond what we already know about deep-sea biology and ecology in general (e.g. Gage and Tyler, 1991; Herring, 2002) there is a historical lack of systematic and comprehensive sampling campaigns at hadal depths, and on issues specific to organisms at very high pressure. The problem stems from a historical mismatch of opportunity and method. The 1950s provided a rare opportunity to compile the best marine scientists of the age and embark on lengthy and expensive voyages to sample as many trenches as possible from around the world. Such investments in sea-going activity are no longer commonplace despite the fact we now have advanced technology and methodology, and a much better understanding of statistical methodology and sampling design. Playing 'catch up' with regards to narrowing the gap in knowledge between the hadal zone and other marine environments is set to be a long drawn out game, despite the recent resurgence in interest.

In an ideal world, this book would end on an all encompassing and in-depth chapter on the ecology of the hadal zone, but in reality this is still a long way off. We do, however, have a good feel for many of the ecological trends that are likely to drive what we observe at the great depths. For example, the trench fore arcs are likely to host chemosynthetic communities, the extent and importance of which are still unknown (Blankenship-Williams and Levin, 2009). There is likely to be a significant effect of food accumulating along the trench axes but this still lacks unequivocal evidence, despite multiple anecdotal or limited observations (Danovaro et al., 2003; Glud et al., 2013). Hadal speciation and endemism is likely driven by combinations of hydrostatic pressure, topography, isolation and disturbance but these have yet to be disentangled across multiple taxa (Wolff, 1960, 1970; Belyaev, 1989). Similarly, the trench communities may differ as a result of proximity to land mass, latitude, individual trench age, inter-trench connectivity, or frequency and magnitude of seismic activity (Belyaev, 1989; Oguri et al., 2013; Watling et al., 2013), which are all likely but yet to be confidently proven. Other more rudimentary ecological trends such as the examination of species diversity with depth or with area, or the effects of transition zones between the trenches and neighbouring abyssal plains are currently tantalising (Jamieson et al., 2011a; Fujii et al., 2013; Kitahashi et al., 2013) but are severely hampered by a lack of consistent and quantitative sampling. While all of the above have some degree of historical data on which to build on in the future, there has not been nor are there any immediate plans to perform long-term monitoring of these ecosystems. This is worrying on two counts. First, the trenches are isolated ecosystems inhabited by largely endemic communities; therefore, large changes or perturbations could have a significant effect on the entire trench (Angel, 1982). Second, we live in an age where climatic changes are ever more evident and the effects are being witnessed at ever-increasing depths (Ruhl and Smith, 2004; Ruhl *et al.*, 2008; Smith *et al.*, 2008). The hadal zone is one of the few marine ecosystems whereby there are no historical long-term datasets on which to steer conservation initiatives or environmental policy in the future.

Chapter 11 documents many of these issues in the hope of providing a baseline on which to build upon as hadal sampling increases. Many of the trends and relationships described hereafter are derived from individual groups, individual species, individual trenches and so on. Therefore, they do not necessarily represent an all encompassing trend in 'hadal ecology' per se. Furthermore, studies examining bathymetric, topographic, area- or depth-related effects on community structure are derived from similarly limited sources, albeit the best sources that are currently available.

On a more contemporary note, we are at a turning point in hadal science (Jamieson and Fujii, 2011) and Chapter 12 explores current perspectives in regards to exploitation and conservation, anthropocentric interactions with the physical instabilities of the trenches and the public perceptions of what they, and the life therein, represent. In an era of responsible exploitation for the biotechnology industry and the quest for understanding life on Earth at its most extreme, the trenches can offer positives to mankind beyond simple curiosity and fascination of the unknown. This perhaps might go some way in counteracting the notoriously negative aspects of trenches such as earthquakes. If future scientific challenges are overcome, we stand a better chance of providing sufficient stewardship of the oceans, thus addressing the shortfall in ecological understanding and long-term sustainability of the deepest parts of the oceans.

11 Ecology and evolution

11.1 Antiquity

The antiquity or age of the hadal community has been a contentious subject and under discussion since the first hadal organisms were recovered (Belyaev, 1989). To understand the invasion of the trenches both the historical oceanography and geology of the deep sea and trenches must be considered. The history of the physical deep-sea environment is characterised by extreme variability in temperature, oxygen and circulation (McClain and Hardy, 2010). The deep sea used to be much warmer than it is today as it has cooled by approximately 14–15°C since the Eocene/Palaeocene boundary (55 Ma), following minor warming in the Late Cretaceous and a similar cool period at the Eocene/Oligocene boundary (34 Ma; Waelbroeck *et al.*, 2001). Furthermore, deep-ocean circulation has alternated between two types of ocean; one driven by high-latitude deep water formation (thermohaline, THC), and one driven by salinity-induced stratification at low latitudes (halothermal, HTC). The former resulted in cold, oxygenated deep water and the latter in warm, saline deep water which reduced global circulation (Rogers, 2000; McClain and Hardy, 2010). THC conditions have existed since the Eocene–Oligocene transition, and the HTC conditions occurred back to the Triassic. During this period, deep-water anoxic events were both frequent and extensive (Jacobs and Lindberg, 1998; Rogers, 2000; Waelbroeck *et al.*, 2001; Takashima *et al.*, 2006), with the most severe events associated with rapid THC–HTC transitions in the mid-Cretaceous, and at the Permian/Triassic and Ordovician/Silurian boundaries (Horne, 1999).

Belyaev (1989) details the arguments on the origin and age of deep-sea and trench fauna from around the 1950s and highlights that there was no consensus between leading experts at the time. The arguments fell into two categories: 'the deep-sea fauna should mainly be considered young' (e.g. Bruun, 1956b) or should be considered as 'refuges that have maintained slightly altered, ancient, archaic forms' (e.g. Zenkevitch and Birstein, 1953) of which both arguments were considered as opposite and mutually exclusive. It is, however, now thought that both arguments are partially right and that the origins of the deep-sea fauna are centred around the 'extinction and replacement' hypothesis where the large, widespread anoxic events led to *near*-complete extinctions followed by the radiation of shallow-water species into deeper waters (Rogers, 2000). This suggests that many deep-sea clades are relatively young, dating back to the Eocene–Oligocene boundary, while many clades had in fact survived the anoxic events

(Raupach *et al.*, 2009; McClain and Hardy, 2010). Vulnerable taxa may have been wiped out by these catastrophic anoxic events producing a more resistant (eurythermic and eurybathic; Wolff, 1960) deep-sea community that may have also encouraged allopatric speciation rather than complete extinction (Wilson, 1999; Rogers, 2000).

These two groups can be categorised into 'ancient' and 'secondary' fauna (Andriashev, 1953; Zenkevitch and Birstein, 1953). The ancient and secondary forms differ in respect to their vertical distribution. In most secondary deep-sea fauna the number of species rapidly decreases with depth whereas in ancient forms the number of species increases with depth in the bathyal zone and decreases towards the abyssal–hadal boundary. Examples of this are the secondary Sipunculoidea and the ancient Siboglinidae (formerly Pogonophora) (Zenkevitch and Birstein, 1953), the ancient Neotanidae and the secondary isopod genera *Macrostylis* and *Storthyngura* (Wolff, 1960) and fish families such as the ancient Macrouridae and Ophidiidae and the secondary Zoarcidae and Liparidae (Andriashev, 1953). Furthermore, Andriashev (1953) states that the secondary fish species are more stenobathic than the relatively eurybathic ancient species, and in the latter, there is a tendency for panoceanic horizontal ranges.

Emiliani (1961) found that the surface waters at high latitude and abyssal–hadal waters were around 14°C in the Upper Cretaceous and had since cooled by 12°C during the last 75 million years. The effect on the abyssal and hadal community may have been significant, although probably very gradual. It was in fact the work of Emiliani (1961) that Bruun (1956b) and Wolff (1960) cited as suggestive evidence that Tertiary abyssal fauna may have been largely killed by the temperature change, and hence most modern abyssal and hadal fauna are relatively young.

Belyaev (1989) concluded that the majority of available hadal data indicated that the modern trench communities originate from groups (at the taxon rank of families and orders) that were formed no earlier than the Mesozoic (251–65.5 Ma), and in many cases, in the Cenozoic (65.5 Ma to present) but also highlight that in the overwhelming majority of cases there is no paleontological chronicle to trace and examine the evolutionary timelines of the trench fauna. This suggests that the hadal community is largely of the secondary invasion whereas he also acknowledges that the wider deep-sea communities largely originate in the Paleozoic (i.e. ancient). Belyaev (1989) also concluded that the abyssal fauna of the neighbouring regions to each trench is the likely source for settlement of the trenches as representatives of the ancient deep-sea groups that were formed and evolved in the abyssal zone should have played a significant role. This is thought to be a result of the lengthy existence at great depths under hydrostatic pressures of several hundred atmospheres, and thus the ancient abyssal fauna were pre-adapted to the colonisation of even greater depths. Wolff (1960) also reiterates this thought that the hadal fauna is derived from the abyssal zone in that endemic families, genera and some species may be considered as relics of a pre-glacial abyssal (and hadal) fauna. This sentiment is further supported by Belyaev (1989) in the idea that the subduction process converts the abyssal plains (and communities therein) into hadal trenches and communities over hundreds of thousands or even millions of years, the duration of which is comparable to the duration of species formation, and sometimes even evolutionary changes to higher taxonomic levels (genera, rarely families). Birstein (1958) noted that the abyssal and hadal Malacostraca are predominately of ancient origin with seemingly endemic ancient genera and families. Wolff (1960) pointed out that this is not in accordance with the thought that the trenches in the current formation are relatively young (formed in their modern form during the Cenozoic period) but could have, however, migrated from the abyssal zone after their formation followed by an extinction at abyssal depths.

11.2   Speciation and endemism

Allopatric speciation occurs as a result of geographic isolation between populations (Hoskin *et al.*, 2005). In areas or habitats with limited or an absence of gene flow, reproductive isolation arises gradually and incidentally as a result of mutation, genetic drift and the indirect effects of natural selection driving local adaptation (Dobzhansky, 1951). The hadal zone represents an ecosystem which should, in theory, promote allopatric speciation as each trench is isolated from one another, often over great distances and partitioned by very differing habitat type (abyssal plains). Since the early taxonomic work of the *Vitjaz* and *Galathea* expeditions, the hadal community as a whole has been frequently reported as being highly endemic, whereas diversity within each trench is relatively low. This is highly analogous with mountain or island bio-geography in terrestrial ecology or seamount and vent systems in marine ecology where an entire ecosystem is represented not by any large continuous habitat but rather by a myriad of individual habitats. With individual trenches being so deep and abruptly punctuating the otherwise vast open and flat abyssal plains the conditions are ideal for driving allopatric speciation.

Allopatric speciation is supported by the findings of different species of *Hirondellea* found in each trench. For example, *Hirondellea gigas* is the single dominant amphipod species at hadal depths in the northwest Pacific trenches (Kuril–Kamchatka, Japan, Izu-Bonin, Mariana, Yap, Palau and Philippine Trenches; Kamenskaya, 1981; France, 1994), *H. dubia* in the southwest Pacific trenches (Kermadec and Tonga Trenches; Dahl, 1959; Blankenship *et al.*, 2006) and three species of *Hirondellea* inhabit the Peru–Chile Trench (Perrone *et al.*, 2002; Fujii *et al.*, 2013; Kilgallen, in press). Furthermore, despite the amphipod *Eurythenes gryllus* being one of the most geographically cosmopolitan species in the deep sea it is known to be morphologically variant depending on habitat type (Ingram and Hessler, 1983; Thurston *et al.*, 2002). Moreover, the HADEEP sampling campaign in the Peru–Chile Trench revealed three distinct morphological forms (HADEEP, unpublished). The two most abundant forms were stratified between abyssal (4602 to 6173 m) and hadal (6173 to 8074 m) depths, whereas the rarest of the three were found between 4602 and 5329 m (Fig. 11.1). The morphological variation occurs in the structure of the pereonies and pleonites, the shape of coxa 2 and the first and second gnathopods.

Population genetic studies on *E. gryllus* from the central North Pacific basin has shown that populations are stratified by depth; in this case one population spread across

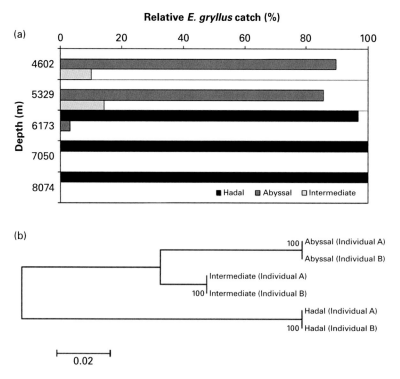

Figure 11.1 (a) Percentage of each morphological form of *Eurythenes gryllus* as a percentage of total *E. gyllus* catch at each depth. (b) Maximum likelihood phylogeny of six *E. gryllus* individuals from hadal, abyssal and intermediate depths, based upon DNA sequence differences across 243 base pairs (b.p.) of the mitochondrial 16S ribosomal RNA locus. Values at nodes represent bootstrap support indices (based upon 1000 iterations).

the summits of guyots and another throughout the adjoining basins (Bucklin *et al.*, 1987). In the Peru–Chile Trench, the 'abyssal' and 'hadal' forms are two distinctively different populations as both are found to contain males, females and juveniles spanning all instars. The abyssal populations were found to be genetically similar to other abyssal populations of the Pacific at similar depths (HADEEP, unpublished). However, the hadal populations were found to be quite divergent, even more so than the guyot population described by Bucklin *et al.* (1987). The abyssal samples were taken in the Milne-Edwards sector of the Peru–Chile Trench and the hadal samples were taken from both the Milne-Edwards and Atacama sectors and therefore it appears that topography does not solely drive allopatric speciation but depth (or rather pressure) is a significant driver.

France and Kocher (1996) discussed the importance of temperature as a controlling ecological factor in the deep sea (with references to Wilson and Hessler, 1987; France, 1994), and to *E. gryllus* in particular, whereby once isolated, such a selective regime may lead to genetic differentiation and speciation (Palumbi, 1994). When comparing bathyal to abyssal populations, typical bottom temperatures are sufficiently distinct to prompt such conclusions, however, such comparisons between abyssal and hadal populations do not. The temperatures experienced at abyssal depths are very similar to those from underlying trenches, in fact the minimum bottom temperatures are typically found at ~4500 m, thereafter the temperature rises and at full ocean depth, it equals those in the upper abyssal and lower bathyal (Jamieson *et al.*, 2010); a result of adiabatic heating (Bryden, 1973). Therefore, the temperature experienced by the abyssal and hadal populations of *E. gryllus* is extremely similar. Oxygen as a controlling factor has also been proposed as an ecological factor promoting allopatric speciation (White, 1987; France and Kocher, 1996). However, like the case of temperature, the oxygen regimes exhibited in the bathyal zone contrast with the abyssal zone, whereas there is no evidence to suggest any such differences in oxygen concentration between abyssal and hadal depths (Belyaev, 1989). Salinity can also be discounted on the grounds it does not fluctuate significantly beyond ~3000 m. The presence of the third divergent *E. gryllus* form, found at the abyssal depths of the Milne-Edwards sector and found only in the lower instars remains subject to further study, but the important issue arising in the case of *E. gryllus* is that allopatric speciation appears to occur as a result of depth alone and not distance isolation.

The story is further complicated in that *E. gryllus* is thought to be stenothermic, favouring colder waters (Thurston *et al.*, 2002). Whilst the central North Pacific and Peru–Chile Trench basin studies show three distinct vertically stratified populations (guyots/seamounts, abyssal plains and hadal trenches), there has yet be a finding of *E. gryllus* in the slightly colder Kermadec and Tonga Trench beyond 6200 m (Blankenship *et al.*, 2006; Jamieson *et al.*, 2011a). This may be explained by the conclusion of Fujii *et al.* (2013) that pressure alone does not necessarily account for these bathymetric patterns and that food supply also contributes significantly (the Kermadec and Tonga Trenches are far more oligotrophic that the Peru–Chile Trench).

Holothurians of the genus *Elpidia* (Elasipodida) are typical and often highly abundant representatives of trench communities. Gebruk and Rogacheva (unpublished data) examined the phylogeny within the genus *Elpidia* based on a matrix comprising 20 morphological characters coded for 22 terminal taxa, including 21 species of *Elpidia* and the outgroup taxon, *Psychroplanes convex*. A strict consensus tree was obtained for four equally parsimonious trees (length = 33 steps; Fig. 11.2).

Their results show that all species endemic to trenches, group into a well-derived clade on the phylogenetic tree. Within this clade, supported by one clear apomorphy, the species *E. kurilensis*, *E. birsteini* and *E. longicirrata* group together confirming the assumption of Belyaev (1975) who designated several species groups in the genus based mainly on the spicule morphology. This group is the strongest supported on the consensus tree after the clade including the Arctic species of *Elpidia*. At the same time the species *E. hanseni* also occurring in the Kuril–Kamchatka and the Izu-Bonin Trenches remains outside the northwest Pacific trench clade. Another trench clade was formed by the species *E. ninae* and *E. lata*, both of which are known from the South Sandwich Trench. These results demonstrated that the trench species in the genus *Elpidia* are mostly derived morphologically, and evolutionary relationships among species occurring in the same or nearby trenches can be different: some species are closely related whilst others evolve separately.

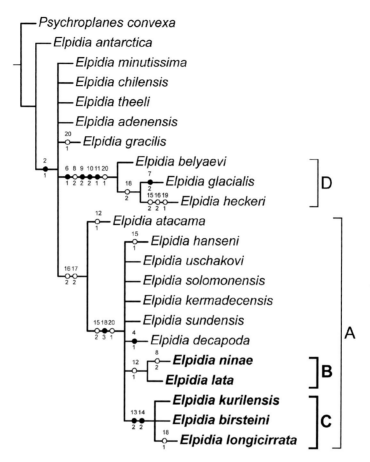

Figure 11.2 Phylogenetic tree of *Elpidia* based on 20 morphological characters. Clade A – trench species; B – South Sandwich Trench species; C – species from the northwest Pacific trenches; D – Arctic species. Numbers correspond to coded characters. Unique characters shown in black circles. Figure courtesy of Andrey Gebruk, P.P. Shirshov Institute of Oceanology, Russia.

Species endemism at hadal depths was estimated by Belyaev (1989) at 56.4% with the hado-pelagic fauna being somewhat lower, averaging 41% (Table 11.1). Of the 56% of endemic species, 95% of those were only found in one trench and of the non-endemic species, 22% were found at abyssal depths in the vicinity of the trench, suggesting the trench fauna originated from the surrounding abyssal province (as suggested by Wolff, 1960). Endemism at genus level was found to be far lower at just 10%. As perhaps expected, the degree of endemism for all benthic organisms increases with depth and is found to be lowest across the abyssal–hadal boundary (6000–7000 m), where a higher proportion of characteristic abyssal species also inhabit. The greatest percentage of endemic species were found at the very deepest points and thus restricted to some of the deepest trenches that exceed 10 000 m (Mariana, Tonga and Philippine). Endemism at these depths ranges from 86 to 100%, although the other two deep trenches (Kuril–Kamchatka and Kermadec) have only 50 and 59% endemism, respectively.

Table 11.1 Percentage of species endemism in each of the trenches and troughs as estimated by Belyaev (1989).

Trench	% endemism	Trench	% endemism
Aleutian	42	Tonga	100
Kuril–Kamchatka	50	Kermadec	59
Japan	53	Peru–Chile	23–50*
Izu-Bonin	48	Banda	43
Volcano	54	Hjort	20
Mariana	100	Java	71
Yap	81	South Sandwich	37
Palau	77	Romanche	60
Philippine	86	Puerto-Rico	50
Ryukyu	72	Cayman	47
Bougainville	71	Pacific Troughs	28
New Hebrides	60	Atlantic Troughs	20
Total			**56.4**

* Indicates values for the northern (Milne-Edwards) and Southern (Atacama) sectors, respectively.

Table 11.2 Summary of the types of dissemination of hadal species based on the *Galathea* and *Vitjaz* expeditions, derived from Belyaev (1989).

Type of dissemination	Percentage of hadal species (%)
Endemic hadal species	56.4
Found only in one trench	4.7
Found in two or multiple but neighbouring trenches	6.4
Found in two or multiple but distant trenches	3
Exclusive to oceanic region in vicinity of trench(es)	22
Known from several regions on one ocean	4.6
Known from two oceans	11
Known from three oceans	6

Belyaev (1989) calculated endemism beyond depth >6000 m in order to broaden the perspective. He calculated that only 4.7% of hadal species were found solely in one trench, 6.4% of species were found in multiple but neighbouring trenches and 3% were known from two or more (but distant) trenches (Table 11.2). A much higher percentage of species, 22%, are found exclusively in one oceanic region in the vicinity of trenches, which in turn prompted the classification of hadal provinces by UNESCO (2009) and Watling *et al.* (2013). These values are likely to change in the future as more sampling is undertaken, particularly in the lesser known or studied trenches. Perhaps an even more enlightening avenue of research would concentrate on sampling the adjacent abyssal plains or continental slopes and rises as well as the trenches and in doing so, greater connectivity between species may be further detected in time.

Belyaev (1989) also indicates a remarkable difference between the trenches at temperate and tropical latitudes (Fig. 11.3). Trenches from temperate latitudes in the

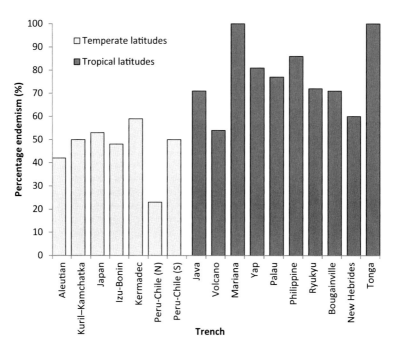

Figure 11.3 Percentage of species endemism in trenches at tropical and temperate latitudes.

Pacific Ocean (Aleutian, Kuril–Kamchatka, Japan, Izu-Bonin, Kermadec and Peru–Chile) showed a far lower degree of endemism than the trenches in Pacific tropical latitudes (Volcano, Mariana, Yap, Palau, Ryukyu, Philippine, Bougainville, New Hebrides and Tonga), with endemism estimated at 42–59% and 54–100%, respectively. These estimates did, however, include the non-Pacific Java Trench and omitted the tropical Banda Trench on account of their apparently young age and thus their relatively low degree of endemism (43%). Despite noticing this stark latitudinal trend, Belyaev was unable to explain why this would be the case. In other more isolated trenches such as the South Sandwich Trench, low degrees of endemism (37%) were thought to be explained by the cold sub-Antarctic locality, known to be inhabited by otherwise deeper species at uncharacteristically shallower depths.

The degree of endemism at species level somewhat reflects the maximum depth at which the group has been found (Table 11.3). For example, the highest percentage of endemism in the benthic fauna is found in the Isopoda, Amphipoda, Gastropoda, Bivalvia and Holothuroidea, all of which are known to inhabit the entire bathymetric range of the hadal zone and exceed 60% endemism. Other groups, such as the Echuiroidea and Tanaidacea, are not as common at the deeper sites and are 30% and 40% endemic, respectively. Polychaeta endemism is also lower (40%), despite their existence at full ocean depth. This may be a result of the fact that they are less likely to be recovered in trawls compared to the other groups. At a genus level, gastropod endemism is slightly lower, around 10–15% for most, although 26% of the 41 Gastropoda genera are endemic.

Table 11.3 Percentage endemism at species and genus level for most major hadal groups, based on *Vitjaz* and *Galathea* expedition data, derived from Belyaev (1989).

Group	Species		Genera	
	Total	% endemic	Total	% endemic
Polychaeta	73	40	50	14
Echuiroidea	13	30	10	0
Tanaidacea	63	40	15	7
Isopoda	122	63	34	9
Amphipoda	35	78	28	11
Gastropoda	56	68	41	26
Bivalvia	47	68	33	15
Holothuroidea	56	69	20	10
Other Echinodermata	53	49	30	7
Siboglinidae	29	76	10	10
Other groups	123	43	93	9
Total	**660**	**56.4**	**364**	**10**

11.3   Community structure

The decrease in species diversity with depth is a well-known trend throughout the deep-sea environment and was also shown to extend to the hadal zone from the earliest investigations into diversity (Wolff, 1960). Understanding the ecological transition of faunal distributions within and between other depth-related habitats, such as the bathyal and abyssal zones, has been discussed in great detail, yet the hadal zone is often omitted from consideration, despite accounting for 45% of the ocean depth range (e.g. Carney, 2005; Rex *et al.*, 2005; Smith *et al.*, 2008; Levin and Dayton, 2009). This omission has provided the impetus for the investigation of individual trenches based not only on observations made from within the trench itself, but also of the transition of faunal distributions from the neighbouring abyssal plains into the trenches (Jamieson *et al.*, 2011a). By examining the abyssal–hadal transition zone, linkages between trench fauna and neighbouring abyssal communities can determine whether the hadal zone comprises an ecocline or ecotone (van der Maarel, 1990). An ecotone represents a narrow transition (an abrupt or rapid distinction of species composition) between two defined habitats; an ecocline refers to a broader area of transition across an environmental gradient (Jenik, 1992; Attrill and Rundle, 2002), in this case, depth.

Using necrophagous amphipods as the target species, Jamieson *et al.* (2011a) statistically examined the compositional change of species across the abyssal–hadal boundary of the Kermadec Trench and found an ecotone between depths of <6007 m and >6890 m, indicating that there is an ecologically distinct, bait-attending fauna within this trench. The exact causes of this shift in composition were unclear as pressure increases monotonically and thus seems unlikely to prompt a sudden shift in the community. One possible explanation for the pattern observed could be related to the

Table 11.4 Deployment data, environmental data and amphipod diversity data of the HADEEP Peru–Chile Trench baited trap sampling. Percentage of catch is shown in parentheses. Modified from Fujii *et al.* (2013).

	Depth (m)				
	4602	5329	6173	7050	8074
Station	SO209/11	SO209/03	SO209/19	SO209/35	SO209/48
Date	030910	010910	050910	100910	130910
Latitude	06° 12.42' S	04° 27.02' S	07° 48.04' S	17° 25.47' S	23° 22.47' S
Longitude	81° 40.13' W	81° 54.72' W	81° 17.01' W	73° 37.01' W	71° 19.97' W
Bottom time (hh:mm)	20:26	11:09	18:40	22:51	20:25
Temperature (°C)	1.80	1.87	1.98	2.07	2.25
Salinity (ppt)	34.69	34.69	34.69	34.69	34.69
Amphipod species					
Abyssorchomene chevreuxi	313 (45.7)	24 (10.5)	44 (33.3)	–	–
Abyssorchomene distinctus	34 (5.0)	1 (0.4)	–	–	–
Eurythenes gryllus	254 (37.1)	21 (9.2)	32 (24.2)	261 (80.6)	54 (32.2)
Paralicella caperesca	72 (10.5)	174 (76.3)	43 (32.6)	–	–
Paralicella tenuipes	–	5 (2.2)	7 (5.3)	14 (4.3)	–
Tectovalopsis sp. (nov.?)	1 (0.1)	–	–	–	–
Valettietta sp.	11(1.6)	–	–	–	–
Princaxelia sp. (nov. ?)	–	3 (1.3)	–	–	–
Hirondellea sp. 1	–	–	4 (3.0)	–	–
Hirondellea sp. 2	–	–	2 (1.5)	15 (4.6)	104 (62.3)
Hirondellea sp. 3	–	–	–	33 (10.2)	–
Tryphosella sp.	–	–	–	1 (0.3)	–
aff. *Pseudorchomene* sp. nov.	–	–	–	–	9 (5.4)
Number of individuals	**685**	**228**	**132**	**324**	**167**
Number of species	**6**	**6**	**6**	**5**	**3**

obvious change in seafloor topography that occurs at ~6400 m. This depth juncture marks the separation between samples taken from a relatively shallow, sloping, plain-type seafloor to those taken on the slopes of the trench. It is, therefore, probable that the deposition of food within these two habitats is quite distinct or possibly that there are physiological issues regarding routine pressure changes of species that inhabit flat or sloping habitats. Furthermore, if stark changes in topography drive the changes in community composition, then composition is likely to vary between trenches, where the change may be shallower or deeper depending on the trench.

To investigate similarities in community structure between trenches that are distinctly isolated from one another Fujii *et al.* (2013), once again, studied necrophagous amphipod samples taken from baited traps from five stations across the abyssal and hadal zones of the Peru–Chile Trench (southeast Pacific Ocean) and from seven in the Kermadec Trench (southwest Pacific Ocean) from depths of 4602–8074 m and 4329–7966 m, respectively (Tables 11.4 and 11.5).

These data were combined to investigate the species composition and structure of the amphipod communities in the two South Pacific trenches, which are isolated by

Table 11.5 Deployment data, environmental data and amphipod diversity data of the HADEEP Kermadec Trench baited trap sampling. Percentage of catch is shown in parentheses. Modified from Fujii *et al.* (2013).

	Depth (m)						
	4329	5173	6000	6007	6890	7561	7966
Station	K0910–8	K0910–2	K0910–6	KT1a	KT2a	K0910–7	KT3a
Latitude	36° 45.31' S	36° 31.02' S	36° 10.07' S	26° 43.94' S	26° 48.73' S	35° 45.10' S	26° 54.96' S
Longitude	179° 11.52' W	179° 12.03' W	179° 00.27' W	175° 11.33' W	175° 18.10' W	178° 52.55' W	175° 30.73' W
Bottom time (hh:mm)	12:10	09:30	12:41	17:28	12:16	13:33	46:57
Temperature (°C)	1.06	1.09	1.17	1.16	1.31	1.40	1.46
Salinity (ppt)	34.70	34.69	34.69	–	–	34.69	–
Amphipod species							
Paralicella tenuipes	1 (4.5)	18 (2.6)	–	–	–	–	–
Paralicella caperesca	12 (54.5)	620 (88.3)	5 (22.7)	78 (4.9)	–	–	–
Cyclocaris tahitensis	–	–	–	2 (0.1)	–	–	–
Eurythenes gryllus	3 (13.6)	7 (1.0)	1 (4.5)	2 (0.1)	–	–	–
Rhachotropis sp.	–	4 (0.6)	–	–	–	–	–
Hirondellea dubia	–	–	2 (9.1)	2 (0.1)	127 (92.7)	279 (99.6)	361 (100)
Paracallisoma sp.	1 (4.5)	1 (0.1)	–	–	–	–	–
Scopelocheirus schellenbergi	–	–	–	1 (0.1)	10 (7.3)	–	–
Abyssorchomene chevreuxi	–	13 (1.9)	–	–	–	–	–
Abyssorchomene distinctus	2 (9.1)	1 (0.1)	–	–	–	–	–
Abyssorchomene musculosus	3 (13.6)	1 (0.1)	–	–	–	–	–
Orchomenella gerulicorbis	–	37 (5.3)	14 (63.6)	1471 (93.3)	–	1 (0.4)	–
Tryphosella sp.	–	–	–	1 (0.1)	–	–	–
Valettietta anacantha	–	–	–	20 (1.3)	–	–	–
Number of individuals	**22**	**702**	**22**	**1577**	**137**	**280**	**361**
Number of species	**6**	**9**	**4**	**8**	**2**	**2**	**1**

10 000 km of abyssal plain. Amphipod diversity was found to decrease significantly with increasing depth across all the sampling stations, but the structure diverged markedly between the two hadal trench communities. Four distinctive community groups were identified (Fig. 11.4) and their relationships were examined using six environmental variables (latitude, longitude, hydrostatic pressure, primary productivity, temperature, sediment characteristics), of which depth (hydrostatic pressure) and longitudinal (geographic isolation) gradients were found to best explain the observed trends. The composition of the abyssal community was dominated by typical cosmopolitan

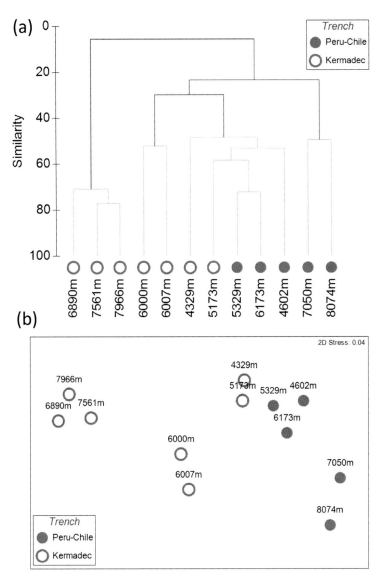

Figure 11.4 Analysis of hadal amphipod species composition from the Kermadec and Peru–Chile Trenches. (a) Dendrogram based on fourth-root transformation, Bray–Curtis similarity and group-average clustering. Four main groups of sites (black branches) were identified based on the similarity profile (SIMPROF) permutation test (p < 0.05). (b) Non-metric multi-dimensional scaling (nMDS) ordination, based on fourth-root transformation and Bray–Curtis similarity. Modified from Fujii *et al.* (2013).

species belonging to the genera *Paralicella*, *Abyssorchomene* and *Eurythenes* (Schulenberger and Hessler, 1974; Dahl, 1979; Thurston, 1990). The abyssal group had a relatively high degree of similarity irrespective of location, suggesting a high degree of connectivity across the vast stretches of the Pacific abyssal plain that connects the two trenches. There appeared to be very few physical impediments to dispersal between the two regions at these depths, since the abyssal areas are almost continuous along the longitudinal gradient between the edges of the Peru–Chile and Kermadec Trenches.

At the deeper hadal sites, there was no similarity between the communities from each trench, suggesting that the deeper trench fauna are either physically isolated and/or the environmental conditions experienced within each trench are sufficiently varied to account for the faunal differences. The hadal Kermadec Trench sites (6890–7966 m) were dominated by *Hirondellea dubia* and the hadal Peru–Chile Trench sites (7050–8074 m) were characterised by *E. gryllus* and three undescribed *Hirondellea* species. The shift from these abyssal genera to the hadal genus *Hirondellea* is also typical of other trench environments (Hessler *et al.*, 1978; Blankenship *et al.*, 2006; Jamieson *et al.*, 2011a).

Fujii *et al.* (2013) concluded that the overlying surface productivities at the trenches studied were in stark contrast to one another; the long-term averages of surface primary production rates at the Peru–Chile Trench ranged from 859.4 to 2144.5 mg $C\,m^{-2}\,d^{-1}$, compared to 261.5 to 554.4 mg $C\,m^{-2}\,d^{-1}$ in the Kermadec Trench. It was concluded, therefore, that the environmental forcing exerted on the amphipod community structure by the pressure and longitudinal gradients was likely to be further exacerbated by the supply of surface-derived food to the trenches. In order to truly test the effects of isolation, further comprehensive studies will be required from other vastly isolated trenches underlying various environmental settings, as well as from trench clusters at a provincial level for comparison. The combined vertical and horizontal isolation is likely to result in allopatric speciation (France and Kocher, 1996; Doebeli and Dieckmann, 2003).

The community structure of another hadal group, the harpacticoid copepods, has also been examined in the context of shifts from bathyal to hadal depths in the Kuril–Kamchatka Trench (Kitahashi *et al.*, 2013) and in comparison between the Kuril–Kamchatka and Ryukyu Trenches (Kitahashi *et al.*, 2012).

The former study focused on the genus diversity and community composition of the harpacticoids and their relationship with environmental factors across the large depth range of 490 to 7090 m. Using a sediment corer, 15 stations that culminated in two hadal stations were sampled down the Kuril–Kamchatka Trench slope, east of Hokkaido. The density of copepods did not decrease with depth and peaked at 1000 m, whereas, copepod diversity showed a unimodal pattern with depth, with peaks at intermediate depths (Fig. 11.5). Kitahashi *et al.* (2013) concluded that the general relationship between depth and diversity described for macro- and megafauna could be extended to meiofauna across all depth ranges. They did not, however, identify a regulating factor, such as food availability, for the observed patterns of diversity.

Across bathyal to hadal depths, the community composition was found to change gradually. Furthermore, comparison of the assemblages between the abyssal plain (5570 m), trench slope (1060–5730 m) and trench floor (7000 m and 7090 m) suggested that the trench floor or 'hadal' community was quite different from those found on the trench slope and abyssal plain; the dissimilarity values between the trench floor and the other stations were considerably larger than those between the three slope zones (Fig. 11.6). Statistical analyses suggested that depth, or certain factors associated with

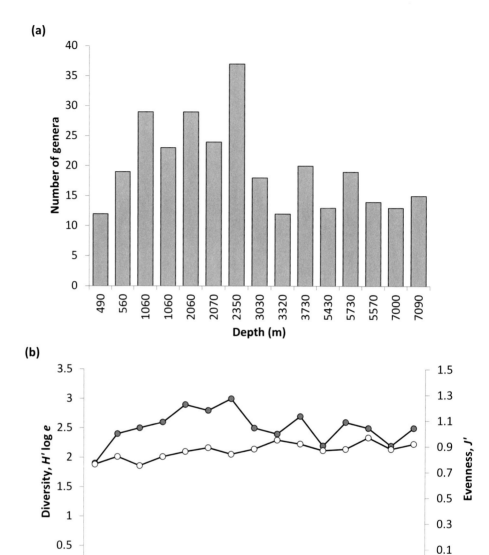

Figure 11.5 (a) Number of genera of harpacticoid copepods across the bathyal to hadal depth range (490–7090 m) and (b) Shannon–Wiener diversity index (H' log e) and evenness (J'). Derived from Kitahashi *et al.* (2013).

depth such as food availability and/or seasonal variation thereof, affects harpacticoid assemblages in and around the Kuril–Kamchatka Trench.

In the Ryukyu Trench, south of Japan, Kitahashi *et al.* (2012) found high average dissimilarities in the harpacticoid assemblages inhabiting the abyssal plain (4910–5710 m),

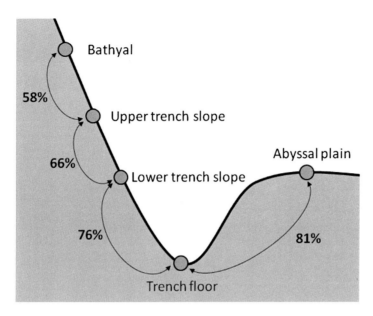

Figure 11.6 Dissimilarity values in harpacticoid copepod community composition between topographical settings in the Kuril–Kamchatka Trench (490–7090 m). Modified from Kitahashi *et al.* (2013).

trench slope (1290–5330 m) and trench floor (6340–7150 m), indicating that the assemblage structures differ substantially between these topographic settings at the family level. In comparison with the Kuril–Kamchatka Trench, they suggested that the hadal harpacticoid assemblage reflects a transition zone between the slope and the abyssal plain in this region and that the composition of the assemblages was influenced by the quantity of organic matter in and around the Ryukyu Trench, while sediment properties play a key role in and around the Kuril–Kamchatka Trench. Direct comparisons of the respective assemblages revealed that the average dissimilarities between the two trenches and between the two surrounding abyssal plains were higher than those between the adjacent slopes (Fig. 11.7). This result suggests that connectivity between regions is difficult for benthic organisms such as the harpacticoids and this is probably due to the presence of topographical barriers around trenches.

11.4 Vertical zonation

Although the total number of hadal species decreases with depth and the percentage of endemic species increases with depth, there are very few species that span the entire bathymetric range of the hadal zone; 0.5% or six species. Based on the ~1200 species listed in the HADEEP hadal species database, the largest percentage depth range is <100 m (38.8%). This is almost certainly because in some cases, individuals were caught in a single trawl that happened to span a few tens of metres in depth (in addition,

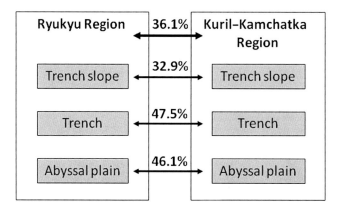

Figure 11.7 Average dissimilarity values in harpacticoid copepod community composition between the regions in and around the Ryukyu and Kuril–Kamchatka Trench (490–7090 m). Modified from Kitahashi *et al.* (2012).

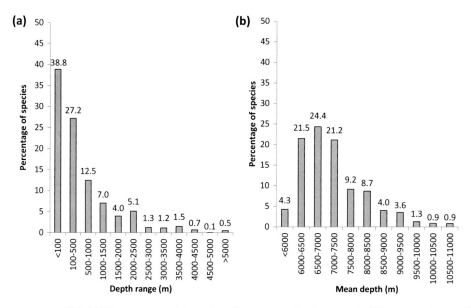

Figure 11.8 (a) The percentage of species with apparent depth ranges in 500 m strata. The <100 category is likely to be solitary or rare findings from trawl that hauled a few tens of metres depth. (b) The mean depth of all hadal species. Data are derived from ~1200 records in the HADEEP database.

rare species sampled from a single locality were not included, i.e. depth range = 0 m). Beyond this, there is an exponential decrease in the percentage of species with increasing depth ranges (Fig. 11.8). However, the mean depth of species is highest between 6000 and 7500 m and it decreases with increasing depth thereafter. The number of species with a mean depth greater than 9000 m is low probably because so few trenches reach this depth and there are a far greater number of trenches with bottom depths of between 6000 and 7500 m (almost all). This result could also reflect the fact that a combination of these shallower depths account for the largest surface area of the trenches, coupled with the abyssal–hadal transition zone. The values for bathymetric ranges are potentially skewed by a group of 10 unidentified scyphozoans of the Coronatae family, whose ranges are recorded as 6000 to 10 000 m (mean depths = 8000 m, range = 4000 m). These range recordings are misleading and are likely to have resulted from the use of vertical plankton hauls that are unable to record the exact depth of capture and thus probably indicated an overly large range. The other, more confident large depth ranges (>4000 m) comprise mostly arthopods (three ostracods, three amphipods) and two holothurians. The largest recorded bathymetric ranges are for the two holothurians; *Peniagone azorica* (Elpidiidae) known from 2640–8300 m, with the deepest finding from the Kermadec Trench, and *Prototrochus bruuni* (Myriotrochidae) known from 6487–10 687 m in seven trenches but, as Belyaev (1989) suggested, the ranges of these species require revision.

The bathymetric trends at the species level, particularly at the deepest depths, are likely to change dramatically as more sampling is undertaken in the future. In areas such as the Mariana Trench, there is a propensity to target the deepest point, in this case Challenger Deep, thus data from the surrounding area are lacking. Furthermore, Belyaev (1989) highlights several species with extraordinary depth ranges, where the maximum and minimum are separated by broad hiatuses. On the back of this, he suggested that certain species may require a thorough systematic revision which could result in the finding that several species have more stenobathic ranges than previously thought. Furthermore, there are many species recorded that have not yet been described (Table 11.6) and this may or may not have altered the ranges of other known species.

The maximum depth of different classification groups known from the trenches also declines with depth and not just in terms of the number of species (Table 11.6; Fig. 11.9). There are, however, some groups that are frequently found that are well known with relatively robust identifications and that have been identified at full ocean depth. These are the amphipods, holothurians, isopods, gastropods and polychaetes. By examination of these groups' bathymetric ranges, it is quite apparent that the amphipods have consistently larger depth ranges than the other groups; nearly double. This may be because the amphipods are a bentho-pelagic and free-swimming group, while the others are truly benthic and, thus, are intrinsically linked to the seafloor. The larger depth range of theses amphipods does, however, prompt further questions as to why some groups are capable of such large bathymetric disseminations when others are either smaller, or limited to shallower depths. It has been shown that it is highly likely that fish are limited to depths of <8000 m by hydrostatic pressure-induced biochemical stress (Yancey *et al.*, in press). While this justification appears to explain the absence of fish and possibly decapods (Laxson *et al.*, 2011) from full ocean depth, it does not yet account for so many other species or classification groups that are not absent, nor does it explain why some, like the amphipods, are so extraordinarily pressure tolerant. As future

Table 11.6 Each phylum and class of hadal organism currently known to inhabit the hadal zone with the total number of known species (+ number of undescribed species or species not registered on WORMS) with the corresponding maximum known depth for the order. Given the large number of undescribed species, these species numbers should be interpreted as indicative only. Also, there are currently no listings for hadal Nematoda.

Phylum	Class	Number of species (+ number of unknown species)	Maximum known depth (m)
Foraminifera	Polythalamea	100 (+15)	10 924
Porifera	Hexactinellida	5 (+3)	8540
	Demospongiae	6 (+ 2)	9990
Cnidaria	Hydrozoa	5 (+ 8)	8700
	Scyphozoa	1 (+ 2)	10 000
	Anthozoa	13 (+ 5)	10 730
Annelidea	Polychaeta	122 (+ 42)	10 730
Echiura	Echiuroidea	13 (+ 2)	10 210
Mollusca	Gastropoda	60 (+ 25)	10 730
	Bivalvia	70 (+ 31)	10 730
	Scaphopoda	3 (+ 2)	7657
	Polyplacophora	2 (+ 1)	7657
	Monoplacophora	3 (+ 1)	6354
Sipuncula	Sipunculidea	7 (+ 1)	6860
Bryozoa	Gymnolaemata	0 (+ 2)	8830
Echinodermata	Crinoidea	0 (+ 8)	9735
	Asteroidea	17 (+ 7)	9990
	Ophiuroidea	16 (+ 8)	8662
	Holothuroidea	51 (+ 9)	10 730
	Echinoidea	8 (+ 2)	7340
Arthropoda	Copepoda	27 (+5)	10 000
	Cirripedia	6 (+3)	7880
	Ostracoda	9 (+5)	9500
	Mysidacea	2 (+10)	8720
	Cumacea	6 (+10)	8042
	Tanaidacea	63 (+10)	9174
	Isopoda	94 (+39)	10 730
	Amphipoda	63 (+14)	10 994
	Decapoda	1 (+1*)	7703
	Acariformes	1	6850
	Pantopoda	8 (+1)	7370

* Indicates species known only from *in situ* observations.

sampling is undertaken, the answers to questions like these will help to further explain the zonation and degree of eurybathy at hadal depths.

11.5   Relationships with area and depth

Applying ecological theory to trench communities beyond individual groups, such as the readily recoverable Amphipoda, is at best difficult, due to a lack of comprehensive sampling at a sufficiently high bathymetric resolution. However, there are enough

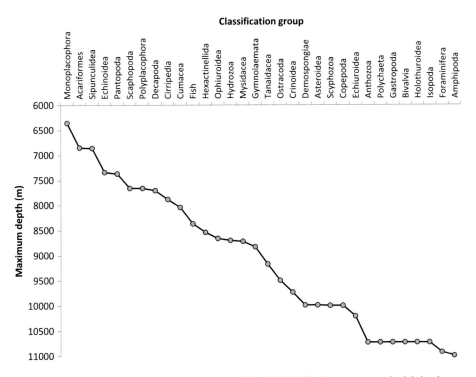

Figure 11.9 Decline in the maximum known depth for all classification groups at hadal depths (except bacteria and nematodes).

available data from some trenches to enable the investigation of some rudimentary ecological patterns. For example, in the Kermadec Trench, 14 baited camera deployments have been undertaken (6000–9281 m; unpublished data, HADEEP; Jamieson *et al.*, 2011a, 2013), 18 baited trap deployments (6097–9856 m; unpublished data, HADEEP; Blankenship *et al.*, 2006; Jamieson *et al.*, 2011a) and 12 trawls (5950–10 015 m; Belyaev, 1989). In addition, all previously known species in the area that were recorded from greater than 4000 m were summarised by Lörz *et al.* (2012), providing a total of 194 recorded species. Using GIS, the Kermadec Trench was stratified into 500 m depth bins and the horizontal abyssal boundaries set at the 4000 m contour on the west flank and the 6000 m contour on the east flank (to coincide with the abrupt change in slope). By combining the diversity data with the topography and bathymetric data, a simple species area relationship (SAR) was tested.

The SAR dictates that as habitat area increases, species diversity (or species richness) also increases (Arrhenius, 1921). This prediction implies that the largest habitats, i.e. the shallowest depth bin of the trench, should contain a higher diversity than smaller habitats, i.e. the deepest point. However, in the Kermadec Trench case study, the highest numbers of species were not sampled from the depth bin with the largest habitat area (6000–6500 m), but from 6500–7000 m (Fig. 11.10).

Currently there is mixed support for SAR where changes in habitat area occur along terrestrial elevational gradients (mountains); some studies have shown strong effects

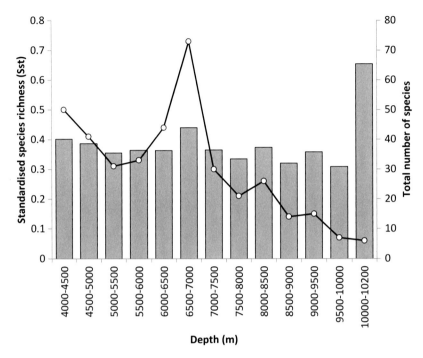

Figure 11.10 Species diversity as a function of depth (black line) and standardised species area relationship (grey bars). Data taken from 194 known species in the Kermadec Trench. The increase at 6500–7000 m shows potential mid-domain effect where abyssal and hadal communities overlap.

and others have shown evidence to the contrary (McCain, 2009, 2010). This lack of a consistent relationship is likely to be due, in part, to variations in the heterogeneity of the topography (the presence of plateaux, escarpments, ridges, etc.), as is also exhibited in trenches and that may occur with changes with elevation/depth. Habitat heterogeneity influences species diversity (the greater number and/or complexity of habitats, the more species a given area will support), and has been found to have a greater influence on diversity patterns than depth in submarine canyons (Schlacher *et al.*, 2007). So, in order to better resolve the diversity–depth relationship for the Kermadec Trench and to elucidate possible drivers for the pattern, the measures of diversity can be standardised using the simple power law commonly used to describe the relationship between species richness and habitat area ($Sst = \ln(S)/\ln(A)$, where S is the number of species and A is the habitat area for a particular depth bin; Connor and McCoy, 1979). Interestingly, standardised species richness does not decrease with increasing depth, rather, it is relatively similar across the bathymetric gradient, apart from a slight elevation for the 6500–7000 m depth bin (Fig. 11.10). The increased richness at 6500–7000 m could be explained by variations in topography (see above) that occur as the seafloor slope increases as the trench itself begins or by insufficiently replicated sampling. Alternatively, it might also be explained by the mid-domain effect (MDE), which assumes that spatial boundaries cause more overlap of species' ranges towards the centre of an area, where species with large- to medium-sized ranges must overlap (see Pineda, 1993). In the Kermadec Trench region, the MDE would, therefore, predict maximum diversity at the abyssal–hadal transition and/or at the midpoint depth of the hadal zone. There is, however, little support for MDE as a sole predictor along bathymetric gradients in the ocean, particularly among macro-invertebrates (McClain and Etter, 2005; Kendall and Haedrich, 2006). Although it is possible that the slight elevation in standardised species richness observed for the 6500–7000 m depth bin could reflect the MDE and represent further evidence of a distinct ecological boundary (an ecotone) between the abyssal and hadal fauna (Jamieson et al., 2011a). Ideally, future sampling in the trench would be conducted, at a fine scale, across the abyssal–hadal transition zone in order to confirm the presence of an ecotone and the possible influence of the MDE on the species diversity–depth relationship. Future sampling effort will also be required to disentangle the species–area relationship from the species–depth relationship, in order to clarify the potential drivers of the latter.

Another interesting point to highlight regarding vertical zonation is that although there are few species that span the entire hadal zone, the hadal zone is a collective of 47 individual habitats (27 subduction trenches, 6 trench faults and 13 troughs) that exceed 6500 m. Most of these areas are not 'full ocean depth', as only five trenches reach depths exceeding 10 000 m. Moreover, given that the area of the seafloor in a trench decreases drastically with depth, the ability to survive at depths of over 10 000 m does not offer a significant increase in available habitat. This point can also be reversed to say that orders, genera or even species with a maximum depth that is far shallower than full ocean depth can still occupy vast majorities of a trench and in some cases, even the entire trench, if it is shallow enough. Again, using the Kermadec Trench as a case study, an examination of the maximum depth of each classification group known to inhabit the trench gives the impression, albeit correct, that only four groups can survive at the deepest point (Fig. 11.11). However, there are twice as many groups with depth ranges between 8000 and 8500 m; which equates to an inhabitable area of the trench of nearly 90%. This point highlights that there are many species or orders that will undoubtedly play a very significant role in the trench ecosystem, despite their absence from the very small but deepest point. One could then question the meaningfulness of studies that only sample the deepest point, as it is not only the smallest area but the area with the lowest diversity and thus the least representative area of a trench. It is analogous with the studies of mountain ecology, whereby investigations made solely on the summit would not provide a meaningful overview of the ecology of the mountain.

By examining the maximum depth of each of the 32 classification groups against the maximum depth of each trench, trench fault and trough, it becomes clear that many groups can span many hadal habitats in their entirety (Table 11.7). For example, the shallower trench faults and troughs can be entirely inhabited by 22 (69%) and 20 (63%) groups, respectively. Furthermore, 22 of these groups can fully extend to the bottom of over 50% of the subduction trenches. Taking into account the total 47 distinct hadal habitats, 24 groups can inhabit the entire depth range of more than 50% of them and

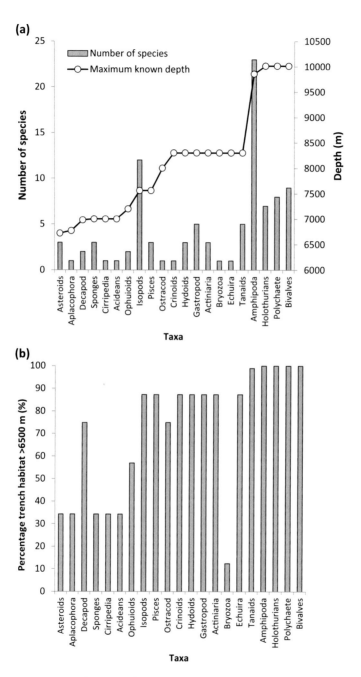

Figure 11.11 (a) Number of species and maximum know depth for each group in the Kermadec Trench. (b) The respective area of the Kermadec Trench potentially occupied by each group.

Table 11.7 The number of hadal trenches ($n = 27$), trench faults ($n = 6$) and troughs ($n = 13$) which each classification group can span the entire depth. Percentages in parentheses. The maximum depth shown is the maximum known depth and are thus in many cases likely to be higher.

Group	Maximum depth (m)	Trenches	Trench faults	Troughs	All
Monoplacophora	6354	1 (4)	0 (0)	0 (0)	1 (2)
Acariformes	6850	4 (15)	1 (17)	1 (8)	6 (13)
Sipunculidea	6860	5 (15)	2 (17)	6 (46)	11 (23)
Echinoidea	7340	8 (30)	3 (50)	9 (69)	20 (43)
Pantopoda	7370	8 (30)	4 (50)	10 (69)	21 (43)
Scaphopoda	7657	9 (33)	5 (50)	11 (69)	21 (45)
Polyplacophora	7657	10 (33)	6 (50)	12 (69)	22 (45)
Decapoda	7703	10 (37)	7 (50)	13 (69)	22 (47)
Cirripedia	7880	11 (37)	4 (67)	11 (85)	25 (53)
Cumacea	8042	12 (44)	5 (67)	12 (85)	27 (57)
Pisces	8370	14 (52)	6 (100)	12 (92)	32 (68)
Hexactinellida	8540	16 (59)	6 (100)	13 (92)	35 (74)
Ophiuroidea	8662	17 (63)	6 (100)	13 (100)	36 (77)
Hydrozoa	8700	17 (63)	6 (100)	13 (100)	36 (77)
Mysidacea	8720	17 (63)	6 (100)	13 (100)	36 (77)
Gymnolaemata	8830	18 (67)	6 (100)	13 (100)	37 (79)
Tanaidacea	9174	21 (78)	6 (100)	13 (100)	40 (85)
Ostracoda	9500	21 (78)	6 (100)	13 (100)	40 (85)
Crinoidea	9735	22 (81)	6 (100)	13 (100)	41 (87)
Demospongiae	9990	22 (81)	6 (100)	13 (100)	41 (87)
Asteroidea	9990	22 (81)	6 (100)	13 (100)	41 (87)
Scyphozoa	10000	22 (81)	6 (100)	13 (100)	41 (87)
Copepoda	10000	22 (81)	6 (100)	13 (100)	41 (87)
Echiuroidea	10210	23 (85)	6 (100)	13 (100)	42 (89)
Anthozoa	10730	25 (93)	6 (100)	13 (100)	42 (94)
Polychaeta	10730	25 (93)	6 (100)	13 (100)	42 (94)
Gastropoda	10730	25 (93)	6 (100)	13 (100)	42 (94)
Bivalvia	10730	25 (93)	6 (100)	13 (100)	42 (94)
Holothuroidea	10730	25 (93)	6 (100)	13 (100)	42 (94)
Isopoda	10730	25 (93)	6 (100)	13 (100)	42 (94)
Polythalamea	10924	27 (100)	6 (100)	13 (100)	47 (100)
Amphipoda	10994	27 (100)	6 (100)	13 (100)	47 (100)

20 groups can inhabit over 75% of them. Also, the maximum depths for each group are still maximum *known* depth and the values are, therefore, likely to be higher.

Access to the deepest point in a trench is significant if the trench resource accumulation hypothesis is correct (Jamieson *et al.*, 2010). If indeed there is an enhanced volume of food availability at the deepest point on the trench axis (as suggested by Glud *et al.*, 2013), then in many trenches, this enrichment is available to a wider range of organisms than those that are typically reported from the deepest places in the deepest trenches. This, in turn, would alter any generalisation regarding the community structure across the bathymetric range of the 'hadal zone' and again favour a focus on individual trench ecology.

11.6   Habitat heterogeneity

In 1966, Belyaev suggested dividing the hadal zone into three subzones: (1) the upper, 6000–7000 m (the abyssal and hadal transition zone), (2) the middle, 7000–8500 m and (3) the lower; greater than 8500 m. He also stressed that the boundaries of these subzones are, to a certain degree, conditional depending on trench identity. From the more recent studies, it appears that these suggestions are highly valid and bring into question how best to subcategorise the trenches into more meaningful and convenient habitats. The abyssal–hadal transition zone does appear to be very real, as indicated by Jamieson *et al.* (2011a) and Fujii *et al.* (2013). However, immediately beyond this transition zone is, perhaps, the most 'representative' trench community at 7000–8500 m as this encompasses the largest area in most of the trenches and is beyond the influence of the abyssal community. The lower hadal zone also seems relevant, as many studies suggest that the deepest points, the small and often unique 'ponds' are not necessarily representative of the wider trench habitat.

This conclusion comes 60 years on from the debate concerning the official coining of the term 'deeps' (Wiseman and Ovey, 1953). During the debate, the British National Committee on Ocean Bottom Features came to the conclusion that the deepest point of a trench or a 'deep' was defined from morphological standpoints and thus such an area was relatively unimportant and should remain unnamed, and the term 'deep' should fall into abeyance (Wiseman and Ovey, 1954). However, the exploration of the Challenger Deep and other 'deeps' within the Mariana Trench shortly after showed that the 'deeps' are often highly unique, important and well defined, warranting at least an ecological subzone to differentiate between these areas and the wider trench environment.

There are, however, more factors to consider when classifying the hadal zone beyond simple 'trenches, trench faults and troughs' or indeed solely by community-based depth strata. The internal heterogeneity of the trenches should also be considered. For example, the trenches comprise two slopes: the continental slope, or fore arc, and the oceanic slope. These two slopes are likely to differ greatly, as the fore arc is, theoretically, an area that hosts many chemosynthetic seep sites (Blankenship-Williams and Levin, 2009). The geological nature of the fore arc is such that the presence of seep sites is highly likely and supported by various observations (Fujikura *et al.*, 1999; Fujiwara *et al.*, 2001). Furthermore, the geology of this area results in much steeper slopes, often with rocky outcrops, escarpments and steep walls. These types of substrata and seafloor conditions, combined with an increased chemosynthetic-based community may well host a community that differs greatly from those at equal depths on the soft sediment of the oceanic slopes.

This entire scenario is complicated further when adding the influence of food supply. The surface-derived input of POM descends into the trench regardless of depth, slope or seafloor setting. If the trenches do accumulate this food towards the axis, then the effects of the community should differ by proximity to the axis, regardless of the slope or depth that they inhabit. The quantity of food will also depend on the biogeographic province (Longhurst *et al.*, 1995), or 'hadal province' (Watling *et al.*, 2013) or proximity to land mass. Other additional factors such as the quantity and type of plant and wood debris input (Bruun, 1957) that can support specific groups (Wolff, 1976), or even trenches that underlie significant migration routes of large marine mammals enhancing the probability of whale-falls and associated fauna (e.g. Smith and Baco, 2003; Glover *et al.*, 2013) may also turn out to be relevant. The complexity does not end there. Further variation may occur on larger scales such as the temperate-tropical latitude variation noted by Belyaev (1989), the age of isolation of the trenches or the degree of partitioning of trenches in close geographic proximity.

12 Current perspectives

The interesting history of discovery, geological aspects, environmental characteristics, ecology and biology associated with the deepest ecosystem on Earth have been acknowledged by scientists for some time and will continue to affect humankind in the future, in both positive and negative ways. For example, the very nature of the ultra-deep regions of the oceans once fuelled the perception that these trenches would make ideal dumping grounds for pharmaceutical and radioactive waste products (Osterberg *et al.*, 1963; Peele *et al.*, 1981; Lee and Arnold, 1983). Thankfully, this idea is no longer commonplace and has been completely overruled by the contemporary perspectives on conservation and sustainability that now apply to all marine environments.

On a grander scale, recent hypotheses suggest that the subduction process is, in part, involved in the disposal of atmospheric carbon, some of which is derived from human activities (Nozaki and Ohta, 1993). While the significance of this contribution to the planet remains unresolved, the plate subduction process impacts upon the everyday life of the human populations living in close proximity to trenches, often in highly negative ways. Ultimately, trench sites are the origin of the devastating earthquakes and tsunamis that have the power to deliver mass death and destruction without warning.

On a more positive note, one of the truly enriching aspects of the hadal zone lies in its status as one of the final frontiers; places on our planet that are still waiting to be explored. Curiosity is an intrinsic part of our human nature and manifests itself in our desire to explore the unknown and in our quest for knowledge concerning the new and fascinating organisms with which we share this planet. One of the members of the 2012 *Deepsea Challenger* expedition to the Mariana Trench, Dr Joe MacInnis, summed up humankind's desire to explain the unexplained in his statement, 'Exploration is a force that gives us meaning. It is driven by our curiosity to know what lies beyond the horizon.'

From a less romantic point of view, we do live in an era where we face an uncertain climatic future and the significance of the role that the oceans play in regulating our climate has never been in doubt. Even before the environmental challenges that face us today were realised, John F. Kennedy predicted in 1961, 'knowledge of the oceans is more than a matter of curiosity. Our very survival may hinge upon it.' In recent years we have seen an ever-increasing body of evidence arise showing that nowhere in the ocean, including the hadal zone, is exempt from a varying or changing climate. Perhaps even more worrying is the fact that the hadal zone represents a large marine ecosystem that we still know very little about and, subsequently, we have no wealth of historical information from which to draw conclusions or make projections for future policy. Therefore, rapid exploration and understanding of the trenches is an ever more pertinent, yet still fascinating, undertaking.

12.1   Exploitation and conservation

Litter, perhaps, has the most conspicuous negative impact on the marine environment. Highly durable and slowly degrading plastic discards are particularly responsible and, unfortunately, they account for the large majority of marine litter (Laist, 1987; Spengler and Costa, 2008). The annual global production of plastic products is estimated at 230 million tons, of which >10% ends up in the oceans (Thompson, 2006). The disposal of solid waste at sea was prohibited in 1988 (Annexe V, MARPOL Convention), yet there are still many instances where even remote environments, such as the deep sea, are contaminated with litter (Galgani *et al.*, 2000; Barnes, 2002; Bergmann and Klages, 2012). Some of the deepest litter ever encountered was observed at 7216 m in the Ryukyu Trench off Japan (Ramirez-Llodra *et al.*, 2011). During the 2009 HROV *Nereus* dive to Challenger Deep, an anecdotal report of a raincoat on the seafloor at over 10 900 m was circulated (Lee, 2012). These examples highlight that the presence of litter in our seas is a full ocean depth problem. Although some forms of litter are recognisable, e.g. bottles and bags, there is mounting evidence that describes a drastic decrease in the size of the litter found. The larger items degrade or erode into 'mermaids' tears' which are approximately 5 mm in diameter and smaller still 'microplastics' (sand grain-sized particles). Both of these types of degraded litter are becoming more evident, even in deep-sea environments (Ramirez-Llodra *et al.*, 2011; Bergmann and Klages, 2012).

There are certain topographical settings such as submarine canyons and troughs that have a tendency to accumulate litter objects ('debris traps') (Galgani *et al.*, 2000). Therefore, the trenches, by their very nature, are likely to accumulate and trap any debris that descends into them. In addition, the problem is exacerbated because the trenches tend to mirror the coastline of continental land masses where the litter originates and, unlike submarine canyons, trenches are closed systems so any trapped items cannot be flushed out into the neighbouring abyssal plain and dispersed. To date, there have been no studies undertaken relating to the presence of litter at hadal depths. However, it is likely that litter-focused investigations will be incorporated into future deep-submersible work as common practices since it can be said with some certainty that the hadal zone is not exempt from the scourge of discarded plastic.

Perhaps the most significant example of irresponsible introduction of anthropogenic material into the hadal zone was the dumping of pharmaceutical waste in the 1970s (mainly antibiotics). During this time, the Puerto-Rico Trench in the North Atlantic was one of the main waste disposal sites because the Puerto Rico government gave tax advantages to the pharmaceutical industry. The subsequent waste material was dumped in the trench approximately 40 miles north of the island, at a depth of 6000 m (Simpson *et al.*, 1981). The figures are astonishing; between 1973 and 1978, >387 000 tons of waste material was discarded in the trench; this is equivalent to 880 Boeing 747s (Ramirez-Llodra *et al.*, 2011). The particular pharmaceutical waste products that were dumped were shown to be acutely toxic to many marine invertebrates (Nicol *et al.*, 1978).

By the early 1980s, this dumping practice ceased but research undertaken at the disposal site found self-evident changes in the marine microbial community (Peele *et al.*, 1981). The abundance of some previously common bacteria, such as *Pseudomonas* spp., decreased drastically at the disposal site over a 3-year study, while others, such as *Staphylococcus*, increased in numbers (Grimes *et al.*, 1984). Furthermore, larger organisms such amphipods (*Ampithoe valida*) experienced chronic toxicity in response to the waste material (Lee and Arnold, 1983).

Around the same era, it was suggested in all seriousness that the trenches could be a suitable dumping ground for nuclear waste. The 'out of sight, out of mind' perception of trenches prompted the emergence of several key papers that highlighted the unsuitability of trenches for such purposes. Angel (1982), in a paper entitled 'Ocean trench conservation', pointed out the simple fact that the isolated and apparently highly endemic trench communities are enclosed and that even a demonstration exercise gone awry could have detrimental effects on an entire trench community. Furthermore, so little was known about the hadal communities at the time that monitoring any of the effects of waste disposal was futile. Yayanos and Nevenzel (1978) reported that not everything out of sight is out of mind forever. By analysing the lipids of various hadal amphipods they showed that a mass contamination-induced extinction would cause contaminated particles to rise to surface waters by binding to the lipids within dead crustaceans. They estimated that, potentially, it would take a dead and ascending amphipod crustacean 1 week or certainly less than 1 year to reach the sea surface from 5000 m below sea level. This 'rising particle hypothesis' may explain how the largest ever specimen of the 'supergiant' deep-sea amphipod, *Alicella gigantea*, was recovered from the regurgitated stomach contents of an albatross in Hawai'i (Harrison *et al.*, 1983; Barnard and Ingram, 1986).

Aside from the direct dumping of nuclear waste, contamination from nuclear weapons testing is also evident in the deep-sea floor, in the form of radionuclides (Tyler, 1995). Radioactive elements from weapons testing were found in deposit-feeding holothurians at depths of 5000 m (Osterberg *et al.*, 1963), presumably accumulated from surface-derived phytodetritus; the process through which this occurred was not known at the time.

Although the industrial practice of dumping nuclear waste products in the trenches never actually materialised, there is a solitary case where it may have occurred. Onboard the ill-fated *Apollo 13*, during its mission to the moon in 1970 was a SNAP-27 radioisotope thermoelectric generator (RTG) that was supposed to remain on the moon after the mission. The RTG contained approximately 3.9 kg of plutonium 238. Once the mission had been aborted, the plutonium was brought back to Earth. The lunar module was purposely burnt up in the Earth's atmosphere and the RTG was deliberately jettisoned over the southwest Pacific where it reportedly survived re-entry and allegedly landed in the Tonga Trench at a depth of 6000–9000 m. Here, the RTG should remain radioactive for several thousand years, however, atmospheric and oceanic monitoring showed no evidence of a release of nuclear fuel (Furlong and Wahlquist, 1999).

A more recent study of trench sediment at 7261 and 7553 m in the Japan Trench revealed that 4 months after the 2011 Tōhoku-Oki earthquake, surface sediment was found to contain ^{134}Cs from the Fukushima Dai-ichi nuclear disaster. The radioactive ^{134}Cs was believed to have been rapidly deposited in the sediment by a concurrent spring bloom of phytoplankton, supplemented by successive sediment disturbances (Oguri et al., 2013). The sinking rate of the phytodetritus and radioactive material was estimated at 78 and 64 m per day and this was comparable to the speed at which fallout material reached the bottom of the Black Sea following the Chernobyl accident (Buesseler et al., 1990). These studies highlight very clearly that contamination from human activities can reach some of the deepest trench communities extremely quickly and, as such, these communities are not exempt from such disasters.

The efforts to protect the trench environment in the 1980s were a direct result of various plans to use them as dumping grounds, which, with the exception of the Puerto-Rico Trench, were implemented before any major detrimental activities took place. The conservation efforts were simply preventative and it was not until January 2009 that the first trench was officially declared 'a marine reserve' by former US President George W. Bush who created the Mariana Trench Marine National Monument (MTMNM, Presidential Proclamation 8335; Tosatto, 2009; Fig. 12.1). It became the largest marine reserve under the authority of the Antiquities Act of 1906, which protects areas of historic or scientific significance. The MTMNM consists of approximately 95 216 square miles of submerged lands and waters of the Mariana Archipelago and it is managed in three units: the 'islands unit', the waters and submerged lands of the three northernmost Mariana Islands; the 'volcanic unit', the submerged lands within 1 nm of 21 designated volcanic sites; and the 'trench unit', the submerged lands extending from the northern limit of the Exclusive Economic Zone (EEZ) of the USA in the Common-wealth of the Northern Mariana Islands (CNMI) to the southern limit of the EEZ of the USA in the Territory of Guam. Interestingly, the US EEZ in the area encompasses almost all the Mariana Trench but not its southernmost tip, where Challenger Deep is located.

The trenches are now considered to be unique areas where negative practices, such as dumping of waste products, should not occur. Furthermore, due to their great depth, industrial extraction of hydrocarbons or mineral resources from trenches is not com-mercially viable and it seems likely that the hadal zone is, for the foreseeable future, exempt from direct anthropogenic impacts. However, issues regarding a changing climate do pose an issue for the entire marine environment, including the hadal zone. The trenches, like all deep-sea environments, are intrinsically linked to the surface waters through the vertical POC flux. Climate change and other human activities, such as ocean fertilisation, will alter the patterns of this surface-derived input of food to the deep sea (Smith et al., 2006). Changes in this downward injection of organic matter will substantially alter the structure, function and biodiversity of the trench ecosystems, and, thus, it is imperative that global assessments of the environmental impacts of global

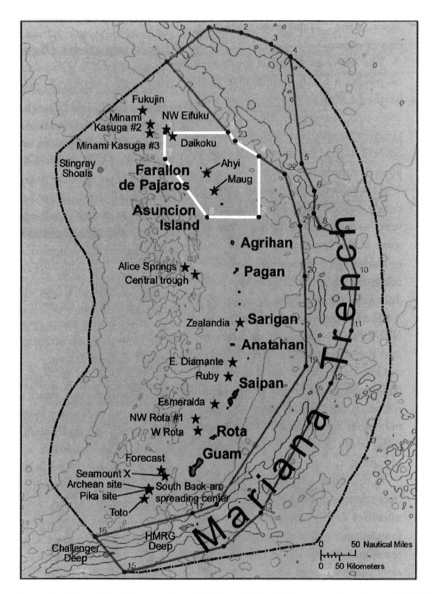

Figure 12.1 Map of the Mariana Trench Marine National Monument showing the EEZ (dashed line), the island unit boundary (white line), the trench unit boundary (solid grey line) and active hydrothermal vents (stars). Image courtesy of Samantha Brooke, NOAA Fisheries Marine National Monuments Program.

warming and ocean fertilisation account for the entire marine environment including the deepest points (Smith et al., 2008).

It has been speculated that the effects of rising atmospheric pCO_2 and climate change will affect the deep seafloor (Ruhl and Smith, 2004; Smith et al., 2008). Warming trends in atmospheric and upper ocean temperatures (attributed to anthropogenic influence) and ocean stratification, combined with a reduction in upwelling, could potentially shift pelagic ecosystems from diatom- and large zooplankton-dominated assemblages with higher export efficiencies to picoplankton- and microzooplankton-dominated assemblages with lower export efficiencies (Smith *et al.*, 2008, 2009). Global warming has been predicted to intensify stratification and reduce vertical mixing which, in turn, will enhance variability in primary production and alter carbon export flux to the deep sea (Smith *et al.*, 2008). These shifts in surface community structure are likely to cause a decrease in overall primary production and thus reduce the efficiency of organic carbon export from the photosynthetically active euphotic zone to the deep sea, causing a substantial reduction in POC flux to the trenches and neighbouring abyssal plains. This process will have detrimental effects on processes such as sediment community oxygen consumption (SCOC), bioturbation intensities and biomass and body sizes of invertebrate taxa, among many others. Furthermore, a reduction in the quality of the POC that reaches the deep sea (in the fatty acid composition, for example, caused by the diatoms to picoplankton shift) will alter the nutritional quality of the food and may, in turn, alter the reproductive success of some species (e.g. Hudson *et al.*, 2004).

Global analysis of ocean temperatures has shown that deep water is warming at an alarming rate (Balmaseda *et al.*, 2013). Although volcanic eruptions and El Niño events were identified as sharp cooling events in the analysis, these events punctuate the trend of long-term ocean warming. While the upper 300 m ocean heat content (OHC) appears to have stabilised despite global ocean warming, the heat is being absorbed in the deeper ocean and it is thought that in the last decade, ~30% of the warming has occurred below 700 m.

The published effects of warming on deep-sea communities are not simply based on theoretical modelling because it has already been shown that broad, biogeographic patterns in abyssal macrofauna community structure can change over contemporary timescales with changes in sea surface conditions (Ruhl *et al.*, 2008). Results from a 10-year study in the abyssal northeast Pacific Ocean at 4100 m found that climate-driven variations in food availability were linked to total metazoan macrofauna abundance, phyla composition, rank-abundance distributions and remineralisation over seasonal and interannual scales. It is these apparent links between climate, the upper ocean and deep-sea biogeochemistry that highlight the worrying vulnerability of all deep-sea ecosystems including the trenches, and indicate that the ocean must be considered in its entirety when determining the long-term carbon storage capacity of the ocean and the effects of a changing climate.

Warming trends in atmospheric and surface layers have occurred over the past four decades (Smith *et al.*, 2009) and our understanding of these trends in the deep sea is fragmented, due to the limited number of coherent long-term monitoring campaigns required to address questions at sufficient scales to understand the effects of a changing climate (Ruhl *et al.*, 2011). Long-term datasets from the abyssal deep sea are particularly rare, and long-term data are only available from a few locations globally (Smith *et al.*, 2009). Equivalent datasets from the hadal zone are entirely nonexistent. Therefore, the long-term stability of these rare and potentially highly endemic, hadal communities requires urgent investigation.

With conservation initiatives now in place and the drive for a greater scientific presence in the hadal zone, one further potential impact on these largely pristine environments remains; the impact of scientific endeavour.

In the aftermath of the highly publicised *Deepsea Challenger* manned submersible dive to Challenger Deep in 2012, various comments and articles were circulated regarding the exciting technological advancements, the ever-improving submarine capabilities and the broadening scientific research capabilities that are currently paving the way for new and unprecedented access to the deepest places on Earth (e.g. Burton, 2012; Lutz and Falkowski, 2012). However, Hartmann and Levin (2012) highlighted that this all means that the deepest oceans are no longer beyond the reach of human activities. As an example, they reported on the finding of bovine DNA in the stomach contents of hadal amphipods from the Tonga Trench (Blankenship and Levin, 2007), which were presumably from ship galley discards. The finding of the raincoat at Challenger Deep (Lee, 2012) also highlights human proximity to the deep ocean. Furthermore, scientists have highlighted that from now and in the future, it is important to consider conserving these pristine environments whilst undertaking scientific endeavours. An ever-increasing presence of scientific instrumentation may well lead to an accumulation of accidentally lost equipment, jettisoned ballast weights and potentially the introduction of foreign bacteria with it. Such accumulations at a specific point of interest, such as at Challenger Deep, could, worryingly, become analogous with the anthropogenic debris currently residing at extraordinary high altitudes on Mount Everest (Karan and Cotton Mather, 1985; Panzeri *et al.*, 2013).

12.2   Living in the shadow of a trench

By far the greatest impact of the hadal zone on humankind is centred on the seismic instability of the trenches themselves. These instabilities present themselves in unpredictable and often devastating earthquakes, often followed by, perhaps even more devastating, tsunamis. Although the magnitude of an earthquake triggered in the subduction zones can be extraordinarily large, it is often the resulting tsunami that delivers the greatest damage and loss of life. In recent years, the devastation wreaked by seismic instability has been unforgettably demonstrated in the 2004 Sumatra–Andaman earthquake (Java Trench), the 2010 Chilean Couquenes earthquake (Peru–Chile Trench) and the 2011 Tōhoku-Oki (Japan Trench).

The Sumatra–Andaman earthquake occurred on 26 December 2004 off the west coast of Sumatra, with a magnitude of M_w 9.1–9.3 (Lay *et al.*, 2005). The earthquake was triggered at the Java Trench, where the Indian plate is subducted beneath the Burma plate. The resulting tsunami was devastating. It generated waves of up to 30 m high that swept across the Indian Ocean coastlines killing over 280 000 people in 14 countries. It was the third largest earthquake ever recorded and lasted for an exceptionally long 8.3–10 min.

On 27 February 2010, a large earthquake (M_w 8.8) occurred off the coast of central Chile, to the northeast of Concepción (Chile's second-largest city). At the time, this was

Table 12.1 Summary of every major earthquake involving multiple fatalities in Japan in the last 100 years. The average magnitude is M_w 7.5, total fatalities = 180 468, and * indicates a tsunami.

Date	Magnitude (M_w)	Name	Death toll
Sept. 1923	8.3	Great Kantō earthquake	142 800
Mar. 1927	7.6	Kita Tango earthquake	3020
Nov. 1930	7.3	North Izu earthquake	272
Mar. 1933	8.4	Sanriku earthquake	3000*
Sept. 1943	7.2	Tottori earthquake	1083
Dec. 1944	8.1	Tōnankai earthquake	1223*
Jan. 1945	6.8	Mikawa earthquake	1180
Dec. 1946	8.1	Nankaidō earthquake	1362
Jun. 1948	7.1	Fukui earthquake	3769
Mar. 1952	8.1	Hokkaidō earthquake	28
Jun. 1964	7.6	Niigata earthquake	26
May. 1968	8.2	Tokachi earthquake	52*
May. 1974	6.5	Izu Peninsula earthquake	25
Jun. 1978	7.7	Miyagi earthquake	28
Jul. 1993	7.7	Hokkaidō earthquake	202
Dec. 1994	7.7	Offshore Sanriku earthquake	3
Jan. 1995	7.2	Great Hansin earthquake	6434
Mar. 2001	6.7	Geiyo earthquake	2
Oct. 2005	6.9	Chūetsu earthquake	40*
Jul. 2007	6.6	Chūetsu offshore earthquake	11
Jun. 2008	6.9	Iwate-Miyage Nairiku earthquake	12
Mar. 2011	9.0	Tōhoku-Oki earthquake	15 883*
Apr. 2011	7.1	Miyage aftershock	4
Apr. 2011	7.1	Fukushima aftershock	6
Dec. 2012	7.3	Kamaishi earthquake	3

the fifth largest earthquake recorded worldwide since 1900 and included over 100 aftershocks of magnitude 5.0 or greater following the initial event (Beittel and Margesson, 2010). The earthquake and subsequent tsunami, which struck Chile's coast roughly 20 min after the earthquake, devastated many coastal areas. The official death toll was over 500 and it is though that around 200 000 homes were badly damaged or destroyed. Estimates suggest as many as 2 million people may have been affected by the earthquake.

The 2011 Tōhoku-Oki earthquake off Japan (M_w 9.0) is believed to have been caused by a fault rupture extending to a shallow part of the subduction zone at the Japan Trench, in the vicinity of 38.322° N 142.369° E (Fujiwara *et al.*, 2011). The event resulted in approximately 20 000 dead or missing people and the tsunami inundated a very large area of about 560 km², covering over 35 cities along the coast of northeast Japan (Ando *et al.*, 2012). The Japan Meteorological Agency observed 666 aftershocks that exceeded M_w 5.0 (Oguri *et al.*, 2013). Even a 'seismic-ready' nation like Japan was still devastated by this earthquake and tsunami, albeit an extraordinarily large one, but history has shown time and again that this region must be prepared for every seismic eventuality (Table 12.1).

Studies focusing on the reasons why so many people died in the Tōhoku-Oki event were held in the following months and concluded that: (1) The figures for the predicted earthquake magnitudes and hazards in northeastern Japan that had been assessed and publicised by the government were significantly smaller than the actual earthquake. (2) The first tsunami warnings were underestimated compared with the actual tsunami heights. (3) Previous overestimated tsunami height predictions influenced the behaviour of the residents. (4) Some local residents believed that with the presence of a breakwater only slight flooding would occur. (6) Many people did not understand how tsunamis are created and thus many did not make the connection or take appropriate action (Ando *et al.*, 2012). They concluded that many deaths resulted unnecessarily because current technology and earthquake science underestimated tsunami heights, warning systems failed and breakwaters were not strong or high enough. This is very surprising given that Japan has arguably the best early-warning systems and tsunami evacuation procedures in the world, yet globally, the loss of life from earthquakes continues to rise, despite increasingly sophisticated methods of estimating seismic risk (Bilham, 2013).

Globally, earthquakes with extraordinarily large death tolls have increased with world population and obey a nonstationary Poisson distribution where rate is proportional to population (Holzer and Savage, 2013). Despite more than a century of seismic-resistant engineering (Tobriner, 2006) and the increasingly sophisticated warning systems, the past decade has been the most fatal if the exceptional Shanxi earthquake in 1556 is ignored (Bilham, 2013). Holzer and Savage (2013) predict that the number of earthquakes with death tolls in the tens of thousands will increase in the twenty-first century to 8.7 ± 3.3 for earthquakes with fatalities of over $100\,000$ and 20.5 ± 4.3 for earthquakes fatalities of over $50\,000$ from 4 and 7, respectively, as observed in the twentieth century, if world population reaches 10.1 billion by 2100. They also estimate that global fatalities in the twenty-first century will be over 2.5 million if the average post-1900 death toll for catastrophic earthquakes (193 000) is assumed. Since 2000, earthquakes have claimed the lives of 630 000 people and the cumulative cost of recent earthquakes has exceeded $300 billion, largely due to reconstruction costs in the industrial nations (Bilham, 2013). An unprecedented death toll exceeding 1 million is now possible in a single earthquake, should it occur near one of the world's megacities (Bilham, 2009). A prime example of this risk would be the city of Tokyo, where the population of the metro area currently exceeds 35 million at a mean density of 2629 people per square kilometre. This risk is further exacerbated because Tokyo is situated at a triple plate junction (between the Japan, Izu-Bonin and Ryukyu Trenches). That being said, Tokyo does, however, have some of the world's most sophisticated seismic-resistant architecture.

Beyond the destruction of infrastructure and direct fatalities, the impact on humans in the aftermath of these events affects a whole manner of everyday things, from declines in health (Daito *et al.*, 2012), declines in house prices (Naoi *et al.*, 2009), to even a decline in the number of males born in close proximity to the areas worst affected by the disaster (Catalano *et al.*, 2013). The latter study found a 2.2% reduction in male births, thought to be a result of men producing less testosterone, reducing the quality of male sperm in a time of high stress.

The vulnerability of the nations that lie in close proximity to subduction zones is also highly variable. While developed nations, such as Japan, are seemingly very advanced when it comes to warning systems and evacuation procedures, many less developed nations lack systems, procedures and seismic-resistant infrastructure. Bilham (2013) attributed the pronounced loss of life in the developing nations to three factors: poverty, corruption and ignorance. Another reason for the vulnerability of some nations lies with the fact that seismic-resistant construction is inadvertently restricted to wealthy or civil segments of the community and is, therefore, either unobtainable or irrelevant to the most vulnerable segment of the public. Nations with coastal borders that lie in close proximity to the trenches, particularly those around the Pacific Rim, must be aware of the potential risks of seismic instability and the devastation that can ensue quickly and sometimes without warning. Living in the shadow of the hadal zone is a risky business.

Of course, not all earthquakes originate directly in hadal trenches, some originate from faulting, shallower subduction and seismic events at the plate interior, but a vast proportion of these statistics will be attributed to the devastating effects of hadal trenches. In fact, England and Jackson (2011) note that the largest earthquake-induced death tolls were not a result of the very high magnitude events ($M_w > 8$), but rather caused by relatively modest events ($Mw < 7.5$), specifically those originating in plate interiors. In recent years, the devastation caused by the larger ($M_w > 8$) plate boundary earthquakes (i.e. hadal trenches) has been expensive but has not resulted in significant loss of life, if it were not for the effects of the resulting tsunami; for example, the 2011 Tōhoku-Oki earthquake (Japan Trench) and Sumatra–Andaman earthquake or 'Boxing Day tsunami' (Java Trench).

In order to truly understand the energy involved in high-magnitude earthquakes it is important to consider the following: the Sumatra–Andaman earthquake caused a shift of mass and a sufficiently massive release of energy to slightly alter the Earth's rotation. Theoretical models suggest the earthquake shortened the length of a day by 2.68 micro-seconds, due to a decrease in the oblateness of the Earth (Cook-Anderson and Beasley, 2005). Similarly, popular media at the time reported that the Tōhoku-Oki earthquake had shifted the Earth's axis by estimates of between 10 and 25 cm, leading to the shortening of a day by 1.8 microseconds due to the redistribution of the Earth's mass.

12.3   Public perception

On a more positive note, humanity has always been fascinated by the animals that survive in extreme and hostile environments such as the trenches (Larsen and Shimomura, 2007a). The intrinsic curiosity of the public is now easily exploited using on-line analytical tools to examine website statistics. For example, the impact on human curiosity of events that have been reported in the media, such as those related to trenches, can be identified with keywords using websites such as Google Trends (www.google.co.uk/trends/). These web services can provide data on the relative search volume of particular words or phrases. The impacts of trench-related stories are particularly evident when examining the search volume patterns of the phrases

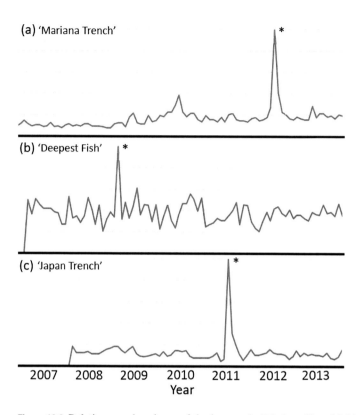

Figure 12.2 Relative search volume of the keywords 'Mariana Trench' (a), 'Deepest fish' (b) and 'Japan Trench' (c) from 2007 to 2013, derived from Google Trends. The peaks in each graph represent a media event associated with the hadal zone (marked *), where in (a) is James Cameron's dive to Challenger Deep in March 2012, (b) is the media release of the deepest fish ever filmed alive; 7703 m in the Japan Trench, October 2008 and (c) is in the aftermath of the Tōhoku-Oki earthquake in Japan, March 2011.

'Mariana Trench', 'Deepest fish' and 'Japan Trench' (Fig. 12.2). Large peaks in search patterns for these phrases are clearly seen during and in the aftermath of James Cameron's dive to Challenger Deep (March 2012), after the announcement of the deepest fish ever filmed alive (October 2008) and following the Tōhoku-Oki earthquake (March 2011). The public's general interest in the subject of trenches has also be monitored on other internet sites such as www.youtube.com, whereby the 2008 'deepest fish ever filmed alive' video received 5 179 620 views over 5 years. Also, the news item, 'Supergiant amphipod found in the Kermadec Trench at 7000 m' received 1 443 057 hits on the BBC news website (www.bbc.co.uk/news/) in the first 24 h, and over 2 million in the first 2 days. Furthermore, both of these stories, and James Cameron's Challenger Deep dive, received extraordinarily large international media coverage, which again is indicative of a real interest in trenches by the inquisitive general public.

There are also indicators that general interest in this subject goes beyond passive news browsing and that large fractions of internet users are searching on-line

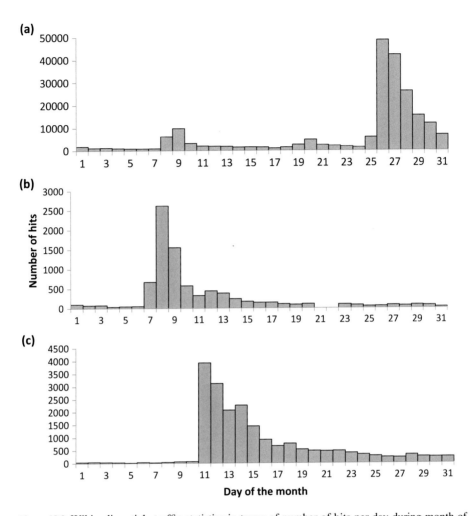

Figure 12.3 Wikipedia article traffic statistics in terms of number of hits per day during month of (a) James Cameron's Mariana Trench dive, 28 March 2012 (Wiki page: Mariana Trench), (b) 'Deepest fish ever seen alive', HADEEP media release, 8 October 2008 (Wiki page: Snailfish) and (c) the Japanese Tōhoku-Oki earthquake, 11 March 2011 (Wiki page: Japan Trench).

encyclopaedias, such as Wikipedia, to find out further information. Wikipedia article traffic statistics can be evaluated at www.stats.grok.se, and for the three stories mentioned above, similar peaks in activity are apparent following the news item or event (Fig. 12.3). The James Cameron dive saw searches for 'Mariana Trench' or 'Challenger Deep' soar from daily hits of 1000–3000 to over 48 000 on the day of the dive. Likewise, the deepest fish story saw the 'snailfish' page increase from approximately 100 hits per day to over 2600, and following the 2011 Japanese earthquake, the 'Japan Trench' page went from a daily hit rate of around 25–50 to nearly 4000. Other stories, such as the 'Supergiant amphipod in the Kermadec Trench' story, saw the 'amphipod' page's hit rate increase from 50–75 per day to over 1000. In the case of the supergiants or the deepest fish events, the encyclopaedia pages were not specific to the news article but are rather a generic entry for amphipods and snailfish, respectively. Hits on these pages indicated that the users' curiosity was sufficiently stimulated to actively seek more information around the general subject of trenches or trench fauna and, thus, the initial story promoted further education of the public on this particular subject. It can, therefore, be construed that the stories from extreme environments, such as from the hadal trenches, are sufficiently relevant and charismatic to incite curiosity in the general subject area and to stimulate a broader interest and independent learning by the general public.

While the response by the public to news stories regarding deep-sea exploration are generally overwhelmingly positive, there are always instances that prompt diverse opinions. A good example is James Cameron's Challenger Deep dive. In June 2013, *National Geographic* ran the 'New explorers' cover story, featuring Cameron himself on the cover. The main article was largely dedicated to his account of the *Deepsea Challenger* dive to the Mariana Trench. However, in the October 2013 edition, several contrasting letters from the public were published in response to the article. The first letter was from a man who, in 1967, was on the USS *Meeker County* transiting from Guam to Vietnam, and when told they were passing over the deepest place on Earth, he decided to throw a hammer over the side of the ship. This correspondence was written in jest and asked if Cameron had found it, and if so, could they return it to the US Navy. One should wonder why the first instinct in such a scenario should be a reckless disregard for such a pristine environment and whether the same reaction would occur in other terrestrial frontier environments? The second letter praises Cameron for all he has achieved in his career including, from *Deepsea Challenger*, 'finding previously unknown micro-organisms' that could 'shed light on the origin of life as we know it'. It is particularly worrying that someone would come to this conclusion based on the report of the dive, and shows that a high profile such as this can detract from years of scientific research. While the first *manned* mission to Challenger Deep did not contribute any scientific finds (Jamieson and Yancey, 2012), Cameron's solitary dive was also rather limited relative to many other studies but gave the impression that Challenger Deep was unexplored and poorly understood, despite the wealth of already published research from the same area spanning over 30 years, for example, Yayanos *et al.* (1981), Kato *et al.* (1997, 1998), Takami *et al.* (1997), Nogi and Kato (1999), Fang *et al.* (2000), Akimoto *et al.* (2001), Todo *et al.* (2005), Pathom-aree *et al.* (2006), Gooday *et al.* (2008), Kitazato *et al.* (2009) and Kobayashi *et al.* (2012), as well as four bottom trawls deeper than 10 000 m by the *Vitjaz* expeditions (Belyaev, 1989), multiple visitations by the ROV *Kaikō* (Takagawa *et al.*, 1997; Mikagawa and Aoki, 2001), HROV *Nereus* (Bowen *et al.*, 2009b), biogeochemical landers (Glud *et al.*, 2013) and oceanographic systems (Taira *et al.*, 2004, 2005). Furthermore, the biotechnology potential of natural products from Challenger Deep sediments are already being examined (Abdel-Mageed *et al.*, 2010). This highlights the danger that high-profile and high-impact events such as the *Deepsea Challenger* can overshadow and perhaps undermine the scientific efforts already achieved. The pursuit of sensational exploration stories does 'raise awareness of exploration', but there is a danger of hindering rather than responsibly promoting dissemination to the public about scientific fact, albeit often less publically digestible.

The third letter to *National Geographic* criticises the fact that the submersible had to leave 1072 lbs (486 kg) of steel ballast behind, and that Cameron 'should have left the trench as it was', whilst the fourth correspondence was tremendously enthusiastic and having read the article was left with a feeling of wanting more. These four correspondents represent diverse and mostly valid opinions and reactions to the same story, which is probably why this selection was picked for publication, but above all else, it is extremely encouraging to see people talking about the deepest places on Earth than not at all.

12.4   Life in extreme environments

Extreme environments are defined as 'environmental parameters showing values permanently close to the lower or upper limits known for life in its various forms' (Rothschild and Mancinelli, 2001; Amils *et al.*, 2007). The study of life in extreme environments is an area of science that has exploded during the past decade, with several reviews and books having been published on extremophiles (Madigan and Marrs, 1997; Horikoshi and Grant, 1998; Horikoshi *et al.*, 2011), and the launch of concerted funding programmes, such as the US National Science Foundation and NASA's programmes in Life in Extreme Environments, Exobiology and Astrobiology, and the European Union's Biotechnology of Extremophiles and Extremophiles as Cell Factories (Aguilar *et al.*, 1998). In 2003, the European Science Foundation (ESF) initiated a new research support activity, 'Investigating Life in Extreme Environments' (ILEE). The main conclusion that emerged from the first interdisciplinary workshop in 2005 was the need for a more coordinated approach to improve future opportunities for funding research in this area. This led to the 2008–10 CAREX project (Coordination Action for Research activities on life in EXtreme environments; Ellis-Evans and Walter, 2008). CAREX tackled the issue of enhancing coordination of 'life in extreme environments research' in Europe and included 60 European and non-European partners from 24 countries. Its approach to 'life in extreme environment research' covered microbes, plants and animals evolving in various marine, polar, terrestrial extreme environments as well as in outer space. Among many other marine and non-marine ecosystems, the hadal zone was selected by CAREX as a model environment against the specific feature of high pressure (CAREX, 2011). However, it should have also been placed under the specific features of 'oligotrophic' (as per seeps and vents), extreme variability (as were seamounts) and perhaps a new category of 'seismic instability'. The categories of the hadal extreme environment as defined by CAREX are shown in Table 12.2.

The study of life in extreme conditions, such as at high pressure, also has significance for understanding life in the emerging scientific area of the 'deep-biosphere' (Jørgensen and D'Hondt, 2006; Huber *et al.*, 2007). Deep-biosphere is the sub-seafloor habitat where prokaryote life has been found to survive hundreds of metres below the seafloor (mbsf). It has been suggested that the sub-seafloor biosphere may contain two-thirds of Earth's total prokaryotic biomass (Whitman, *et al.*, 1998), although these estimates are somewhat

Table 12.2 Selected model marine environments and their specific features as identified by CAREX (2011), including the hadal zone as contrasted against other marine environments. This has been adapted from CAREX (2011) by adding * where the trenches should be considered oligotrophic, especially as seeps and vents are already deemed so, **extreme variability is undoubtedly as high as seeps, vents and seamounts with respect to endemism, food supply, size and environmental setting and *** the addition of seismic instability as a specific feature (Oguri et al., 2013).

	Acidity	Alkalinity	Salinity	Anoxia/hypoxia	Extreme temperatures	Oligotrophic environments	High pressure	High radiation	Toxic compounds	Extreme variability	Irregular energy supply	Seismic instability
Cold seeps				X	X	X	X		X	X	X	
Hydrothermal vents	X	X		X	X	X	X		X	X	X	
Hypersaline lakes	X		X	X	X		X	X		X	?	
Seamounts										X	?	
Antarctic continental slopes					X							
Canyons							X			?	X	
Hadal trenches						*	X			**		***
Oxygen minimum zones				X						X		
Coastal saline lakes			X	X					X			
Inland saline lakes			X	X					X			
Intertidal zones				X					X		X	

contentious (Jørgensen, 2012). Roussel et al. (2008) provided evidence for living prokaryotic cells in sediments 1626 mbsf that were 46–111 million years old and living at 60° to 100°C. This study also suggested that Archaea, capable of anaerobic oxidation of methane, and novel members of the high-temperature Thermococcales (*Pyrococcus* and *Thermococcus*), can dominate deep and hot sediments where there are thermogenic energy sources. The extent of the deep sub-seafloor biosphere is still in debate, for example, Roussel et al. (2008) versus Hinrichs and Inagaki (2012), but the important point is that even in an age where technology has finally caught up with the desire to go deeper, we find that life exists even deeper still. If we continue to study the most extreme environments where life is sustained, then the more we will understand about life on our planet.

Understanding the evolution and existence of life under extreme environmental parameters will aid in determining the boundaries of where life can exist. This may lead to an understanding of organismal properties that have evolved under particular environmental stressors, and this may, in turn, aid our understanding of the ecology and evolution of life on Earth and potentially that of extraterrestrial life also (Allwood et al., 2013). The discovery of extreme environments and the organisms that inhabit them has made the search for extraterrestrial life more plausible (Rothschild and Mancinelli, 2001), and even advocated the possibility of panspermia (the transport of life from one planet to another; Nicholson et al., 2000; Wickramasinghe et al., 2013).

12.5   Bioprospecting and biotechnology

The discovery of extremophiles has also sparked great interest from the biotechnology industry. The marine environment is currently emerging as a hotbed of microbial diversity that has rarely been exploited for biotechnological gain (so-called 'blue biotechnology'; DeSilva, 2004), despite preliminary work showing huge potential (Aertsen et al., 2009; Blunt et al., 2009; Fang and Kato, 2010). Extreme deep-sea habitats such as the hadal trenches, as well as the polar regions, O_2 minimum zones and hypersaline pools are also considered to be likely repositories of previously unknown, novel biocompounds of potential importance in medicine and biotechnology (Rittman and McCarty, 2001). However, the commercial potential of bioprospecting in the *deep sea* has rarely been recognised, and is certainly far from being realised (Abe and Horikoshi, 2001; Haefner, 2003).

The ocean's biotechnological potential is only now being recognised, with the global market currently estimated at US$2.4 billion and with a predicted annual growth of 10% (Allen and Jaspars, 2009). The European Commission describes it as 'one of the most exciting technology sectors', and the Institute of Marine Engineering, Science, and Technology (IMarEST) describes the sea as a 'biotechnological frontier waiting to be explored' with 'potential for marine biotechnological products to be used as [among others] anticancer agents' (Anon, 2007).

The high-pressure and low-temperature environments within the hadal trenches are also likely to be repositories of novel biocompounds that are currently unknown to modern medicine. Recent advances in exploration and analytical techniques have discovered a myriad of bacterial communities that have evolved novel bioactive compounds through their physiological adaptations to environmental stressors. The properties and potential applications of these compounds have rarely been fully appreciated (Allen and Jaspars, 2009). Research on marine natural products (MNPs) has evolved into a multidisciplinary international collaboration between scientists from pharmacology, chemical ecology, biosynthesis, molecular biology, genomics, metabolomics, chemical biology and chemical genomics, under the umbrella terminology of 'bioprospecting'.

Although bioprospecting may be perceived at face value as deep-sea exploitation and thus has negative connotations, it is still in its infancy and there are opportunities to pursue the search for novel, natural products at hadal depths responsibly, opportunities that may herald many positive outcomes for human life and well-being. At present, there are about 20 (shallower) marine natural products in advanced clinical development as anticancer drugs and many more are under testing for the treatment of pain and various neurodegenerative disorders, as well as tuberculosis, AIDS, malaria and many other diseases (Mayer et al., 2010; Querellou et al., 2010).

All extremophiles such as thermophiles, psychrophiles, acidophiles, alkaliphiles and, of course, peizophiles have been, or are, of great potential interest to the biotechnology industry (Simonato et al., 2006). Possible biotechnology applications for piezophiles have proven slow to develop as a result of cultivation difficulties. A number of potential routes of exploration are described in Simonato *et al.* (2006), such as the ability of deep-sea bacteria to synthesise chemical compounds such as omega-3 polyunsaturated fatty acids. These PUFAs are considered useful in decreasing the risk of cardiovascular disease (Nichols *et al.*, 1993). Similarly, for the food industry, enzymes involved in the biosynthesis of these chemicals could be transferred into more suitable organisms in order to obtain increased quantities of these compounds.

The number of reported compounds that have been isolated from hadal trench organisms so far is <10 (Arnison *et al.*, 2013), and recently, 12 more compounds have been isolated from a pressure-tolerant bacterium found in Mariana Trench sediment (Abdel-Magreed *et al.*, 2010). Recent evidence shows that piezotolerant bacteria from Mariana Trench sediments produce biologically active and unusual secondary metabolites that have very strong anti-trypanosomal activity (M. Jaspars, unpublished). The metabolic activity of such deep-trench organisms was shown to change significantly under high hydrostatic pressures. The exploration of extreme environments with high hydrostatic pressures and low temperatures is likely to yield microorganisms from clades that are far removed from those found in other environments, thus increasing the possibility of discovering new chemical entities with potent and selective bioactivity.

12.6   Future challenges

Closer to home, our understanding of the distribution of marine biodiversity is a crucial first step towards the effective and sustainable management of marine ecosystems (Webb *et al.*, 2010). A marine conservation paradox currently exists: we need to assess and protect the marine environment with the utmost urgency, but we still know relatively little about it (Holt, 2010). This problem is particularly pertinent in the hadal zone, where fundamental understanding of its ecology is lagging far behind that of the more readily accessible coastal and shallower zones. 'Global marine conservation' must do just this; conserve the global marine environment which extends from the air–sea interface to the deepest ocean trench. In the past, ocean research has tended to focus on shallower habitats that were perceived to have a greater direct influence on day-to-day human endeavours. Sadly, there is still an anthropocentric opinion that the deep sea is a remote and enigmatic environment, far removed from everyday human activities.

To redress this issue, recent efforts by international networks such as the Census of Marine Life (www.coml.org) have substantially advanced our knowledge of the marine diversity of specific regions and habitats (Snelgrove, 2010). Yet, despite the 10-year long project, including projects on abyssal plains, in the Arctic, Antarctic, on continental margins and shelves, at coral reefs, mid-ocean ridges, seamounts, hydrothermal vents and seeps, it did not include the hadal trenches and, therefore, the knowledge gap between the trenches and the rest of the ocean is ever widening.

The research effort regarding the hadal zone has, however, increased in recent decades. An on-line search for peer-reviewed scientific papers on Thomson Reuters 'Web of Knowledge' journal search engine (http://wok.mimas.ac.uk/) with the term

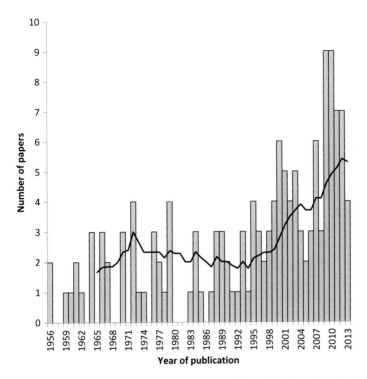

Figure 12.4 The number of peer-reviewed scientific publications found using the search word 'hadal' on Thomson Reuters 'Web of Knowledge' (accessed September 2013) from 1956 to 2013. The black line represents a moving 10-year average.

'hadal' listed 143 papers between 1956 and 2013 (Fig. 12.4). In the first 40 years (1956 to 1996), 61 papers were published on hadal biology (43% of the total). In the last 10 years (2002 to 2012), a further 58 'hadal' publications have emerged, accounting for 41% of all hadal papers ever published. The maximum number of hadal papers published was in 2009 and 2010 (nine papers each). While this increase in research activity is encouraging, it is still greatly lagging behind similar work in the shallower depths ranges, for example, the same search using the term 'abyssal' produces 1359 papers, 'hydrothermal vent' produces 1605 papers, 'seamounts' produces 1275 papers, and 'continental slope' and 'continental shelf' produces 2528 and 5026 papers, respectively. These values were derived in September 2013.

Webb *et al.* (2010) compiled a list of ~7 million georeferenced records of marine organisms, recorded in the Ocean Biogeographic Information System (OBIS) in order to provide a graphical summary of the three-dimensional distribution of recorded global marine biodiversity. This exercise demonstrated a clear under-sampling of deep oceans, and, in particular, the deep pelagic zone. This work highlighted the significance of under-sampling the deep pelagic ocean in terms of its extraordinary large volume and this concern became the focus of their conclusions. However, the hadal zone was somewhat overlooked, presumably on account of its relatively low area coverage despite the enormous depth range that it encompasses.

The challenges for the immediate future are two-fold. First, there is a technology and access challenge. The challenge being to develop low cost, compact and innovative methods by which to access the greatest depths in order to perform multidisciplinary observational and experimental tasks, including long-term monitoring (Jamieson and Fujii, 2011). This comes at a time where the reliance on large deep-submergence platforms, such as ROVs, HOVs and AUVs, may dwindle due to the pressure of financial strain (Monastersky, 2012) on engineers and designers to develop tools for full ocean depth that have been previously restricted to shallower waters.

The second challenge is to alter the perception of the scientific community in order to ensure that the deepest parts of the world are included in future research programmes and on an equal par with the other marine environments. This is necessary, not just to narrow the knowledge gap between hadal and other zones, but also to encourage a more holistic approach to marine science, especially given the climate-related changes happening currently in the atmosphere and surface waters and the cascading effects these may have on the underlying habitats. Furthermore, depth-related trends in diversity, biology, physiology and ecology, among many others, are likely to be heavily influenced by the incorporation of the deepest 45% of the world's marine environment.

These technological and psychological challenges need to be urgently met in order to enable the comprehensive sampling of multiple trenches with sufficient resolution, bathymetric coverage and replication to enable global generalisations, not just about marine life, but about all life on Earth; from the upper atmosphere and high elevations to the deepest trench and the deep-biosphere beyond. We now live in an age where technology is at a level where few, truly unexplored frontiers remain. With increasing technological developments, exploration, education and the overarching fundamental drive to push the limits of human endeavour, our understanding and appreciation of these deep environments will hopefully become a reality in the not so distant future.

Ocean exploration gives mankind a sense of human progress and heritage. It provides the experience and knowledge necessary to undertake stewardship of the ocean and its resources, and thus sets a course for future generations to navigate. What lies ahead is still unknown. Whatever it is, however, will be influenced by what is found through tomorrow's exploration – and, will likely be different than today's predictions! (Anon, 1998).

Appendix

List of all known species from hadal depths showing maximum and minimum depth of capture from in, and in the vicinity of, the trenches. Those marked * are pelagic and those marked ** were identified from *in situ* photography. NPT = North Pacific Troughs.

PHYLUM: FORAMINIFERA			

CLASS: POLYTHALAMEA
ORDER: Allogromiida
FAMILY: Allogromida incertae sedis

Conicotheca nigrans	Mariana	10 896	10 896
FAMILY: Allogromiidae			
Nodellum aculeata	Mariana	10 896	10 896
Nodellum membranaceum	NPT	2140	7224
Resigella bilocularis	Mariana	10 896	10 896
Resigella laevis	Mariana	10 896	10 896
Xenothekella elongata	Kuril–Kamchatka	9220	9380
FAMILY: Ammodiscidae			
Ammodiscus consonus	Kuril–Kamchatka	4710	9050
Ammodiscus profundissimus	NW Pacific, NPT	3400	9220
Ammolagena clavata	Aleutian	68	7660
Glomospira gordialis	Kuril–Kamchatka	2507	9050
Turritellella shoneana	Japan	7225	7225
Usbekistania charoides profunda	Aleutian, Ryukyu	2532	6520
FAMILY: Astorhizidae			
Pelosina cylindrica	Kermadec	3429	6240
Pelosina rotundata	NPT	3400	6070
Pelosina variabilis	Aleutian, Kurile–Kamchatka	1760	6980
Pelosphaera trunca	NPT	6070	6070
FAMILY: Botellinidae			
Protobotellina pacifica	Aleutian, Kurile–Kamchatka	2000	8430
FAMILY: Dendrophryidae			
Dendrophrya abyssalica	NPT	5510	6060
Dendrophrya kermadecensis	Kermadec	8950	10 002
FAMILY: Hyperamminidae			
Hyperammina echinata	Kuril–Kamchatka, Ryukyu	2930	9580
Hyperammina elongata	Aleutian	2048	6980
Hyperammina imbecilla	NPT	2200	6072
Hyperammina kermadecensis	Kermadec	8950	10 002

Hyperammina zenkevichi	Kuril–Kamchatka	5060	9540
Saccorhiza praealta	Kuril–Kamchatka	4120	6870
Saccorhiza ramosa	Kuril–Kamchatka, NPT	1739	6072
Saccorhiza zankevichi	Kuril–Kamchatka	6700	9540
FAMILY: Normaninidae			
Normanina elongata	Kuril–Kamchatka, NPT	2890	7180
Normanina fruticosa	Aleutian, Kurile–Kamchatka	4330	7180
Normanina ultrabyssalica	Kermadec, Tonga	8950	10 687
FAMILY: Polyaccaminidae			
Saccamminis incrusatum	Kuril–Kamchatka, Bougainville	3360	8006
FAMILY: Psamminidae			
Psammina planata	Kuril–Kamchatka	6860	7320
FAMILY: Psammosphaeridae			
Psammosphaera orbiculata	NPT	2532	6070
Sorosphaera abyssorum	Volcano, Tonga	2582	10 687
FAMILY: Rhabdamminidae			
Bathysiphon lanosum	NPT	500	6240
Psammosiphonella bougainwillica	Bougainville	6800	8006
Psammosiphonella rustica	Japan, NPT	2680	6150
Rhabdammina abyssorum	Mariana	10 924	10 924
Rhabdammina abyssorum	Kuril–Kamchatka, NPT	2020	6860
Rhabdammina bougainwillica	Bougainville	9022	9022
Rhabdammina inaudita	Kuril–Kamchatka, NPT	2000	6260
Rhabdammina recondita	Kuril–Kamchatka, NPT	3429	6880
FAMILY: Rhizamminidae			
Rhizammina algaeformis	Unknown	1015	6240
Rhizammina alta	Aleutian, Kurile–Kamchatka	5050	6520
Rhizammina transversa	NPT	6020	6070
FAMILY: Saccamminidae			
Lagenammina alta	Kuril–Kamchatka, NPT	1724	6860
Lagenammina minuta	Kuril–Kamchatka	8220	9220
Lagenammina difflugiformis	Mariana	10 924	10 924
Tholosina irregularis	Japan, Bougainville, Kermadec	6070	10 002
Thurammina albicans	Peru–Chile	1800	7720
Thurammina corrugata	Peru–Chile	2140	7720
FAMILY: Stannomida			
Stannophyllum granularium	Kuril–Kamchatka	6215	6675
	Japan	6116	6116
Stannophyllum mollum	Japan	6380	6380
FAMILY: Syringamminidae			
Aschemonella delicata	NPT	3420	6070
Aschemonella ramuliformis	Kermadec	2998	8950
Aschemonella scabra	Kuril–Kamchatka, NPT	2760	7180
Ocultammina profunda	Izu-Bonin	8260	8260
ORDER: Litoulida			
FAMILY: Hormosinidae			
Hormosina globulifera	Kuril–Kamchatka, NPT	10 924	10 924
FAMILY: Reophacidae			
Leptohalysis kaikoi	Mariana	10 896	10 896
FAMILY: Ammosphaeroidinidae			
Adercotryma glomerata abyssorum	Kuril–Kamchatka, NPT	2000	7351
Cribrostomoides nitidum abyssalicus	Kuril–Kamchatka, NPT	2000	6250

Cribrostomoides profundum	Aleutian, Kuril–Kamchatka, Izu-Bonin, NPT	2380	6240
Cribrostomoides rotulatum	Aleutian, Kurile–Kamchatka, NPT	2508	7266
Cystammina pauciloculata	Ryukyu, NPT	252	6200
Recurvoidatus parcus	Kuril–Kamchatka, Volcano, NPT	4105	8087
Recurvoidatus trochamminiforme trochamminiformis	Kuril–Kamchatka, NPT	2726	6740
Recurvoidatus ultraabyssalicus	Kuril–Kamchatka, Bougainville	6700	7678
Recurvoidella bradyi	NPT	1732	6050
Recurvoides contortus gurgitis	Kuril–Kamchatka, NPT	1500	6740
Recurvoides mutilus	Japan, Volcano, Bougainville	7225	8380
FAMILY: Discamminidae			
Ammoscalaria tenuimargo	Japan	7225	7225
FAMILY: Haplophragmoididae			
Cribrostomellus apertus	Kuril–Kamchatka	8220	9580
Haplophragmoides pulicosus	Kuril–Kamchatka, NPT	2611	6740
Labrospira canariensis profunda	Kuril–Kamchatka	1739	6250
Veleroninoides scitulus	Unknown	1450	6006
FAMILY: Hormosinellidae			
Hormosinella distans distans	Kuril–Kamchatka	1134	7660
Hormosinella ovicula oviculus	Unknown	1620	7225
Subreophax aduncus	Unknown	2515	7225
FAMILY: Hormosinidae			
Hormosina normani	Aleutian, NPT	1620	7266
Hormosinella guttifera	Mariana	10 924	10 924
Pseudonodosinella rubra	NPT	3540	6120
FAMILY: Lituolidae			
Ammobaculites echinatus echinatus	Kuril–Kamchatka, NPT	2414	7316
Ammobaculites filiformis	Kuril–Kamchatka	1669	6180
Ammobaculites microformis	Kuril–Kamchatka	5080	6250
Ammobaculites sp.	Aleutian	640	6520
Eratidus foliaceus	Kuril–Kamchatka	1015	6250
FAMILY: Prolixoplectidae			
Karrerulina apicularis	Ryukyu, NPT	2862	7225
FAMILY: Reophacidae			
Hormosinoides perpastus	NPT	6070	6070
Pseudonodosinella bacillaris	Aleutian, Kuril–Kamchatka, NPT	2853	7500
Reophax gaussicus	Aleutian, Kuril–Kamchatka, Japan, Izu-Bonin, NPT	4580	7180
Reophax nodulosus		2561	9540
Reophax dentaliniformis	Aleutian, Kuril–Kamchatka, NPT	2140	9220
Reophax echinatus	NPT	4920	6070
Reophax excentricus	NPT	1739	6250
Reophax pesciculus	Kuril–Kamchatka	8220	9580
FAMILY: Spiroplectamminidae			
Astrorhizinulla aetheria	Kuril–Kamchatka	4610	6250
Morulaeplecta sp.	Aleutian	1950	6980
Psammosiphonella beata	Aleutian, Kuril–Kamchatka, Bougainville, NPT	1887	9540

Spiroplectammina subcylindrica	Unknown	2000	6810
FAMILY: Trochamminoidae			
Trochamminoides lituotubus	Unknown	750	6250
ORDER: Loftusiida			
FAMILY: Cyclamminidae			
Cyclamina cancellata cancellata	Kuril–Kamchatka, NPT	2750	6200
Cyclamina subtrullissata	Unknown	2770	6240
Cyclamina trullissata	NPT	3000	6200
FAMILY: Globotextularidae			
Globotextularia anceps	NPT	2507	6065
Tritaxis nana	NPT	1550	6070
ORDER: Miliolida			
FAMILY: Cornuspiroidinae			
Cornuspiroides striolatus	Kuril–Kamchatka	2197	6240
FAMILY: Hauerinidae			
Involvohauerina globularis	Japan, NPT	5030	6150
Miliolinella laeva	New Hebrides	2048	7225
FAMILY: Spiroloculinidae			
Pseudospirilina abyssalica	New Hebrides	4930	6927
ORDER Textulariidae			
FAMILY: Eggerellidae			
Eggerella bradyi	Kuril–Kamchatka	1748	6250
FAMILY: Textulariidae			
Textularina sp.	Mariana	10 896	10 896
ORDER: Trochamminida			
FAMILY: Conotrochamminidae			
Conotrochammina abyssorum	Kuril–Kamchatka, NPT	2507	7300
FAMILY: Trochamminidae			
Trochammina abyssorum	Kuril–Kamchatka	3314	9220
Trochammina alta	Unknown	713	6008
Trochammina macroformis	Kuril–Kamchatka, NPT	5017	6860
Trochammina subglabra	Kuril–Kamchatka	1800	6250

PHYLUM: PORIFERA

CLASS: HEXACTINELLIDA
ORDER: Amphidiscophora
FAMILY: Hyalonematidae

Hyalonema apertum	Kuril–Kamchatka	6090	6235
Hyalonema sp.	Kuril–Kamchatka	6860	6860
ORDER: Haxasterophora			
FAMILY: Caulphacidae			
Caulophacus hadalis	Kermadec	6660	6770
Caulophacus latus latifolium	Kuril–Kamchatka	6090	6710
Caulophacus sp. sp.	Japan	6156	6207
FAMILY: Euplectellidae			
Holascus undulatus	Aleutian	6296	6328
FAMILY: Rossellidae			
Bathydoris fimbriatus	Kuril–Kamchatka	6090	6135
Hyalospomgidae?	Volcano	8530	8540
	South Sandwich	6766	7216
	San Cristobal	5650	6070

CLASS: DEMOSPONGIAE
ORDER: Poecilosclerida
FAMILY: Chondrocladiidae

Chondrochadia concresens	Kuril–Kamchatka	6090	8660
FAMILY: Cladorhizidae			
Asbestopluma occidentalis	Kermadec	6960	7000
	Kuril–Kamchatka	7265	8840
Asbestopluma wolffi	Kermadec	6620	6730
	Kuril–Kamchatka	6675	8120
Asbestopluma sp. sp.	Kuril–Kamchatka	6860	6860
	Palau	8440	9990
Cladorhiza septemdentalis	Kuril–Kamchatka	7265	7295
Cladorhizidae sp. sp.	Izu-Bonin	6770	6890
	Mariana	8215	8225
	Yap	7230	7280
	Philippine	7000	7170
	Palau	7000	7880
	New Britain	7875	7921
Abyssocladia bruuni	Bougainville	6920	7567
FAMILY: Esperiopsidae			
Esperiopsis plumosa	Kuril–Kamchatka	6860	6860

PHYLUM: CNIDARIA

CLASS: HYDROZOA
ORDER: Anthoatheceta
FAMILY: Corymorphidae

Branchiocerianthus imperator	New Hebrides	6758	6776
Branchiocerianthus sp.	Kermadec	6260	6260
ORDER: Leptothecata			
FAMILY: Lafoeidae			
Cryptolaria sp.	Kuril–Kamchatka	6860	6860
FAMILY: Hebellidae			
Hailiphonia galathea	Kermadec	8210	8300
FAMILY: Aglaopheniidae			
Aglaopenia tenuissima	Kermadec	6660	6770
Aglaopenia galathea	Java	6900	7000
Aglaopenia sp.	Cayman	6300	6300
ORDER: Leptomedusae			
FAMILY: Mitroconidae?			
Leptomedusae sp.	New Britain	8258	8260
ORDER: Anthomedusae			
FAMILY: Anthomedusidae?			
Anthomedusae sp.	New Britain	8258	8260
ORDER: Trachymedusae			
FAMILY: Rhoplanematidae			
Crossota sp.	Palau	8021	8042
	New Hebrides	6758	6776
Voragonema profundicula*	Kuril–Kamchatka	6800	8700
Trachymedusae sp.	New Britain	7057	7075

CLASS: SCYPHOZOA
ORDER: Coronatae
FAMILY: Atorellidae

Stephanoscyphus simplex	Banda	6490	6650
	Kermadec	6180	7000
Stephanoscyphus sp. sp.*	Aleutian	6000	10 000
	Kuril–Kamchatka	6000	10 000
	Japan	6000	10 000
	Izu-Bonin	6000	10 000
	Ryukyu	6000	10 000
	Philippine	6000	10 000
	New Hebrides	6000	10 000
	Palau	6000	10 000
	Puerto-Rico	6000	10 000
	Cayman	6000	10 000

ORDER: Semaeostomeae
FAMILY: Ulmaridae

Ulmaridae sp.	Bougainville	7847	8662

CLASS: ANTHOZOA
ORDER: Alcyonacea
FAMILY: Alcyonaira

Calvulariidae	New Britain	7057	8260
	Bougainville	7847	8662
	New Hebrides	6758	6776

FAMILY: Primnoidae

Primnoella sp.	Palau	8021	8042

ORDER: Pennatularia
FAMILY: Kophobelemnonidae

Kophobelemnon biflorum	Kuril–Kamchatka	6090	6135
Kophobelemnon molanderi	South Sandwich	6052	6150
	San Cristobal	5650	6070

FAMILY: Umbellulidae

Umbellula lindashi	Palau	6100	6100
	South Sandwich	6052	6150
	San Cristobal	5650	6070
Umbellula magniflora	Kuril–Kamchatka	6090	6135
	Palau	6040	6240
Umbellula thomsoni	Kuril–Kamchatka	6090	6235
	Palau	6040	6240
Umbellula sp. sp.	Kermadec	6180	6730
	Palau	6006	6260
Pannatularia?	New Britain	7875	7921

ORDER: Hexacorallia
FAMILY: Actinosolidae

Bathydactulus kroghi	Kermadec	8210	8230
Hadalanthus knudseni	Kermadec	6660	6770

FAMILY: Bathyphelliidae

Daontesia mielchei	Banda	7250	7290

FAMILY: Edwardsiidae

Paredwardsia lemchei	Java	7160	7160

ORDER: Actinaria
FAMILY: Galatheanthemidae

Galatheanthemum hadale	Philippine	9820	10210
	Kuril–Kamchatka	7210	7230
	Cayman	5800	6500
	Puerto-Rico	5749	8130
Galatheanthemum profundale	Kermadec	6180	8300
Galatheanthemum sp. n.	Mariana	10 170	10 730
Galatheanthemidae sp. sp.	Aleutian	6965	7250
	Kuril–Kamchatka	6090	7295
	Japan	6156	7370
	Ryukyu	6660	6670
	Izu-Bonin	6770	9735
	Volcano	6780	6785
	Mariana	8215	8225
	Yap	8560	8720
	Palau	7420	9750
	Bougainville	6920	8662
	New Hebrides	6758	6776
	Tonga	8950	9020
	Kermadec	8928	9174
	Puerto-Rico	7500	8143
	Cayman	5800	6500
	San Cristobal	5650	6070

FAMILY: Actiniidae

Paractis sp.	Cayman	6740	6780

ORDER: Antipatharia
FAMILY: Unknown

Bathypathes patula	Aleutian	7200	7200
	Kuril–Kamchatka	8175	8840

ORDER: Sceleractina
FAMILY: Fungiidae

Fungiacyathys symmetricus	Aleutian	6296	6328
	Kuril–Kamchatka	6090	6135

PHYLUM: PLATYHELMINTHES

CLASS: TURBELLARIA
ORDER: Polycalidida
FAMILY: Unknown

Unknown	Kuril–Kamchatka	7265	9335
Unknown	Peru–Chile	6000	6354
Unknown	Aleutian	7298	7298

PHYLUM: GASTROTRICHA

CLASS: UNKNOWN
ORDER: Unknown
FAMILY: Unknown

Unknown	Peru–Chile	6000	6354

PHYLUM: ANNELIDA			

CLASS: POLYCHAETA
ORDER: Eunicida
FAMILY: Dorvilleidae

Ophryotrocha hadalis	Aleutian	7298	7298
FAMILY: Lumbrineridae			
Lumbrineris abyssorum	South Sandwich	6052	6150
Lumbrineris sp. sp.	Aleutian	6960	7250
	Japan	6156	7587
Paraninoe fusca	Kermadec	6620	7000
Paraninoe harmani	Aleutian	7250	7250
	Japan	6156	7587
	Kuril–Kamchatka	6475	8100
FAMILY: Onuphiidae			
Onuphis ehlersi	Kuril–Kamchatka	6090	6135
	Aleutian	6296	6328
Paraonuphis ulraabyssalis	Volcano	6330	6330
	Philippine	6290	6330
ORDER Phyllodocidae			
FAMILY: Pilargidae			
Sigambra (Ancistrosyllis) constricta	Banda	6580	6580
FAMILY: Hesionidae			
Hesionidae sp.	Bougainville	8980	9043
FAMILY: Nephtyidae			
Micronephthys abranchiata	Izu-Bonin	8800	8830
	Kermadec	8928	9174
Nephthys elamelata	Kermadec	6180	7000
Nephthyidae gen. sp.	Bougainville	8980	9043
FAMILY: Nereididae			
Ceratocephale loveni pacifica	Japan	6600	6670
Nereis profundi	Banda	7250	7290
Nereis caymanensis	Cayman	5800	6850
Nereis sp. sp.	Philippine	8080	8400
	Java	6935	7060
Nereidae gen. sp.	Japan	7265	7587
FAMILY: Aphroditidae			
Laetmonice benthaliana	South Sandwich	6766	6875
FAMILY: Goniadidae			
Bathyglycinide longisetosa	South Sandwich	7218	7934
FAMILY: Phyllodicidae			
Eulalia sandwichensis	South Sandwich	6052	7218
Eulalia sigeiformis	Aleutian	7246	7246
Vitjazia dogiela	Kuril–Kamchatka	6150	8100
	Japan	7190	7587
FAMILY: Polynoidae			
Bathyeditia berkeleyi	Aleutian	6965	7000
	Ryukyu	6810	7450
	Philippine	7420	7880
Bathyeliasona abyssicola	Aleutian	7286	7286
	Bougainville	6920	8006

Bathyeliasona kirkegaardi	Aleutian	6925	7250
	Izu-Bonin	6770	6890
	Volcano	6330	6330
	Palau	7000	7170
	Philippine	7420	7880
	Ryukyu	7440	7450
	Kermadec	6620	7000
	Banda	7250	7290
	Java	7130	7160
Bathykermadeca hadalis	Kermadec	6660	8300
	Yap	8560	8720
	Japan	7350	7370
	Banda	7250	7290
	Philippine	10 160	10 210
Bathykurila zenkevitchi	Kuril–Kamchatka	8100	8135
	Japan	6600	6670
	Philippine	8080	8400
Bathylevensteinia bicornis	Tonga	9735	9875
Bathymariana zebra	Ryukyu	7440	7450
Bathymoorea aff. *renotubulata*	South Sandwich	6052	6150
Macellicephala alia	Palau	7970	8035
Macellicephala mirabilis	Puerto-Rico	7625	7900
Macellicephala tricornis	South Sandwich	7200	8116
Macellicephala violaca	Aleutian	7250	7250
	Kuril–Kamchatka	6135	9530
	Japan	7370	7370
Macellicephaloides grandicirra	Kuril–Kamchatka	8100	9500
Macellicephaloides improvisa	Kuril–Kamchatka	8035	8120
Macellicephaloides sandwichiensis	South Sandwich	7200	7934
Macellicephaloides uschakovi	Kuril–Kamchatka	8035	8120
Macellicephaloides verrucosa	Kuril–Kamchatka	7210	8015
	Japan	6156	6207
Macellicephaloides villosa	Japan	7350	7370
Macellicephaloides vitjazi	Kuril–Kamchatka	7210	8430
Macellicephaloides sp.	Mariana	7990	10 710
Polynoidae sp.	Kuril–Kamchatka	8175	9335
	Japan	6475	7370
	Izu-Bonin	6770	9735
	Volcano	6780	6785
	Mariana	10 170	10 730
	Palau	8021	8042
	Philippine	7610	9990
	New Britain	7875	8260
	Bougainville	7847	8662
	New Hebrides	6758	8930
	Tonga	10 415	10 687
	Java	6820	6850

FAMILY: Sigalionidae

Leonira quatrefagesi	South Sandwich	6050	6150

ORDER: Drilomorpha
FAMILY: Capitallidae

Notomastus latericeus	South Sandwich	6875	7216
Notomastus sp. sp.	Aleutian	6550	6550
	Kuril–Kamchatka	6860	8660
	Japan	6600	7587
	Kermadec	8210	8300
	Java	6820	6850
Capitellidae sp. sp.	Aleutian	6410	7246
	Japan	7190	7190

ORDER: Terebellida
FAMILY: Faveliopsidae

Laubieriopsis (Fauveliopsis) brevis	Kuril–Kamchatka	6835	6835
	Palau	6200	6240
	South Sandwich	6052	6150
Fauveliopsis challengeriae	Kuril–Kamchatka	6090	6135

FAMILY: Flabelligeridae

Ilyphagus (brada) irenaia	Kuril–Kamchatka	6860	6860
Ilyphagus (brada) sp.	South Sandwich	6052	7934
Ilyphagus bythincola	Java	6730	6850
Ilyphagus wyvillei	San Cristobal	5650	6070
Flabelligeridae sp. n.	Aleutian	6965	7000
	Japan	6600	7587

FAMILY: Maldanidae

Maldanella harai	Kermadec	6620	6720
Maldanella japonica	Japan	6156	6840
Notoproctus oculatus antarcticus	South Sandwich	6766	6875
Notoproctus sp.	Japan	6156	6207
Petaloproctus	Banda	7250	7290
Maldanidae sp. sp.	Kuril–Kamchatka	6860	7230
	Japan	6380	6380

FAMILY: Opheliidae

Ammotrypane galatheae	Banda	7250	7290
Ammotrypane sp.	Aleutian	7250	7250
	Japan	7370	7587
	Bougainville	8980	9043
Kesun abyssorum	Aleutian	6960	7250
	Japan	6156	7370
	Kuril–Kamchatka	6080	8430
	Izu-Bonin	8800	9735
	Philippine	8080	8400
	Bougainville	6920	8006
	Kermadec	6960	8300
	Java	6740	6850
	Ryukyu	6330	7600
	South Sandwich	6052	8116
	San Cristobal	5650	6070
Kesun fuscus	Japan	7565	7587
Travisia profundi	Aleutian	7250	7250
	Kuril–Kamchatka	6090	7230

	Japan	6156	7190
	Banda	6490	7290
Travisia sp.	Japan	7460	7557
FAMILY: Cirratulidae			
Chaetozone sp.	Java	6935	7060
Cossura longicirrata	Java	6487	6487
Tharix multifilus	Banda	6580	6580
Tharix sp. sp.	Java	6820	6850
	South Sandwich	7200	8116
Cirratulidae sp. sp.	Aleutian	7246	7246
	Kermadec	8928	10015
FAMILY: Amparetidae			
Amagopsis cirratus	Palau	6200	6240
Amphicteis gunneri	South Sandwich	7686	7686
Amphicteis gunneri japonica	Aleutian	6965	7250
	Kuril–Kamchatka	6475	6571
	Japan	6156	7587
Amphicteis mederi	Kuril–Kamchatka	7210	8430
	Japan	6380	6380
Anobothrus sp.	Kuril–Kamchatka	6860	6860
	Japan	6156	6207
Mellinampharete eoa	Kuril–Kamchatka	6150	6860
Ampharetidae sp. sp.	Aleutian	6296	6296
	Kuril–Kamchatka	7210	7230
	Japan	6156	6207
	Kermadec	6660	6720
	Java	6820	7000
FAMILY: Terebellidae			
Pista sp.	Aleutian	6296	6328
	Japan	6156	6207
Pista mirabilis	Kuril–Kamchatka	6205	6215
	Japan	6156	6207
	Palau	6040	6328
FAMILY: Trichobranchiidae			
Terebellides eurystethus	Kuril–Kamchatka	7210	7230
	Japan	7190	7587
	Kermadec	6660	6770
	Aleutian	6960	6960
ORDER: Sabellida			
FAMILY: Siboglinidae			
Lamellisabella johanssoni	Japan	6156	6207
Birsteinia sp.	Kuril–Kamchatka	9000	9050
Cyclobrachia auriculata	Bougainville	7974	7974
Diplobrachia japonica	Aleutian	7200	7200
	Japan	6600	7587
	Izu-Bonin	8800	8830
Heptabrachia abyssicola	Kuril–Kamchatka	6475	8100
Heptabrachia subtilis	Izu-Bonin	9715	9735
Polybrachia choanata	Kuril–Kamchatka	9000	9050
	Japan	6600	6670
Polybrachia sp. 1	Kuril–Kamchatka	8100	8100

Polybrachia sp. 2	Palau	6324	6328
Zenkevitchiana longissima	Kuril–Kamchatka	8330	9500
Polybrachiidae gen. sp.	Palau	6324	6328
Sclerolinum Javanicum	Java	6820	6850
Sclerolinum sp.	Ryukyu	6810	6810
Siboglinum caulleryi	Kuril–Kamchatka	8100	8100
Siboglinum longimanus	Palau	6324	6328
Siboglinum pusillum	Aleutian	6960	6980
Siboglinum sp. II	Kuril–Kamchatka	9000	9050
Siboglinum sp. IV	Kuril–Kamchatka	8100	8100
Siboglinum sp. VI	Kuril–Kamchatka	9000	9050
Siboglinum sp. VIII	Bougainville	7974	8006
Siboglinum sp. IX	Bougainville	7974	8006
Siboglinum sp. n. 1	Aleutian	6410	6757
Siboglinum sp. n. 2	Japan	7565	7587
Siboglinum sp. n. 3	Java	6841	6841
Siboglinum sp. n. 4	Palau	6324	6328
Siboglinum sp. n. 5	Palau	6324	6328
Spirobrachia beklemishevi	Kuril–Kamchatka	9000	9050
Spirobrachia leospira	South Sandwich	8004	8186
FAMILY: Oweniidae			
Myriochele sp.	Kermadec	6180	8300
Owenia lobopygidiata	Banda	6490	6650
Oweniidae sp. sp.	Aleutian	6296	7246
FAMILY: Scalibregmatidae			
Pseudoscalibregma pallens	Kermadec	8928	9174
Pseudoscalibregma collaris	San Cristobal	5650	6070
Scalibregmidae sp. sp.	Japan	6600	6700
	Kuril–Kamchatka	6860	6860
FAMILY: Sabellidae			
Potamethus filatovae	Aleutian	6328	6960
	Kuril–Kamchatka	7210	7230
	Japan	6380	6700
	Izu-Bonin	9715	9735
Potamethus sp. sp.	Japan	6156	6380
	Izu-Bonin	8530	8735
	Kermadec	6620	8300
	Banda	7280	7280
Potamilla sp.	Kuril–Kamchatka	8100	8100
Sabellidae sp. sp.	Kuril–Kamchatka	7210	7230
	Palau	8021	8042
	New Hebrides	6758	6776
FAMILY: Serpulidae			
Serpulidae sp.	Aleutian	6410	6757
	Izu-Bonin	9715	9735
	Kermadec	6620	6620
ORDER: Spionida			
FAMILY: Poecilochaetidae			
Poecilochaetus vitjazi	Tonga	10 415	10 687
FAMILY: Chaetopteridae			
Phyllochaetopterus	Kuril–Kamchatka	6860	6860

CLASS: OLIGOCHAETA
ORDER: Haplotaxida
FAMILY: Tubificidae

Bathydrilus hadalis	Aleutian	7298	7298

PHYLUM: ECHUIRA

CLASS: ECHIUROIDAE
ORDER: Bonelliida
FAMILY: Bonelliidae

Alomasoma chaetifera	Aleutian	7246	7246
Alomasoma nordpacifica	Aleutians	7246	7246
	Kuril–Kamchatka	6090	8430
	Volcano	7584	7614
Bruunelia bandae	Banda	7250	7290
Hamingia arctica	Palau	7970	8035
	Ryukyu	7440	7450
Ikedella bogorovi	Java	6820	6850
Jakobia birsteini	Kuril–Kamchatka	6475	9730
	Japan	6600	7370
	South Sandwich	7200	7216
Kurchatovus tridentatus	Puerto-Rico	5890	6000
	Cayman	6740	6780
Pseudoiledellia sp.	Philippine	9980	9990
Sluiterina flabellorhynchus	San Cristobal	5650	6070
	Puerto-Rico	6400	6400
Sluiterina vitjazi	Aleutian	6965	7000
Torbenwolffia galatheae	Kermadec	6660	8300
	Japan	8560	8720
	Aleutian	6965	7000
Vitjazema aleutica	Aleutian	7246	7286
Vitjazema planirostris	Philippine	9750	9750
Vitjazema ultraabyssalis	Kuril–Kamchatka	7210	9530
	Izu-Bonin	9715	9735
	Yap	7190	7250
Vitjazema sp. sp.	Philippine	8080	8400
	Philippine	10 150	10 210
	Palau	6200	6240

PHYLUM: ECHINODERMATA

CLASS: ECHINOIDEA
ORDER: Echinothuroida
FAMILY: Echinothuroida

Kamptosoma abyssale	Kuril–Kamchatka	6090	6235

ORDER: Spatangoida
FAMILY: Holasteridae

Rhodocystis rosea	Puerto-Rico	6290	6314
	Cayman	5800	6850

FAMILY: Pourtalesiidae

Ceratophysa ceratopyga valvaecristata	Aleutian	6272	6328

Echinosigra amphora amphora	Aleutian	6272	6282
Echinosigra amphora sp.?	Java	6433	6850
Echinosigra amphora indica	Palau	7000	7170
Pourtalesia heptneri	Banda	7130	7340
Pourtalesia sp. (aff. *debilis*)	South Sandwich	6100	6650
	San Cristobal	5650	6070
	Japan	6156	6207
FAMILY: Urechinidae			
Pilematecinus belyaevi	Cayman	5800	6780
CLASS: OPHIUROIDEA			
ORDER: Ophiurae			
FAMILY: Ophiacanthidae			
Ophiacantha bathybia	Kuril–Kamchatka	6090	6235
Ophiocymbium cavernosum	Izu-Bonin	6065	6850
Ophiocymbium sp. n.	Izu-Bonin	6770	6850
	Philippine	7420	7880
Ophiacanthidae sp.?	New Hebrides	6758	6776
FAMILY: Ophiodermatidae			
Ophiurochaeta sp.	South Sandwich	6052	6150
FAMILY: Ophioleucidae			
Bathylepta pacifica	Bougainville	6920	8006
	New Hebrides	6680	6830
FAMILY: Ophiuridae			
Abyssura brevibranchia	Aleutian	6965	7000
	Kuril–Kamchatka	6675	7295
	Japan	6156	6207
Amphiophuira bullata bullata	Puerto-Rico	5890	6035
Amphiophuira bullata pacifica	Kuril–Kamchatka	6090	6235
	Japan	6096	6390
Amphiophuira convexa	Ryukyu	6065	6810
Amphiophuira vitjazi	Ryukyu	6810	6810
Amphiophuira sp.	South Sandwich	6052	6150
	San Cristobal	5650	6070
Homalophiura madseni	Kuril–Kamchatka	6675	7230
	Japan	6156	6380
Homalophiura aff. *madseni*	Ryukyu	7440	7450
	Palau	7000	7170
	Mariana	7340	7450
Homalophiura sp. nov.	Kuril–Kamchatka	6475	6571
Ophiocten sp. nov.	San Cristobal	5650	6070
Ophiotypa simplex	Java	6477	6487
Ophiura bathybia	Aleutian	6296	6328
	Kuril–Kamchatka	6090	6135
Ophiura indet	Kermadec	6620	6620
Ophiura loveni	Kermadec	6660	6770
	Romanche	6330	7340
Ophiura irrorata irrorata	San Cristobal	5650	6070
Ophiura irrorata polyacantha	South Sandwich	6052	6150
Ophiurolepis sp.	South Sandwich	6052	7216
Perlophiura profundisima	Aluetian	7200	7200

	Kuril–Kamchatka	6795	8135
	Izu-Bonin	6770	6890
	Volcano	6330	6330
Ophiuridae sp. sp.	Bougainville	7847	8662
	New Hebrides	6758	6776

CLASS: ASTEROIDEA
ORDER: Brisingida
FAMILY: Freyellidae

Freyella kurilo-kamchatica	Kuril–Kamchatka	6205	6860
	Japan	6156	6207
Freyella mortenseni	Kermadec	6180	6180
Freyella mutabila	San Cristobal	5650	6070
Freyellidae sp. sp.	New Britain	7875	8260
	Bougainville	7847	8662
	New Hebrides	6758	6776

ORDER: Paxillosida
FAMILY: Porcellansteridae

Albatrossia sp.	Aleutian	6296	6328
Eremicaster pacificua	Kermadec	6620	6620
Eremicaster crassus	Aleutian	6296	6328
Eremicaster vicinus	Aleutian	6296	7246
	Kuril–Kamchatka	6090	6860
	Japan	6700	7340
	Izu-Bonin	6770	6890
	Peru–Chile	6006	6328
	Kermadec	6620	6730
	Java	6433	7000
	South Sandwich	6052	6150
	San Cristobal	5650	6070
Lethmaster rhipidophorus	Ryukyu	6460	7540
Lethmaster rhipidophorus	Philippine	7420	7880
Porcellanaster sp. sp.	Aleutian	6296	6328
	South Sandwich	6052	6150
Styracaster longispinus	Aleutian	6296	6328
Styracaster sp. n.	Cayman	6466	6600
Porcellanasteridae?	New Britain	7057	7078
	New Hebrides	6758	6776

ORDER: Valvatida
FAMILY: Goniasteridae

Litonaster sp.	Palau	8021	8042

FAMILY: Caymanostellidae

Caymanostella spinimarginata	Cayman	6740	6780

FAMILY: Pterasteridae

Hymenaster blegvadi	Kermadec	6660	6770
Hymenaster sp.	Kuril–Kamchatka	6090	8400
	Japan	6380	6380
	Izu-Bonin	6770	6890
	Volcano	7584	7657
	Yap	8560	8720
	Palau	7000	8042

	Philippine	8440	9990
	Bougainville	6920	7657
	New Hebrides	6758	6776
	Romanche	6330	7600
	South Sandwich	6052	6150

CLASS: HOLOTHUROIDEA
ORDER: Apodida
FAMILY: Myriotrochidae

Lepidotrochus kermadecensis	Kermadec	8928	9174
	South Sandwich	6766	7934
Myriotrochus longissimus	Japan	6475	7370
	Philippine	6290	6330
	Palau	7000	7170
Myriotrochus mitus	Kermadec	8928	9174
Prototrochus angulatus	Philippine	7610	9990
Prototrochus bipartitodentatus	South Sandwich	7694	8116
Prototrochus bruuni	Philippine	9360	10 210
	Palau	7970	8035
	Izu-Bonin	8900	9180
	Bougainville	8940	9043
	Tonga	8950	10 687
	Kermadec	8928	10 015
	Java	6487	7060
Prototrochus kurilensis	Kuril–Kamchatka	7795	8430
Prototrochus wolffi	Mariana	8215	8225
	Volcano	8530	8540
	Yap	8560	8720
Prototrochus zenkevitchi *zenkevitchi*	Kuril–Kamchatka	8175	9530
	Japan	7500	7500
	Izu-Bonin	8800	9735
Prototrochus zenkevitchi atlanticus	Romanche	7430	7600
Prototrochus zenkevitchi exiguus	Kuril–Kamchatka	8060	8135
Prototrochus aff. longissimus	South Sandwich	6052	6150
	San Cristobal	5650	6070
Prototrochus sp. n.	Mariana	10 630	10 730
	San Cristobal	5650	6070
Siniotrochus spiculifer	Kuril–Kamchatka	8330	8430

ORDER: Aspidochirotida
FAMILY: Synalactidae

Mesothuria murrayi	Banda	6490	6650
Paroriza grevei	Banda	6490	7290
Pseudostichopus villosus	Kermadec	6660	7000
	New Britain	7875	8260
	Bougainville	6758	6776
Molpadiodemas (P) villosus	Kermadec	6660	7000
Pseudostichopus sp.	Kuril–Kamchatka	8100	8100

ORDER: Elasipoda
FAMILY: Elpidiidae

Amperima naresi	Java	6820	7160
	Palau	6200	6240

Amperima velacula	South Sandwich	6052	6150
	San Cristobal	5650	6070
Ellipinion galatheae	Philippine	9820	10 000
Elpidia atakama	Peru–Chile	7720	8074
Elpidia birsteini	Kuril–Kamchatka	8060	9345
	Kermadec	8185	8400
	Izu-Bonin	8530	8540
Elpidia decapoda	San Cristobal	5650	6070
	South Sandwich	6052	6150
Elpidia hanseni hanseni	Kuril–Kamchatka	8610	9530
Elpidia hanseni idsuboninensis	Izu-Bonin	8800	9735
Elpidia hanseni javanica	Java	6820	6850
Elpidia glacialis kermadecensis	Kermadec	6620	8300
Elpidia kurilensis	Kuril–Kamchatka	6675	8100
	Japan	6156	7587
	Aluetian	6410	6757
Elpidia lata	South Sandwich	8004	8116
Elpidia longicirrata	Kuril–Kamchatka	8035	8345
Elpidia ninae	South Sandwich	6766	7634
Elpidia solomonensis	Bougainville	7847	9043
	New Hebrides	7057	8260
Elpidia sundensis	Java	6433	7160
Elpidia uschakovi	New Hebrides	6680	6830
Elpidia sp. 3	Romanche	7340	7340
Elpidia aff. *minutissima*	Yap	8560	8720
	Palau	7970	8042
Kolga hyalina	Kuril–Kamchatka	6205	6215
	South Sandwich	6052	6150
Peniagona azorica	New Hebrides	7057	7921
	Bougainville	7847	8662
	Kermadec	2640	8300
	Romanche	7100	7300
Peniagona gracilis	Aluetian	6965	7250
	Izu-Bonin	6770	7315
Peniagona herouardi	Puerto-Rico	7950	8100
	South Sandwich	7694	7934
Peniagona incerta	Kuril–Kamchatka	6090	7230
	Japan	6156	6207
	South Sandwich	6052	6875
Peniagona purpurea	Palau	8021	8042
Peniagona vedali	Kermadec	6140	8300
Peniagona sp. sp.	Kuril–Kamchatka	6090	8400
	Japan	7565	7587
	Izu-Bonin	7305	7315
	Mariana	6580	6650
	Peru–Chile	6002	6030
	Romanche	6330	7600
	Puerto-Rico	6290	7960
Scotoplanes globosa	Kermadec	2470	6770
Scotoplanes hanseni	Kuril–Kamchatka	6090	6860

	Japan	6480	6640
	New Britain	7057	7075
	Bougainville	6920	7660
	New Hebrides	6758	6830
Elpidiidae sp. sp.	Philippine	8440	9990
FAMILY: Laetmogonidae			
Apodogaster sp.	Kermadec	4410	6730
FAMILY: Psychropotidae			
Benthodytes sanguinolenta	Banda	6490	7290
Psychropotes verrucosa	Kermadec	3710	6730
	Banda	6490	7290
Psychropotes sp. sp.	Peru–Chile	6260	6260
	Kuril–Kamchatka	6090	6215
Psychropotidae gen. et. sp. n.	Kuril–Kamchatka	9170	9335
FAMILY: Palagothuriidae			
Palgothuria natatrix	New Hebrides	6758	6776
ORDER: Molpadonia			
FAMILY: Gephyrothuriidae			
Hadalothuria wolffi	New Britain	7057	7071
	Bougainville	8780	8940
	New Hebrides	6758	6776
Hadalothuria sp.	Kuril–Kamchatka	9070	9530
FAMILY: Molpadiidae			
Ceraplectana trachyderma	Banda	6490	6650

PHYLUM: MOLLUSCA

CLASS: SCAPHOPODA			
ORDER: Galilida			
FAMILY: Entalinidae			
Costentalina caymanica	Cayman	5900	6780
Costentalina tuscarorae tuscarorae	Japan	6480	6640
Entalinidae sp.	Kuril–Kamchatka	6090	6675
	South Sandwich	6052	6150
FAMILY: Unknown			
Siphonodentalium galatheae	Java	6900	7000
FAMILY: Pulsellidae			
Unknown	Japan	6480	6640
	Bougainville	6920	7657
	Java	6820	6850
	Romanche	6330	6330
	San Cristobal	5650	6070
CLASS: GASTROPODA			
ORDER: Docoglossa			
FAMILY: Bathypeltidae			
Bathypelta pacifica	Japan	8560	8720
FAMILY: Bathysciadiidae			
Bonus petrochenkoi	Kuril–Kamchatka	8240	9530
	Tonga	8950	9020
FAMILY: Propolidiidae			
Propilidium reticulatum	Kuril–Kamchatka	6090	6135

FAMILY: Fissurellidae			
Fissurellidae sp.	Philippine	6290	6300
ORDER: Alata			
FAMILY: Sequenziidae			
Seguenzia sp.	Japan	7190	7250
	Peru–Chile	7000	7170
	Ryukyu	7440	7450
ORDER: Anisobranchia			
FAMILY: Skeneidae			
Skeneidae sp.	Philippine	6290	6330
FAMILY: Trochidae			
Guttula galathea	Kermadec	6660	6770
Trenchia wolffi	Kermadec	6620	6730
Trenchia sp.	Ryukyu	7440	7450
Trochidae sp. sp.	Peru–Chile	7970	8035
	Ryukyu	7335	7340
ORDER: Aspidophora			
FAMILY: Naticidae			
Naticidae	Romanche	6330	6430
ORDER: Hamiglossa			
FAMILY: Buccinidae			
Calliloncha iturupi	Kuril–Kamchatka	8240	8345
Calliloncha solida	Izu-Bonin	6770	6850
Paracalliloncha ultra-abyssal	Kuril–Kamchatka	8035	8120
Tacita arnoldi	Kuril–Kamchatka	6090	6135
Tacita holoserica	Kuril–Kamchatka	6090	6135
	Japan	6480	6640
Tacita zenkevitchi	Peru–Chile	5329	6173
Tacita sp. n.	Kuril–Kamchatka	7210	7230
Tacita sp. sp.	Kuril–Kamchatka	9000	9050
	Japan	7370	7370
Buccinidae sp. sp.	Kuril–Kamchatka	6090	8015
	Japan	6156	6640
Buccinidae sp.?	San Cristobal	5650	6070
FAMILY: Cancellaridae			
Admete bruuni	Kermadec	6660	6770
Cancellariidae sp.	Peru–Chile	7000	7170
	Banda	7335	7340
ORDER: Heterosropha			
FAMILY: Aclididae			
Aclis kermadecensis	Kermadec	8210	8300
FAMILY: Piramidellidae			
Piramidellidae sp. sp.	Japan	7230	7280
	Peru–Chile	7000	7170
ORDER: Homoeostropha			
FAMILY: Eulimidae			
Melanella hadalis	Kermadec	6660	6770
ORDER: Planilabiata			
FAMILY: Bathyphytophilidae			
Bathyphytophilus caribaeus	Cayman	5800	6780

Aenigmabonus kurilo-kamtschaticus	Kuril–Kamchatka	6090	8120
FAMILY: Cocculinidae			
Bandabyssia sp.	Bougainville	6920	7657
Caymanabyssia spina	Cayman	6740	7247
Cocculina sp.	Puerto-Rico	7540	7960
Fedikovella caymanensis	Cayman	6740	7247
Fedikovella sp. n. 1	Puerto-Rico	7950	8100
Fedikovella sp. n. 2	Puerto-Rico	8330	8330
Pseudococculina sp. n. 1	Cayman	6740	6800
Pseudococculina sp. n. 2	Kuril–Kamchatka	6090	6135
Pseudococculina sp. n. 3	Philippine	8080	8400
Pseudococculina sp. n. 4	Java	6820	6850
Macleaniella moskalevi	Puerto-Rico	5179	8595
Amphiplica plutonica	Cayman	6466	7247
ORDER: Toxoglassa			
FAMILY: Turridae			
Abyssocomitas kuirlo-kamchatika	Kuril–Kamchatka	6090	6117
Kurilohadalia brevis	Kuril–Kamchatka	7210	8015
	Japan	6156	6207
Kurilohadalia sysoev	Kuril–Kamchatka	7210	8430
Oenopotella ultrabyssalis	Kuril–Kamchatka	6090	7230
	Japan	6156	6207
Oenopotella aleutica	Aleutian	6965	7000
Pleurotomella cancellata	Izu-Bonin	6770	6850
Tuskaroria ultrabyssalis	Kuril–Kamchatka	7210	7230
Vityazinella multicostata	Kuril–Kamchatka	6090	6135
Xanthodaphne bougainvillensis	Bougainville	6920	7657
Xanthodaphne laevis	Bougainville	7947	8006
Xanthodaphne palauensis	Palau	7000	7170
Xanthodaphne tenuistriata	Izu-Bonin	6770	6850
Gastropoda prosobranchia	Volcano	6330	6330
	Mariana	10 220	10 730
	Philippine	7420	9990
	New Hebrides	6680	6830
	Tonga	10 415	10 687
	Kermadec	9995	10 015
	Romanche	6330	7600
	South Sandwich	6052	8116
	San Cristobal	5660	6070
ORDER: Tectibranchia			
FAMILY: Phylinidae			
Phyline sp. 3	Japan	7565	7587
Phyline sp. 5	Aleutian	6410	6757
Phyline sp. 6	Yap	6820	6850
Phyline sp. 7	Yap	6820	6850
FAMILY: Retusidae			
Volvula sp. 2	Bougainville	7974	8006
FAMILY: Scaphandridae			
Cylichna sp. 4	Izu-Bonin	7305	7315
Tectibranchia sp. sp.	Izu-Bonin	6770	6850

	Yap	7230	7280
	Palau	7000	8035
	South Sandwich	7206	7934
	San Cristobal	5650	6070
CLASS: POLYPLACOPHORA			
ORDER: Cyclopoida			
FAMILY: Chitonophilidae			
Leptochiton vitjazae	Bougainville	6920	7657
	Palau	7000	7100
Leptochiton sp.	New Hebrides	6680	6830
Ferreiraella caribbea	Cayman	6740	6780
CLASS: MONOPLACOPHORA			
ORDER: Tryblidiida			
FAMILY: Neopilinidae			
Rokopella oligotropha	NW Pacific Trough	6065	6079
Vema bacescui	Peru–Chile	5986	6134
Vema ewingi	Peru–Chile	5817	6002
Neopilina sp.	Peru–Chile	1647	6354
CLASS: BIVALVIA			
ORDER: Nuculida			
FAMILY: Ledellidae			
Bathyspinula (bathyspinula) bogorovi	Ryukyu	6810	6810
	Philippine	8080	8400
	Japan	7350	7370
Bathyspinula (bathyspinula) knudseni	Kuril–Kamchatka	6860	6860
Bathyspinula (bathyspinula) latirostris	Japan	7350	7370
Bathyspinula (bathyspinula) oceanica	Aleutian	6296	6328
	Kuril–Kamchatka	6090	6710
	Japan	6165	6207
	Izu-Bonin	7500	7500
	Peru–Chile	6324	6328
Bathyspinula (bathyspinula) thorsoni	Romanche	6330	6430
	Puerto-Rico	6400	6400
Bathyspinula (bathyspinula) vityazi	Aleutian	6965	7250
	Kuril–Kamchatka	6475	9335
	Japan	6660	7587
Ledella (Ledella) crassa	Cayman	5800	6500
Ledella (Mageleda) inopinata	San Cristobal	5650	6070
Ledellina convexirostra	Japan	6600	6670
Ledellina olivecea	Peru–Chile	7720	7720
Spinula sp. sp.	Izu-Bonin	6770	6890
	Mariana	7340	7450
	Yap	7190	7250
	Palau	7000	7170
	Philippine	6290	9990
	Ryukyu	6660	6670
	Romanche	6330	6430
Parayoldiella mediana	Kuril–Kamchatka	7600	7710
	Japan	7350	7587
Parayoldiella angulata	Mariana	8890	8900

	Philippine	8440	8580
Parayoldiella hadalis	Philippine	10 150	10 190
Parayoldiella idsubonini	Izu-Bonin	8800	8900
Parayoldiella inflata	Mariana	7340	7450
	Japan	8560	8720
	Palau	7970	8035
	Philippine	8440	9990
Parayoldiella knudseni	Philippine	9820	10 000
Parayoldiella ultraabyssalis	Kuril–Kamchatka	8355	9530
Parayoldiella sp.	South Sandwich	8004	8116
Parayoldiella sp. sp.	Volcano	6330	8540
	Palau	7000	7170
	Philippine	7610	9750
FAMILY: Malletiidae			
Malletia cuneata	Japan	6156	6207
Malletia sp. n.	South Sandwich	7200	7934
Malletia sp. sp.	Kuril–Kamchatka	6860	7230
	Japan	7565	7587
	Palau	7970	8035
	Philippine	6290	7880
	Ryukyu	6660	6670
Malletia sp. sp.	San Cristobal	5650	6070
FAMILY: Nuculanidae			
Neilonella hadalis	Kermadec	6660	7000
Nielonella sp. sp.	Aleutian	7246	7246
	Kuril–Kamchatka	6475	8430
	San Cristobal	5650	6070
Yoldia kermadecensis	Kermadec	8210	8300
Yoldiella sp. sp.	South Sandwich	6875	6875
Nuculanidae gen. sp. n.	Tonga	10 415	10 687
	Kermadec	8928	9174
	Java	6820	6850
FAMILY: Tindariidae			
Tindaria sp. sp.	Aleutian	6296	7286
	Kuril–Kamchatka	7210	7230
	Izu-Bonin	6770	6890
	Kermadec	6960	7000
ORDER: Lucinida			
FAMILY: Montacutidae			
Montacuta sp. sp.	Banda	7130	7340
Montacutidae sp. sp.	Philippine	6290	8580
FAMILY: Mytilidae			
Dacridium sp.	South Sandwich	6050	6150
FAMILY: Thyasiridae			
Axinopis sp.	Japan	6156	6207
Axinulus aff. *pygmaeus*	Java	6820	6850
Axinulus sp. n. 1	Java	6841	7060
Axinulus sp. n. 2	Tonga	10 415	10 687
Axinulus sp. n. 3	Kermadec	8928	10 015
Axinulus sp. sp.	Aleutian	6460	7285

	Kuril–Kamchatka	6150	9050
	Japan	7350	7587
	Volcano	8530	8540
	Palau	7000	7170
	Bougainville	7974	9043
Maorithyas hadalis	Japan	7326	7434
Axinulus sp. sp.	Kermadec	9995	10 015
Parethyasira kaireiae	Japan	6270	6440

ORDER: Pectinida
FAMILY: Limarlidae

Lima sp. sp.	Kuril–Kamchatka	9000	9050
	Japan	6156	6207
	Izu-Bonin	9715	9735
	Volcano	6135	6135

FAMILY: Pectinidae

Cyclopecten (H) hadalis	Kermadec	6620	7000
Cyclopecten sp.	San Cristobal	5650	6070
Delectopecten randolphi	Japan	6156	6207
	Java	6820	6850
Delectopecten sp.	Kuril–Kamchatka	6860	8100
Propeamussium sp. sp.	Aleutian	6410	7246
	Kuril–Kamchatka	7210	7230
	Bougainville	7974	8006
	New Hebrides	6680	6830
Pectinidae sp.	Romanche	6330	6430

ORDER: Venerida
FAMILY: Pholadidae

Xylophaga grevi	Banda	7250	7290

FAMILY: Pholadidae

Xylophaga hadalis	Kermadec	6660	6770

FAMILY: Teredinidae

Bankia carinata	Banda	7250	7290
Uperotus clavus	Banda	7250	7290

FAMILY: Vesicomyidae

Vesicomya bruuni	Kermadec	6620	9174
Vesicomya profundi	Aleutian	7246	7246
Vesicomya bruuni	Kuril–Kamchatka	7210	9050
Vesicomya sergeevi	Kuril–Kamchatka	6090	9530
Vesicomya sundensis	Java	6820	7000
Vesicomya sp. sp.	Japan	6156	7587
	Volcano	6330	6330
	Mariana	10 700	10 730
	Romanche	6330	7600
Calyptogena sp.	Japan	6270	6440
Abyssogena phaseoliformis	Japan	6270	6440

ORDER: Cuspidariida
FAMILY: Cuspidariidae

Cuspidaria hadalis	Banda	6580	7210
Cuspidariidae sp. sp.	Japan	7565	7587
	Volcano	6330	6330

	Mariana	8215	8225
	Yap	7190	7250
	Palau	7970	8035
	Philippine	6290	9990
	Ryukyu	6660	6670
	Bougainville	7974	8006
	Romanche	7460	7600
ORDER: Verticordiida			
FAMILY: Verticordiidae			
Laevicordia sp.	South Sandwich	6050	6150
Lyonsiella	Kuril–Kamchatka	8175	8840
Polycordia (Angustebranchia) extente	Kuril–Kamchatka	8185	8400
Polycordia (A) maculata	Kuril–Kamchatka	9000	9050
Polycordia (A) rectangulata	Kuril–Kamchatka	8175	9335
Polycordia (A) sp. 1	Kuril–Kamchatka	8610	8660
Polycordia (A) sp. 2	Kuril–Kamchatka	8175	8840
Polycordia (Latebranchia) ovata	Palau	6040	6040
Polycordia sp. 1	South Sandwich	7200	7216
	San Cristobal	5650	6070
Polycordia sp. 2	Philippine	6290	6330

PHYLUM: SIPUNCULA

CLASS: SIPUNCULIDEA			
ORDER: Golfingiida			
FAMILY: Golfingiidae			
Golfingia (Golfingia) anderssoni	Kuril–Kamchatka	6090	6135
Golfingia (Golfingia) muricaudata	Kuril–Kamchatka	6090	6860
	Japan	6156	6214
Golfingia (Nephasoma) improvisa	Kuril–Kamchatka	6090	6135
Golfingia (Nephasoma) minuta	Kuril–Kamchatka	6090	6710
	Japan	6380	6380
	South Sandwich	6052	6150
Golfingia (Nephasoma) schuttei	Kuril–Kamchatka	6090	6235
Golfingia (Nephasoma) sectile	South Sandwich	6052	6150
FAMILY: Phascolionidae			
Phascolion (montuga) lutense	Aleutian	6296	6328
	Kuril–Kamchatka	6090	6860
	Japan	6156	6207
	Java	6820	6850
	South Sandwich	6052	6150
	Aleutian	6296	6328
	Kuril–Kamchatka	6090	6860
	Japan	6600	6670
	San Cristobal	5650	6070
ORDER: Unknown			
FAMILY: Unknown			
Sipuncula sp.	Aleutian	6520	7298
	Japan	6600	6670
	Palau	6040	6229

	PHYLUM: BRYOZOA		

CLASS: GYMNOLAEMATA
ORDER: Cheilostomatida
FAMILY: Bugulidae

Bugula	Kermadec	8210	8300
Kinetoskias?	Java	6487	6487
	Kuril–Kamchatka	6090	8400
	Izu-Bonin	8800	8830
	Peru–Chile	7000	7000
	Romanche	7340	7340

	PHYLUM: CHAETOGNATHA		

CLASS: SAGITTOIDEA
ORDER: Phragmophora
FAMILY: Eukrohniidae

Eukrohnia fowleri*	Kuril–Kamchatka	6000	8700

	PHYLUM: ARTHROPODA		

CLASS: COPEPODA
ORDER: Calanoida
FAMILY: Aetideidae

Batheuchaeta anomala*	Kuril–Kamchatka	960	7040
Batheuchaeta gurjanovae*	Kuril–Kamchatka	1200	8100
Batheuchaeta heptneri*	Kuril–Kamchatka	3680	7100
Batheuchaeta peculiaris*	Kuril–Kamchatka	3250	6540
Batheuchaeta tuberculata*	Kuril–Kamchatka	3000	7040
Pseudochaeta spinata*	Kuril–Kamchatka	5130	6210
Pseudochaeta sp.*	Kuril–Kamchatka	6000	8500
FAMILY: Bathypontiidae			
Zenkevitchiella abyssalis*	Kuril–Kamchatka	6000	8500
FAMILY: Calanidae			
Neocalanus tonsus*	Pacific/Antarctica	6000	8500
FAMILY: Euchaetidae			
Paraeuchaeta plicata*	Kuril–Kamchatka	1440	7390
Paraeuchaeta sp.*	Kuril–Kamchatka	6000	8500
FAMILY: Heterorhabdidae			
Paraheterorhabdus compactus*	All oceans	5850	8500
FAMILY: Lucicutiidae			
Lucicutia anomola*	Arctic/Atlantic/Pacific	3600	6900
Lucicutia biuncata*	Kuril–Kamchatka	1900	3900
Lucicutia cinerea*	Kuril–Kamchatka	3470	3860
Lucicutia curvifurcata*	Kuril–Kamchatka	1500	8150
Lucicutia ushakovi*	Kuril–Kamchatka	1140	8500
Lucicutia sp.*	Kuril–Kamchatka	5020	6140
FAMILY: Metridinidae			
Metridia okhotensis*	Pacific	6000	8500
Metridia similis abyssalis*	Pacific	6000	8500
FAMILY: Phaennidae			
Xanthocalanus pavlovskii*	Pacific	6000	8500

FAMILY: Spinocalanidae

Mimocalanus distinctocephalus *	NW Pacific	6000	8500
Spinocalanus similis profundalis *	NW Pacific	6000	8500

FAMILY: Scolecitrichidae

Parascaphocalanus zenkevitchi *	NW Pacific	6000	8500
Puchinia obtusa *	NW Pacific	6500	8000
Scaphocalanus acutocornis *	NW Pacific	4295	6551
Falsilandrumis bogorovi *	Atlantic/Indian	6000	8500
Scaphocalanus sp.*	Atlantic/Indian	6000	8500
Scolecithrix birsteini major *	Atlantic/Indian	6000	8500
	Atlantic/Indian	6000	8500
Scolecithrixidae gen. sp.*	Atlantic/Indian	6000	8500

CLASS: CIRRIPEDIA
ORDER: Scalpelliformes
FAMILY: Scalpellidae

Annandaleum japonicum	Kuril–Kamchatka	6675	6860
	Japan	6156	6380
	Ryukyu	6810	6810
Meroscalpellum ultraabissicolum	Ryukyu	6660	6670
Neoscalpellum eltaninae	Peru–Chile	6040	6040
Planoscalpellum hexagonum	Peru–Chile	6040	6040
Trianguloscalpellum regium	Kuril–Kamchatka	6090	6135
Weltnerium speculum	San Cristobal	5650	6070
Arcoscalpellinae sp. 1	Izu-Bonin	6770	6850
Arcoscalpellinae sp. 2	Philippine	7420	7880
Scalpellum sp.	Kermadec	6620	7000

CLASS: OSTRACODA
ORDER: Podocopida
FAMILY: Bairdiidae

Bairdiidae sp. n.	Bougainville	6920	7657
Bairdia sp.	Palau	7000	7170

FAMILY: Bythocyprididae

Bythocypris sp.	Philippine	6290	6330
Retibythere scaberrima	Puerto-Rico	7950	8100
Zabythocypris helicina	Peru–Chile	5986	6134

FAMILY: Krithidae

Krithe setosa	Java	6487	6487

FAMILY: Trachyleberidae

Actinocyteris sp.	Palau	6290	6330

ORDER: Halocyprida
FAMILY: Halocyprididae

Archiconchoecia maculata	Kuril–Kamchatka	7280	9500
Archiconchoecia sp. n.	Mariana	4400	8000
	Bougainville	6700	8000
Bathyconchoecia paulula pacifica	Kuril–Kamchatka	5400	9500
Juryoecia abyssalis	Kuril–Kamchatka	4200	8500
	Mariana	4400	8000
	Bougainville	4200	6200
	Kermadec	6200	8000
Paraconchoecia mamillata	Kuril–Kamchatka	4200	8500

Bathyconchoecia sp. n. sp.	Kermadec	5173	5173
Matavargula cf. *adinothrix*	Kermadec	5173	5173
Paraconchoecia vitjazi	Kuril–Kamchatka	7280	9500
CLASS: MALACOSTRACA			
ORDER: Mysida			
FAMILY: Mysidae			
Amblyops magna	Kuril–Kamchatka	6435	7230
Amblyops sp. n. 1	Yap	7190	8720
	Banda	7335	7340
Amblyops sp. n. 2	South Sandwich	7200	7216
Amblyops sp. n. 3	Volcano	6780	6785
Amblyops sp. n.	Japan	7190	7190
Amblyops sp. 4	Kermadec	6265	6709
Birsteiniamysis sp. n.	South Sandwich	7200	7216
Mysimenzies hadalis	Peru	6146	6354
Mysimenzies sp. n. 1	Palau	7970	8035
Mysimenzies sp. n. 2	Ryukyu	7440	7450
Paramblyops sp. n.	Aleutian	7246	7246
	Yap	7190	8720
Mysidacea sp. sp.	Palau	8021	8042
	New Britain	8258	8266
	Bougainville	7847	8662
	New Hebrides	6758	6776
ORDER: Cumacea			
FAMILY: Botodriidae			
Bathycuma sp. n.	Bougainville	6920	7657
Vaunthompsonia aff. *cristata*	Kuril–Kamchatka	6475	6571
FAMILY: Diastylidae			
Makrokylindrus hadalis	Java	7160	7160
Makrokylindrus hystrix	Japan	6380	6450
FAMILY: Lampropidae			
Lamprops sp. n.	Java	6820	6850
FAMILY: Leuconidae			
Leucon sp. nov.	Aleutian	7426	7426
FAMILY: Nannastacidae			
Cumacea sp. sp.	Izu-Bonin	6770	6890
	Volcano	6330	6330
	Marianas	6580	6650
	Palau	8021	8042
	New Britain	7875	7921
	Bougainville	7984	8006
	New Hebrides	6758	6776
	Peru	6324	6328
	South Sandwich	6766	7216
	San Cristobal	5650	6070
ORDER: Tanaidacea			
FAMILY: Apseudidae			
Apseudes galathea	Kermadec	6660	6770
Apseudes zenkevitchi	Kermadec	6065	6065
Apseudidae gen. sp.	Bougainville	6920	7657

FAMILY: Gigantapseudidae

Gigantapseudes adactylus	Philippine	6920	7880

FAMILY: Neotanaidae

Herpotanais kirkegaardi	Kermadec	6960	7000
Neotanais americanus	Izu-Bonin	6770	6850
Neotanais armiger	Peru	5986	6134
Neotanais hadalis	Kermadec	6960	8300
	Bougainville	6920	7657
	Puerto-Rico	8330	8330
Neotanais insignis	Mariana	8215	8225
Neotanais kurchatovi	South Sandwich	7200	7934
Neotanais serratispinosus	Kermadec	6960	8300
	Bougainville	6920	7657
	Aleutian	6520	6520
Neotanais tuberculatus	Kuril–Kamchatka	7265	7295
	Puerto-Rico	6800	7030
Neotanais sp. A	Japan	6380	7460
Neotanais sp. B	Japan	6380	6380
Neotanais sp.	New Hebrides	6680	6830
Neotanais?	New Britain	7875	7921

FAMILY: Paratanaidae

Heterotanoides ornatus	Japan	7370	7370

FAMILY: Pseudotanaidae

Cryptocopoides arctica	Izu-Bonin	6770	6850
Pseudotanais affinis	Izu-Bonin	6770	6890
Pseudotanais nordenskioldi	Izu-Bonin	6770	6890
	Banda	7335	7430
Pseudotanais vitjazi	Japan	7370	7370
Pseudotanais sp.	Kuril–Kamchatka	6675	6710

FAMILY: Taniadae

Protanais sp.	Kuril–Kamchatka	6090	6135

FAMILY: Agathotanaidae

Paragathonais typicus	Izu-Bonin	6770	6890

FAMILY: Anarthotanaidae

Anarthruropsis langi	Kuril–Kamchatka	7795	8015

FAMILY: Leptognathiidae

Colletea cylindrata	Kuril–Kamchatka	6090	6710
Colletea minima	Izu-Bonin	6770	6890
Leptognathia angustocephala	Romanche	6330	7500
Leptognathia armata	Aleutian	6520	6520
	Bougainville	7974	8006
	South Sandwich	6052	6150
Leptognathia birsteini	Bougainville	6920	7657
Leptognathia brevimeris	Kuril–Kamchatka	7265	7295
Leptognathia caudata	Bougainville	7974	8006
Leptognathia dentifera	Kuril–Kamchatka	6225	6225
Leptognathia dissimilis	Puerto-Rico	5890	6000
Leptognathia elegans	Bougainville	7974	8006
Leptognathia gracilis	Kuril–Kamchatka	6675	6710
	Japan	7370	7370

	South Sandwich	6052	7218
Leptognathia greveae	Kuril–Kamchatka	8185	8400
Leptognathia langi	Kuril–Kamchatka	6675	6710
Leptognathia longiremis	Izu-Bonin	6770	6890
	Aleutian	6520	6520
	Philippine	6290	6850
	Java	6433	6475
	Kermadec	8928	9174
Leptognathia microcephela	Volcano	6330	6330
Leptognathia parabranchiata	Palau	7000	7170
Leptognathia paraforcifera	South Sandwich	6052	6150
Leptognathia robusta	Izu-Bonin	6770	6850
Leptognathia sp.	Kuril–Kamchatka	8700	8700
Libanus longicephalus	Izu-Bonin	6770	6890
Tryphlotanais compactus	Kuril–Kamchatka	6090	6135
Tryphlotanais elegans	Volcano	6330	6330
	Izu-Bonin	6770	6850
Tryphlotanais grandis	Kuril–Kamchatka	6090	6135
Tryphlotanais kussakini	Kuril–Kamchatka	6090	6135
Tryphlotanais mucronatus	Kuril–Kamchatka	6675	6710
Tryphlotanais rectus	Volcano	6330	6330
	Japan	7370	7370
Tanadacea sp. sp.	New Hebrides	6758	6776
	New Britain	7875	7921

ORDER: Isopoda
FAMILY: Laptanthuridae

Laptanthura hendii	Banda	6580	6580

FAMILY: Desmosomatidae

Desmosoma similpes	Peru–Chile	5986	6134
Desmosoma tenuipes	Kuril–Kamchatka	6675	6710

FAMILY: Echinothambematidae

Echinothambema sp. n.	Cayman	5800	6850

FAMILY: Munnopsidae

Bathyopsurus nybelini	Puerto-Rico	7265	7900
Betamorpha acuticoxalis	Kuril–Kamchatka	7210	8400
	Japan	6156	7587
Eurycope complanata	Puerto-Rico	6800	7030
Eurycope curtirostris	Kuril–Kamchatka	7210	7230
	Japan	7370	7370
Eurycope eltanje	Peru–Chile	5986	6134
Eurycope galatheae	Kermadec	6960	7000
Eurycope kurchatovi	Puerto-Rico	6800	7030
Eurycope madseni	Kermadec	6960	7000
Eurycope magna	Aleutian	7246	7246
	Kuril–Kamchatka	7210	8345
Eurycope ovata	Japan	7370	7370
Eurycope quadratifrons	Romanche	7200	7200
Eurycope sp. 1	Java	6820	6850
Eurycope sp. 2	Kermadec	8928	9174
Eurycope sp. A	Puerto-Rico	8330	8330

Eurycope sp. 3	Izu-Bonin	6770	6890
Munneurycope curticephala	Kuril–Kamchatka	6675	7230
Munneurycope menziesi	Kermadec	6960	7000
Storthyngurella (Storthyngura) benti	Kermadec	6620	7000
Vanhoeffenura (S) bicornins	Japan	6156	6207
Vanhoeffenura (S) chelta	Kuril–Kamchatka	6090	6860
Rectisura (S) furcata	Kermadec	6620	6770
Rectisura (S) herculea	Aleutian	7246	7246
	Kuril–Kamchatka	6475	9345
	Japan	6700	7703
Storthyngura pulchra kermadecensis?	Kermadec	6620	6730
Rectisura (S) tenuispinis	Kuril–Kamchatka	6205	8430
	Aleutian	7246	7246
	Japan	6700	7370
Storthyngurella (S) zenkevitchi	Romanche	7200	7200
Storthyngurella (S) sp. n. 1	Java	6820	6850
Storthyngurella (S) sp. n. 2	Puerto-Rico	6400	7030
Storthyngurella (S) sp. sp.	Izu-Bonin	6770	6850
	Volcano	6330	6330
Syneurycope sp.	Kermadec	8928	9174
Ilyarachna defecta	Peru–Chile	6073	6281
Ilyarachna kermadecensis	Kermadec	6600	7000
Ilyarachna kussakini	Kuril–Kamchatka	6090	7230
	Japan	6156	7370
Ilyarachna vemae	Peru–Chile	6052	6328
Ilyarachna sp. 1	Kuril–Kamchatka	8330	8430
Ilyarachna sp. 2	Kermadec	8928	9174
Ilyarachna sp. 3	Tonga	10 415	10 687
Ilyarachna sp. 4	Peru–Chile	6324	6348
FAMILY: Haploniscidae			
Haploniscus belyaevi	Kuril–Kamchatka	6090	6225
Haploniscus bruuni	Peru–Chile	5986	6260
Haploniscus gibbernastus	Kuril–Kamchatka	6435	6710
Haploniscus hydroniscoides	Kuril–Kamchatka	6675	8120
	Japan	7370	7370
Haploniscus inermis	Kuril–Kamchatka	8035	8345
Haploniscus intermedius	Kuril–Kamchatka	6090	6135
Haploniscus menziesi	Kuril–Kamchatka	6090	6135
Haploniscus profundicola	Kuril–Kamchatka	6090	7710
	Japan	7370	7370
Haploniscus pygmeus	Romanche	7280	7280
Haploniscus ultraabyssalis	Bougainville	6920	8006
Haploniscus cf. *unicornis*	Puerto-Rico	8330	8330
Haploniscus sp. n. 1	Java	6935	7060
Haploniscus sp. n. 2	Tonga	10 415	10 687
Haploniscus sp. n. 3	Kermadec	8928	9174
Haploniscus sp. n. 4	Puerto-Rico	7430	7430
Haploniscus sp. sp.	Izu-Bonin	6770	6850
	Phillipine	8440	8580
Haploniscus sp.	Japan	7190	7250

Mastigoniscus concavus	Peru–Chile	6073	6281
Mastigoniscus latus	Kuril–Kamchatka	6435	8400
FAMILY: Ischnomedidae			
Haplomesus bervispinus	Kuril–Kamchatka	6090	6135
Haplomesus concinnus	Kuril–Kamchatka	6090	6135
Haplomesus consanguineus	Izu-Bonin	8800	8830
Haplomesus cornutus	Kuril–Kamchatka	6475	6571
Haplomesus gigas	Kuril–Kamchatka	6475	6571
	Japan	6675	8430
Haplomesus profundicola	Kuril–Kamchatka	7265	7295
Haplomesus robustus	Kuril–Kamchatka	6675	6710
Haplomesus thomsoni	Kuril–Kamchatka	6435	6710
Haplomesus sp. sp.	Izu-Bonin	6770	6850
	Volcano	6330	6330
Haplomesus sp. n. A	Puerto-Rico	7430	8330
Haplomesus sp. n. B	Puerto-Rico	7938	7938
Ischnomesus andriashevi	Japan	6156	6207
Ischnomesus bruuni	Kermadec	6960	7000
Ischnomesus elongatus	Bougainville	7974	8006
Ischnomesus sparcki	Kermadec	6660	7000
Ischnomesus sp. B	Cayman	6840	6850
Ischnomesus sp. n.	Puerto-Rico	8330	8330
Stylomesus hexatuberculatus	Kuril–Kamchatka	6050	6135
Stylomesus inermis	Aleutian	6079	6079
Stylomesus menziesi	Kuril–Kamchatka	6090	6135
Stylomesus sp. n.	Kuril–Kamchatka	6675	6710
FAMILY: Janirellidae			
Janirella erostrata	Bougainville	7974	8006
Janirella fusiformis	Japan	6156	6207
Janirella macrura	Kuril–Kamchatka	6435	7230
Janirella quadritubercuata	Kuril–Kamchatka	6150	6150
Janirella sedecimtuberculata	Japan	6350	6450
Janirella spinosa	Kuril–Kamchatka	6435	8430
	Japan	7370	7370
Janirella tuberculata	Japan	6380	6450
Janirella verrucosa	Kuril–Kamchatka	6205	6850
Janirella sp.	Izu-Bonin	6770	6850
FAMILY: Acanthaspidiidae			
Acanthaspidia curtispinosa	South Sandwich	6766	7216
Acanthaspidia iolanthoidea	San Cristobal	5650	6070
Acanthaspidia cf. *decorata*	Puerto-Rico	6400	7030
FAMILY: Macrostylidae			
Macrostylis compactus	Bougainville	6920	7657
Macrostylis curticornis	Kuril–Kamchatka	6225	6225
	Japan	6600	6670
Macrostylis galatheae	Palau	9820	10 000
Macrostylis grandis	Kuril–Kamchatka	7265	7295
Macrostylis hadalis	Banda	7270	7270
Macrostylis longifera	Peru–Chile	5986	6354
Macrostylis ovata	Kuril–Kamchatka	6435	6710

Macrostylis profundissimus	Kuril–Kamchatka	8240	9530
Macrostylis porrecta	Java	6433	6433
Macrostylis vitjazi	Bougainville	6920	7657
Macrostylis zenkevitchi	Kuril–Kamchatka	6090	6135
Macrostylis sp. n.	Puerto-Rico	7950	8100
Macrostylis sp. 1	Mariana	10 630	10 710
Macrostylis sp. 2	Tonga	10 415	10 687
Macrostylis sp. 3	Izu-Bonin	6770	8900
	Volcano	6330	8540
	Mariana	8890	10 730
	Palau	6290	9750
FAMILY: Mesosignidae			
Mesosignum latum	Kuril–Kamchatka	6090	6135
Mesosignum multidens	Peru–Chile	5986	6450
Mesosignum vitjazi	Bougainville	6920	7657
Mesosignum sp.	Peru–Chile	6002	6002
Mesosignum sp. sp.	Izu-Bonin	6770	6890
	Volcano	6330	6330
	Philippine	7420	7880
FAMILY: Munnidae			
Aryballurops japonica	Japan	6380	6450
Munna sp.	Japan	6380	6380
Zoromunna setifrons	Peru–Chile	5986	6134
FAMILY: Nannoniscidae			
Janthura (Austroniscoidea) bougainvillei	Bougainville	6920	9043
Austroniscus acutus	Kuril–Kamchatka	6090	6135
Austroniscus sp. n.	Cayman	6800	6850
Nannoniscus ovatus	Peru–Chile	6321	6328
Nannoniscus perunis	Peru–Chile	5986	6134
Nannoniscus sp.	Peru–Chile	6073	6281
Nannoniscidae gen. n. sp. n. A	Cayman	6800	6800
Nannoniscidae gen. n. sp. n. B	Puerto-Rico	8330	8330
FAMILY: Cirolanidae			
Corolana sp.	Peru–Chile	5986	6134
FAMILY: Antartcturidae			
Chaetarcturus (Antarcturus) abyssalis	Kuril–Kamchatka	6090	6135
Chaetarcturus (Antarcturus) ultraabyssalis	Kuril–Kamchatka	6435	7320
	Japan	7190	7190
Antarcturus zenkevitchi	Kuril–Kamchatka	7370	7370
FAMILY: Arcturidae			
Arcturus primus	Japan	7370	7370
Arcturus sp.	South Sandwich	7200	7216
ORDER: AMPHIPODA			
FAMILY: Stilipedidae			
Alexandrella carinata	Kuril–Kamchatka	7210	7230
FAMILY: Stegocephalidae			
Andaniexis australis	Peru	6324	6328
Andaniexis sp.*	Izu-Bonin	6770	6890
*Andaniexis stylifer**	Bougainville	6500	8500

FAMILY: Andaniexinae

*Andaniexis subabyssi**	Kuril–Kamchatka	6000	8500

FAMILY: Maeridae

Bathyceradocus stephenseni	Bougainville	6920	7652
	Banda	7250	7340
Metaceradocoides vitjazi	Yap	7190	7250
	Japan	6600	7370
	Mariana	8215	8225
	Izu-Bonin	8900	8900

FAMILY: Lysianassidae

Bathyschraderia fragilis	Philippine	7000	9990
Bathyschraderia magnifica	Kermadec	6960	9174
	Tonga	7354	9875
Onesimoides cavimanus	Banda	6490	6650
Orchomene abyssorum	Kermadec	8210	8230
Orchomene sp.	Mariana	10 500	10 500
Galathella galatheae	Kermadec	6960	7000
Unknown (Uristes) gen. n. sp. n.	Tonga	7349	9273
	Kermadec	9104	9104

FAMILY: Tryphosidae

Tryphosella bruuni	Kermadec	6660	6770
Tryphosella sp. 2	Kermadec	6007	6007
	Peru–Chile	7050	7050

FAMILY: Epimeriidae

Epimeria sp. nov.	Japan	6156	6207
	Kuril–Kamchatka	7210	7230

FAMILY: Eurytheneidae

*Eurythenes gryllus***	Peru–Chile	4329	8072
	Izu-Bonin	6770	7850
	Tonga	5155	6252
	Kermadec	4329	6252

FAMILY: Eusiridae

Eusirella longisetosa	Bougainville	8500	8500
Eusirus bathybius	Bougainville	7500	7500
	Philippine	7625	7900
	Puerto-Rico	7625	7900
*Eusirus fragilis**	Tonga	9120	9120
Rhachotropis flemmingi	Kuril–Kamchatka	6090	6135
	Java	6820	7160
Rhachotropis sp. n.	Philippine	7420	7880
Rhachotropsis sp.?	Kermadec	6960	7000

FAMILY: Pardaliscidae

*Halice aculeata**	Izu-Bonin	4000	6500
	Kuril–Kamchatka	4200	8050
	Bougainville	6500	6500
	Tonga	7100	10 500
*Halice quarta**	Kuril–Kamchatka	6000	8500
	Izu-Bonin	8480	9000
	Tonga	9120	9120
	Mariana	10 000	10 000

*Halice rotunda**	Bougainville	4050	8400
	Tonga	9120	9120
*Halice subquarta**	Yap	7190	7250
	Philippine	7420	7880
	Kermadec	9400	9400
	Tonga	10 500	10 500
Pardaliscoides longicaudatus	Kermadec	6180	6180
	Philippine	9820	10 000
Princaxelia abysallis	Kermadec	6620	8300
	Philippine	7420	7880
	Kuril–Kamchatka	6435	9530
	Aleutian	6965	7000
	Yap	7190	8720
	Japan	6380	7370
	Izu-Bonin	6770	8830
	Bougainville	7974	8006
Princaxelia magna	Japan	7190	7250
	Tonga	7354	8411
Princaxelia jamiesoni	Izu-Bonin	9316	9316
	Japan	7703	7703
FAMILY: Phoxocephalidae			
Harpinia abyssalis	Peru	6324	6328
Harpiniopsis spaercki	Banda	6580	7340
Metaphoxus sp.	Japan	7550	7550
FAMILY: Hirondelleidae			
Hirondellea dubia	Kermadec	6000	9400
Hirondellea gigas	Izu-Bonin	6770	8900
	Palau	7970	8035
	Kuril–Kamchatka	7250	9345
	Volcano	8530	8540
	Yap	8560	8720
	Mariana	7218	10 592
	Philippine	8467	10 190
Hirondellea sp. 1	Peru–Chile	6173	6173
Hirondellea sp. 2	Peru–Chile	6173	8072
Hirondellea sp. 3	Peru–Chile	7050	7050
FAMILY: Hyperiopsidae			
*Hyperiopsis anomala**	Tonga	6900	6900
*Hyperiopsis laticarpa**	Kuril–Kamchatka	6000	8500
	Izu-Bonin	8480	8480
	Bougainville	8500	8500
*Parargissa affinis**	Izu-Bonin	6500	6500
	Bougainville	8150	8500
	Bougainville	8150	8500
*Parargissa arquata**	Kuril–Kamchatka	4200	8500
*Protohyperiopsis curticornis**	New Hebrides	7000	7000
*Protohyperiopsis longipes**	Bougainville	8500	8500
FAMILY: Lanceolidae			
*Lanceola clausi gracilis**	Kuril–Kamchatka	4200	8000
	Philippine	6200	6750

*Lanceola sphaerica**	Kuril–Kamchatka	7800	7800
*Metalanceola chevreuxi**	Bougainville	6500	8500
	Kermadec	9400	9400
	Tonga	9100	10 500
FAMILY: Lepechinellidae			
Lepechinella aberrantis	Volcano	6330	6330
FAMILY: Atylidae			
Lepechinella ultraabyssalis	Kuril–Kamchatka	6475	8015
	Japan	7370	7370
Lepechinella vitrea	Yap	7190	7250
Lepechinella wolfii	Kermadec	6660	6770
FAMILY: Liljeborgiidae			
Liljeborgia caeca	Japan	6156	6207
FAMILY: Alicellidae			
Paralicella microps	Japan	6580	6580
	Kuril–Kamchatka	8000	8000
	Izu-Bonin	8480	8480
*Paralicella tenuipes***	Kermadec	4786	7000
	Peru–Chile	6173	7050
	Tonga	7300	7300
Paralicella caperesca	Kermadec	4329	6007
	Peru–Chile	4602	6173
Alicella gigantea	Kermadec	6200	7000
FAMILY: Cyclocaridae			
Cyclocaris tahitensis	Kermadec	6007	6007
FAMILY: Scinidae			
*Scina cheleta**	Kuril–Kamchatka	7750	7750
*Scina wagleri abyssalis**	Kuril–Kamchatka	6000	8500
	Izu-Bonin	8500	8500
	Kermadec	9400	9400
FAMILY: Scopelocheiridae			
*Scopelocheirus hopei (pacifica)**	Kermadec	6960	7000
Scopelocheirus schellenbergi	Kuril–Kamchatka	6000	7000
	Japan	6380	7370
	Java	6935	7060
	Aleutian	6965	7200
	New Hebrides	6680	8000
	Tonga	6252	8723
	Puerto-Rico	7625	7900
FAMILY: Stegocephalidae			
Stegocephalus nipoma	Philippine	6290	6330
Stegocephalus sp. nov.	Kuril–Kamchatka	7600	7710
	Kuril–Kamchatka	7795	8015
Stegocephalus sp. nov. 1	Japan	6380	6380
Steleuthera maremboca	Peru	6324	6380
FAMILY: Uristidae			
Abyssorchomene gerulicorbis	Kermadec	5173	6007
Abyssorchomene chevreuxi	Peru–Chile	6173	6173
FAMILY: Velettiopsidae			
Valettietta anacantha	Kermadec	6007	6007

FAMILY: Vitjazianidae

*Vitjaziana gurjanovae**	Izu-Bonin	4200	8480

CLASS: PYCNOGONIDA
ORDER: Pantopoda
FAMILY: Ascorhynchidae

Acorhynchus birsteini	Palau	6040	6040
Acorhynchus inflatum	San Cristobal	5565	6070

FAMILY: Austrodecidae

Colossendels sp.	Aleutian	6410	6757
Pantopipetta longituberculata	Kuril–Kamchatka	6090	6710
	South Sandwich	6052	6150

FAMILY: Nymphonidae

Heteronymphon profundum	Kuril–Kamchatka	6860	6860
	Japan	6156	6380
Nymphon femorale	Banda	6490	6650
Nymphon longitarse	Japan	7270	7370
Nymphon procerum	Kuril–Kamchatka	6090	6135
Nymphon tripectinatum	Japan	7370	7370

PHYLUM: HEMICHORDATA

CLASS: ENTEROPNEUSTA
ORDER: Enteropneusta
FAMILY: Torquaratiodae?

Enteropneausta sp.	Kuril–Kamchatka	5615	8100
	Aleutian	6520	7250
	South Sandwich	8004	8116
	New Britain	8258	8260
	New Hebrides	6758	6776

PHYLUM: CHORDATA

CLASS: ASCIDIACEA
ORDER: Phlebobranchia
FAMILY: Corellidae

Corellidae sp.?	New Britain	7057	7075
	New Hebrides	6758	6776

FAMILY: Octanemidae

Octacnemus sp.	Kuril–Kamchatka	8185	8400
Situla pelliculosa	Kuril–Kamchatka	7265	8430

ORDER: Stolidobranchia
FAMILY: Hexacrobylidae

Asajirus (hexacrobylus) sp.	Philippine	7420	7880
Asajirus (hexacrobylus?) gen. et. sp. n.	Volcano	6330	6330

FAMILY: Pyuridae

Culeolus murrayi	Japan	6156	6207
Culeolus robustus	Kuril–Kamchatka	7265	7295
Culeolus tenuis	Japan	6156	6207
Culeolus sp. sp.	Kuril–Kamchatka	7210	8015
	Japan	6380	6380

FAMILY: Styelidae

Cnemidocarpa bythia	Kermadec	6180	7000

CLASS: ACTINOPTERYGII
ORDER: Gadiformes
FAMILY: Macrouridae

Coryphaenoides yaquinae	Japan	6160	6945

ORDER: Ophidiiformes
FAMILY: Ophidiidae

Leucicorus atlanticus	Cayman	4580	6800
Unidentified ophidid	Peru–Chile	5329	6173
Abyssobrotula galatheae	Puerto-Rico	8370	8370
	Japan	6480	6640
Holcomycteronus profundissimus	Java	5600	7160
Apagesoma edentatum	Puerto-Rico	5082	8082

FAMILY: Carapidae

Echiodon neotes	Kermadec	8200	8300

ORDER: Scorpaeniformes
FAMILY: Liparidae

Notoliparis antonbruuni	Peru–Chile	6150	6150
*Notoliparis kermadecensis***	Kermadec	6474	7561
Notoliparis sp.**	Peru–Chile	7050	7050
Pseudoliparis ambylstomopsis	Kuril–Kamchatka	7210	7230
	Japan	6945	7703
Pseudoliparis belyaevi	Japan	7565	7587

References

Abe, F. and Horikoshi, K. (2001). The biotechnological potential of piezophiles. *Trends in Biotechnology*, **19**(3), 102–108.

Abdel-Mageed, W.M,. Milne, B.F., Wagner, M. *et al.* (2010). Dermacozines, a new phenazine family from deep-sea dermacocci isolated from Mariana Trench sediment. *Organic and Biomolecular Chemistry*, **8**(10), 2352–2362.

Aertsen, A., Meersman, F., Hendrickx, M.E., Vogel, R.F. and Michiels, C.W. (2009). Biotechnology under high pressure: applications and implications. *Trends in Biotechnology*, **27**(7), 434–441.

Agassiz, A. and Mayer, A.G. (1902). Reports on the scientific results of the expedition to the tropical Pacific in charge of Alexander Agassiz by the US Fish Commission steamer *Albatross* from August 1899 to March 1900. III. *The Medusae, Memoirs of the Museum of Comparative Zoology at Harvard College*, **26**, 139–176.

Aguilar, A., Ingemansson, T. and Magnien, E. (1998). Extremophile microorganisms as cell factories: support from the European Union. *Extremophiles*, **2**, 367–373.

Aguzzi, J., Jamieson, A.J., Fujii, T. *et al.* (2012). Shifting feeding behaviour of deep-sea buccinid gastropods at natural and simulated food falls. *Marine Ecology Progress Series*, **458**, 247–253.

Akimoto, K., Hattori, M., Uematsu, K. and Kato, C. (2001). The deepest living Foraminifera, Challenger Deep, Mariana Trench. *Marine Micropaleontology*, **42**, 95–97.

Albertelli, G., Amaud, P.M., Della Croce, N., Drago, N. and Elefteriou, A. (1992). The deep Mediterranean macrofauna caught by traps and its trophic significance. *Comptes Rendus de l'Academie des Sciences*, **315**(111), 139–144.

Alexander, D.E. (1988). Kinematics of swimming in two species of *Idotea* (Isopoda: Valvifera). *Journal of Experimental Biology*, **138**, 37–49.

Allen, M.J. and Jaspars, M. (2009). Realizing the potential of marine biotechnology: challenges and opportunities. *Industrial Biotechnology*, **5**(2), 77–83.

Allwood, A., Beaty, D., Bass, D. *et al.* (2013). Conference summary: life detection in extraterrestrial samples. *Astrobiology*, **13**(2), 203–216.

Amils, R., Blix, A., Danson, M. *et al.* (2007). Investigating life in extreme environments: a European perspective. European Science Foundation Position Paper.

Amstutz, A. (1951). Sur l'e´volution des structures alpines. *Archive Des Sciences*, **4**, 323–329.

Anderson, M.E., Crabtree, R.E., Carter, H.J., Sulak, K.J. and Richardson, M.D. (1985). Distribution of demersal fishes of the Caribbean Sea found below 2000 meters. *Bulletin of Marine Science*, **37**, 794–807.

Ando, M., Ishida, M., Nishikawa, Y., Mizuki, C. and Hayashi, Y. (2012). What caused a large number of fatalities in the Tohoko earthquake? *Geophysical Research Abstracts*, **14**, EGU2012–5501–1.

Andriashev, A.P. (1953). Archaic deep-sea and secondary deep sea-fishes and their role in zoogeographical analysis. *Essays on the General Problems of Ichthyology*, 58–64. (In Russian; translation on-line.)

Andriashev, A.P. (1955). A new fish of the snailfish family (Pisces, Liparidae) found at a depth of more than 7 kilometers. *Trudy Instituta Okeanologii im. P.P. Shirshova*, **12**, 340–344.

Andriashev, A.P. and Pitruk, D.L. (1998). A review of the ultra-abyssal (hadal) genus *Pseudoliparis* (Scorpaeniformes, Liparidae) with a description of a new species from the Japan Trench. *Voprosy Ikhtiologii*, **33**, 325–330.

Andriashev, A.P. and Stein, D.L. (1998). Review of the snailfish genus *Careproctus* (Liparidae, Scorpaeniformes) in Antarctic and adjacent waters. *Natural History Museum of Los Angeles County Contributions in Science*, **470**, 1–63.

Angel, M.V. (1982). Ocean trench conservation. International Union for Conservation of Nature and Natural Resources. *The Environmentalist*, **2**, 1–17.

Anon. (1998). Executive summary. The legendary ocean: the unexplored frontier. Year of the Ocean Discussion Papers (March 1998). Silver Spring, MD: Office of the Chief Scientist, NOAA, US Department of Commerce. p. L-12.

Anon. (2007) *Investigating the Oceans*. London: House of Commons Science and Technology Select Committee.

Aono, E., Baba, T., Ara, T. *et al.* (2010). Complete genome sequence and comparative analysis of *Shewanella violacea*, a psychrophilic and piezophilic bacterium from deep sea floor sediments. *Molecular BioSystems*, **6**, 1216–1226.

Archer, D.E. (1996). An atlas of the distribution of calcium carbonate in sediments of the deep sea. *Global Biogeochemical Cycles*, **10**, 159–174.

Armstrong, J.D., Bagley, P.M. and Priede, I.G. (1992). Photographic and acoustic tracking observations of the behaviour of the grenadier *Coryphaenoides* (*Nematonurus*) *armatus*, the eel *Synaphobranchus bathybius*, and other abyssal demersal fish in the North Atlantic Ocean. *Marine Biology*, **112**, 535–544.

Arnison, P.G., Bibb, M.J., Bierbaum, G. *et al.* (2013). Ribosomally synthesized and post-translationally modified peptide natural products: overview and recommendations for a universal nomenclature. *Natural Products Report*, **30**(1), 108–160.

Arrhenius, O. (1921). Species and area. *Journal of Ecology*, **9**(1), 95–99.

Arzola, R.G., Wynn, R.B., Lastras, G., Masson, D.G. and Weaver, P.P.E. (2008). Sedimentary features and processes in the Nazaré and Setúbal submarine canyons, west Iberian margin. *Marine Geology*, **250**, 64–88.

Attrill, M.J. and Rundle, S.D. (2002). Ecotone or ecocline: ecological boundaries in estuaries. *Estuarine, Coastal and Shelf Science*, **55**(6), 929–936.

Bacescu, M. (1971). *Mysimenzies hadalis* g. n. sp. n., a benthic mysid of the Peru Trench, found during cruise XI/1965 of R/V *Anton Bruun* (USA). *Revue Roumaine de Biologie (Zoologie)*, **16**(1), 3–8.

Bagley P.M., Priede, I.G., Jamieson, A.J. *et al.* (2005). Lander techniques for deep ocean biological research. *Underwater Technology*, **26**(1), 3–11.

Bailey, D.M. and Priede, I.G. (2002). Predicting fish behaviour in response to abyssal food-falls. *Marine Biology*, **141**(5), 831–840.

Bailey, D.M., King, N.J. and Priede, I.G. (2007). Camera and carcasses: historical and current methods for using artificial food falls to study deep-water animals. *Marine Ecology Progress Series*, **350**, 179 191.

Balmaseda, M.A., Trenberth, K.E. and Källen, E. (2013). Distinctive climate signals in reanalysis of global ocean heat content. *Geophysical Research Letters*, **40**, 1–6.

Barker, B.A.J., Helmond, I., Bax, N.J. *et al.* (1999). A vessel-towed camera platform for surveying seafloor habitats of the continental shelf. *Continental Shelf Research*, **19**, 1161–1170.

Barnard, J.L. (1961). Gammaridean Amphipoda from depths of 400 to 6000 meters. *Galathea Report*, **5**, 23–128.

Barnard, J.L. and Ingram, C.L. (1986). The supergiant amphipod, *Alicella gigantea* Chevreux from the North Pacific Gyre. *Journal of Crustacean Bioogy*, **6**, 825–839.

Barnes, D.K.A. (2002). Biodiversity: invasions by marine life on plastic debris. *Nature*, **416**, 808–809.

Barnett, P.R.O., Watson, J. and Connelly, D. (1984). The multiple corer for taking virtually undisturbed samples from shelf, bathyal and abyssal sediment. *Oceanologica Acta*, **7**, 399–408.

Barradas-Ortiz, C., Briones-Fourzán, P. and Lozano-Álvarez, E. (2003). Seasonal reproduction and feeding ecology of giant isopods *Bathynomus giganteus* from the continental slope of the Yucatán peninsula. *Deep-Sea Research I*, **50**, 495–513.

Barry, J.P. and Hashimoto, J. (2009). Revisiting the Challenger Deep using the ROV *Kaikō*. *Marine Technology Society Journal*, **43**(5), 77–78.

Barry, J.P., Kochevar, R.E. and Baxter, C.H. (1997). The influence of pore-water chemistry and physiology in the distribution of vesicomyid clam at cold seeps in Monterey Bay: implications for patterns of chemosynthetic community organization. *Limnology and Oceanography*, **42**, 318–328.

Bartlett, D.H. (2002). Pressure effects on *in vivo* microbial processes. *Biochimica et Biophysica Acta*, **1595**, 367–381.

Bartlett, D.H. (2009). Microbial life in the trenches. *Marine Technology Society Journal*, **43**(5), 129–131.

Beaulieu, S.E. (2002). Accumulation and fate of phytodetritus on the sea floor. *Oceanography and Marine Biology Annual Review*, **40**, 171–232.

Beittel, J.S. and Margesson, R. (2010). Chile earthquake: US and international response. *Congressional Research Service Report for Congress*, 7–5700.

Belman, B.W. and Gordon, M.S. (1979). Comparative studies on the metabolism of shallow-water and deep-sea marine fishes. 5. Effects of temperature and hydrostatic pressure on oxygen consumption in the mesopelagic *Melanostigma pammelas*. *Marine Biology*, **50**, 275–281.

Belyaev, G.M. (1966). Bottom fauna of the ultra-abyssal depths of the world ocean. *Akademia Nauka SSSR, Trudy Instituta Okeanologii*, **591**, 1–248.

Belyaev, G.M. (1975). New species of holothurians of the genus *Elpidia* from the southern part of Atlantic Ocean. *Trudy Instituta Okeanologii*, **103**, 259–280. (In Russian.)

Belyaev, G.M. (1989). *Deep-sea Ocean Trenches and Their Fauna*. Moscow: Nauka.

Berger, W.H. (1974). Deep-sea sedimentation. In *The Geology of Continental Margins*, ed. C.A. Burke and C.D. Drake. New York: Springer, pp. 213–241.

Berger, W.H. and Wefer, G. (1996). *Late Quaternary Movement of the Angola–Benguela Front: SE Atlantic, and Implications for Advection in the Equatorial Ocean*. Berlin: Springer.

Bergmann, M. and Klages, M. (2012). Increase of litter at the Arctic deep-sea observatory HAUSGARTEN. *Marine Pollution Bulletin*, **64**, 2734–2741.

Bett, B.J., Vanreusal, A., Vincx, M. *et al.* (1994). Sampler bias in the qualitative study of deep-sea meiobenthos. *Marine Ecology Progress Series*, **104**, 197–203.

Bett, B.J., Malzone, M.G., Narayanaswamy, B.E. and Wigham, B.D. (2001). Temporal variability in phytodetritus and megabenthic activity at the seabed in the deep Northeast Atlantic. *Progress in Oceanography*, **50**, 349–368.

Bilham, R. (2009). The seismic future of cities. *Bulletin of Earthquake Engineering*, **7**, 839–887.

Bilham, R. (2013). Societal and observational problems in earthquake risk assessments and their delivery to those most at risk. *Tectonophysics*, **584**, 166–173.

Billett, D.S.M. and Hansen, B. (1982). Abyssal aggregations of *Kolga hyalina* Danielssen and Koren (Echinodermata: Holothurioidea) in the northeast Atlantic Ocean: a preliminary report. *Deep-Sea Research A*, **29**(7), 799–818.

Billett, D.S.M., Lampitt, R.S., Rice, A.L. and Mantoura, R.F.C. (1983). Seasonal sedimentation of phytoplankton to the deep-sea benthos. *Nature*, **302**, 520–522.

Billett, D.S.M., Bett, B.J., Rice, A.L. *et al.* (2001). Long-term change in the megabenthos of the Porcupine Abyssal Plain (NE Atlantic). *Progress in Oceanography*, **50**, 325–348.

Billett, D.S.M., Bett, B.J., Jacobs, C.L., Rouse, I.P. and Wigham, B.D. (2006). Mass deposition of jellyfish in the deep Arabian Sea. *Limnology and Oceanography*, **51**, 2077–2083.

Billett, D.S.M., Bett, B., Reid, W.D.K., Boorman, B. and Priede, I.G. (2010). Long-term change in the abyssal NE Atlantic: the 'Amperima Event' revisited. *Deep-Sea Research II*, **57**(15), 1406–1417.

Birstein, J.A. (1957). Certain peculiarities of the ultra-abyssal fauna as exemplified by the genus *Storthyngura* (Crustacea Isopoda Asellota). *Zoologichesky Zhurnal*, **36**, 961–985. (In Russian with English summary.)

Birstein, J.A. (1958). Deep-sea Malacostraca of the north-western part of the Pacific Ocean, their distribution and relations. *15th International Congress of Zoology (London)*, **5**.

Birstein, J.A. (1963). *Deep-sea Isopod Crustaceans of the Northwestern Pacific Ocean*. Moscow: Institute of Oceanology of the USSR, Nauka. (In Russian with English summary.)

Birstein, J.A. and Tchindonova, J.G. (1958). Glubocovodniie Mysidii severo zapadnoi ciasti Tihogo Okeana (The deep sea Mysidacea from the north-western Pacific Ocean). *Trudy Instituta Okeanologii*, **27**, 258–355. (In Russian.)

Birstein, Y.A. and Vinogradov, M.E. (1955). *Pelagicheskle gammaridy* (Amphipoda: Gammaridea) Kurilo-Kamchatskoi padiny. *Trudy Instituta Okeanologii*, **12**, 210–287. (In Russian.)

Biscaye, P.E. and Anderson, R.F. (1994). Fluxes of particulate matter on the slope of the southern Middle Atlantic Bight: SEEP-II. *Deep-Sea Research II*, **41**, 459–510.

Blankenship, L.E. and Levin, L.A. (2007). Extreme food webs: foraging strategies and diets of scavenging amphipods from the ocean's deepest five kilometres. *Limnology and Oceanography*, **52**(4), 1685–1697.

Blankenship, L.E., Yayanos, A.A., Cadien, D.B. and Levin, L.A. (2006). Vertical zonation patterns of scavenging amphipods from the hadal zone of the Tonga and Kermadec Trenches. *Deep-Sea Research I*, **53**, 48–61.

Blankenship-Williams, L.E. and Levin, L.A. (2009). Living deep: a synopsis of hadal trench ecology. *Marine Technology Society Journal*, **43**(5), 137–143.

Blaxter, J.H.S. (1978). Baroreception. In *Sensory Ecology*, ed. M.A. Ali. New York: Plenum Publishing Corporation, pp. 375–409.

Blaxter, J.H.S. (1980). The effect of hydrostatic pressure on fishes. In *Environmental Physiology of Fishes*, ed. M.A. Ali. New York: Plenum Publishing Corporation, pp. 369–386.

Blunt, J.W., Copp, B.R., Hu, W.P. *et al.* (2009). Marine natural products. *Natural Product Reports*, **26**(2), 170–244.

Bostock, H.C., Hayward, B.W., Neil, H.L., Currie, K.I. and Dunbar, G.B. (2011). Deep-water carbonate concentrations in the southwest Pacific. *Deep-Sea Research I*, **58**, 72–85.

Boulègue, J., Benedetti, E.L., Dron, D., Mariotti, A. and Létolle, R. (1987). Geochemical and biogeochemical observations on the biological communities associated with fluid venting in Nankai Trough and Japan Trench subduction zones. *Earth and Planetary Science Letters*, **83**, 343–355.

Boutan, L. (1900). *La Photographie Sous-marine et les Progrés de la Photographie*. Paris: Schleicher Frères.

Bouvier, E.L. (1908). Crustaces decapodes (peneides) provenant des campagnes de '*Hirondelle*' et de la '*Princess Alice*' (1886–1907). *Resultants des Campagnes Scientifiques Acomplies sur son Yacht Prince Albert I*, **33**, 1–122.

Bowen, A.D., Yoerger, D.R., Taylor, C. *et al.* (2008). The Nereus hybrid underwater robotic vehicle for global ocean science operations to 11,000 m depth. OCEANS '08, IEEE/MTS Conference Proceedings, Quebec.

Bowen, A.D., Yoerger, D.R., Taylor, C. *et al.* (2009a). The Nereus hybrid underwater robotic vehicle. *Underwater Technology*, **28**(3), 79–89.

Bowen, A.D., Yoerger, D.R., Taylor, C. *et al.* (2009b). Field trials of the *Nereus* hybrid underwater robotic vehicle in the Challenger Deep of the Mariana Trench. OCEANS '09, IEEE/MTS Conference Proceedings, Biloxi, MS.

Bowman, J.P., Gosink, J.J., McCammon, S.A. *et al.* (1998). *Colwellia demingiae* sp. nov., *Colwellia hornerae* sp. nov., *Colwellia rossensis* sp. nov. and *Colwellia psychrotropica* sp. nov.: psychrophilic Antarctic species with the ability to synthesize docosahexaenoic acid (22:6o3). *International Journal of Systematic Bacteriology*, **48**, 1171–1180.

Brady, H.B. (1884). Report on the Foraminifera dredged by H.M.S. Challenger during the years 1873–1876. *Reports on the Scientific Results of the Voyage of the H.M.S. Challenger During the Years 1873–1876, Zoology*, **9**, 1–814.

Brandt, A., Malyutina, M., Borowski, C., Schriever, G. and Thiel, H. (2004). Munnopsidid isopod attracted to bait in the DISCOL area, Pacific Ocean. *Mitteilungen Hamburgisches Zoologisches Museum Institut*, **101**, 275–279.

Brehan, M.K., MacDonald, A.G., Jones, G.R. and Cossins, A.R. (1992). Homeoviscous adaptation under pressure: the pressure dependence of membrane order in brain myelin membranes of deep-sea fish. *Biochimica et Biophysica Acta*, **1103**, 317–323.

Britton, J.C. and Morton, B. (1994). Marine carrion and scavengers, *Oceanography and Marine Biology Annual Review*, **32**, 369–434.

Broecker, W.S. (1991). The Great Ocean Conveyer. *Oceanography*, **4**(2), 79–89.

Broecker, W.S. and Peng, T.-H. (1982). *Tracers in the Sea*. Palisades, New York: Eldigio Press.

Broecker, W.S., Takahashi, T. and Stuiver, M. (1980). Hydrography of the central Atlantic, II: waters beneath the two degree discontinuity. *Deep-Sea Research*, **27**(6A), 397–420.

Brown, A. and Thatje, S. (2011). Respiratory response of the deep-sea amphipod *Stephonyx biscayensis* indicates bathymetric range limitation by temperature and hydrostatic pressure. *PLoS ONE*, **6**(12), e28562–[6pp].

Brown, A.D. and Simpson, J. (1972). Water relations of sugar-tolerant yeasts: the role of intracellular polyols. *Journal of General Microbiology*, **72**, 589–591.

Bruun, A.F. (1956a). Animal life of the deep-sea bottom. In *The Galathea Deep Sea Expedition 1950–1952*, ed. A.F. Bruun, S. Greve, H. Mielche and R. Spärk. London: George, Allen and Unwin, pp.149–195.

Bruun, A.F. (1956b). The abyssal fauna: its ecology distribution and origin. *Nature*, **177**, 1105–1108.

Bruun, A.F. (1957). General introduction to the reports and list of deep-sea station. *Galathea Report*, **1**, 7–48.

Bryden, H.L. (1973). New polynomials for thermal expansion, adiabatic temperature gradient and potential temperature of sea water. *Deep-Sea Research*, **20**, 410–408.

Bucklin, A., Wilson, R.R. and Smith, K.L. (1987). Genetic differentiation of seamount and basin populations of the deep-sea amphipod *Eurythenes gryllus*. *Deep-Sea Research*, **34**, 1795–1810.

Buesseler, K.O. and Boyd, P.W. (2009). Shedding light on processes that control particle export and flux attenuation in the twilight zone of the open ocean. *Limnology and Oceanography*, **54**, 1210–1232.

Buesseler, K.O., Livingston, H.D., Honjo, S. *et al.* (1990). Scavenging and particle deposition in the southwestern Black Sea – evidence from Chernobyl radiotracers. *Deep–Sea Research*, **7**, 413–430.

Bühring, S.I. and Christiansen, B. (2001). Lipids in selected abyssal benthopelagic animals: links to the epipelagic zone? *Progress in Oceanography*, **50**, 369–382.

Burton, A. (2012). Way down deep. *Frontiers in Ecology and the Environment*, **10**, 112.

Cairns, S.D., Bayer, F.M. and Fautin, D.G. (2007). *Galatheanthemum profundale* (Anthozoa: Actinaria) in the Western Atlantic. *Bulletin of Marine Science*, **80**(1), 191–200.

Caldeira, K. and Wickett, M.E. (2003). Anthropogenic carbon and ocean pH. *Nature*, **425**, 365.

Campenot, R.B. (1975). The effects of high hydrostatic pressure on transmission at the crustacean neuromuscular junction. *Comparative Biochemistry and Physiology B*, **52**, 133–140.

Canals, M., Puig, P., Durrieu de Madron, X. *et al.* (2006). Flushing submarine canyons. *Nature*, **444**, 354–357.

CAREX (2011). CAREX roadmap for research on life in extreme environment. *CAREX Publication*, **9**, 1–48.

Carey, S.W. (1958). The tectonic approach to continental drift. In *Continental Drift: A Symposium*, ed. S.W. Carey. Hobart: University of Tasmania, pp. 177–363.

Carney, R.S. (2005). Zonation of deep biota on continental margins. *Oceanography and Marine Biology Annual Review*, **43**, 211–278.

Carruthers, J.N. and Lawford, A.L. (1952). The deepest oceanic sounding. *Nature*, **169**, 601–603.

Carter, H.J. (1983). *Apagesoma edentatum*, a new genus and species of Ophidiid fish from the western north Atlantic. *Bulletin of Marine Science*, **33**, 94–101.

Castellini, M.A., Castellini, J.M. and Rivera, P.M. (2001). Adaptations to pressure in the RBC metabolism of diving animals. *Comparative Biochemistry and Physiology A*, **129**, 751–757.

Catalano, R., Yorifuji, T. and Kawachi, I. (2013). Natural selection *in utero*: evidence from the Great East Japan Earthquake. *American Journal of Human Biology*, **25**, 555–559.

Chapelle, G. and Peck, L.S. (1999). Polar gigantism dictated by oxygen availability. *Nature*, **399**, 114–115.

Chapelle, G. and Peck, L.S. (2004). Amphipod crustacean size spectra: new insights in the relationship between size and oxygen. *Oikos*, **106**, 167–175.

Charmasson, S.S. and Calmet, D.P. (1987). Distribution of scavenging Lysianassidae amphipods *Eurythenes gryllus* in the northeast Atlantic: comparison with studies held in the Pacific. *Deep-Sea Research*, **34**(9), 1509–1523.

Chastain, R.A. and Yayanos, A.A. (1991). Ultrastructural changes in an obligatory barophilic marine bacterium after decompression. *Applied and Environmental Microbiology*, **57**(5), 1489–1497.

Chernova, N.V., Stein, D.L. and Andriashev, A.P. (2004). Family Liparidae Scopoli 1777, annotated checklists of fishes. *California Academy of Science*, **31**, 1–82.

Chevreux, E. (1899). Sur deux espèces géantes d'amphipodes provenant des campagnes du yacht Princesse Alice. *Bulletin de la Société of Zoologique de France*, **24**, 152–158.

Childress, J.J. (1971). Respiratory rate and depth of occurrence of midwater animals. *Limnology and Oceanogaphy*, **16**, 104–106.

Childress, J.J. (1977). Effects of pressure, temperature and oxygen on the oxygen-consumption rate of the midwater copepod *Gausia princeps*. *Marine Biology*, **39**, 19–24.

Childress, J.J. (1995). Are there physiological and biochemical adaptations of metabolism in deep-sea animals? *Trends in Ecology and Evolution*, **10**, 30–36.

Childress, J.J. and Fisher, C. (1992). The biology of hydrothermal vent animals; physiology, biochemistry and autotrophic symbioses. *Oceanography and Marine Biology Annual Review*, **30**, 337–442.

Childress, J.J. and Somero, G.N. (1979). Depth-related enzymatic activities in muscle, brain, and heart of deep-living pelagic teleosts. *Marine Biology*, **52**, 273–283.

Childress, J.J. and Thuesen, E.V. (1995). Metabolic potentials of deep-sea fishes: a comparative approach. In *Biochemistry and Molecular Biology of Fishes*, ed. P.W. Hochachka and T.P. Mommsen. Berlin: Elsevier Science, pp. 175–195.

Chiswell, S.M. and Moore, M.I. (1999). Internal tides near the Kermadec Ridge. *Journal of Physical Oceanography*, **29**, 1019–1035.

Christiansen, B. and Diel-Christiansen, D. (1993). Respiration of lysianassoid amphipods in a subarctic fjord and some implications on their feeding ecology. *Sarsia*, **78**, 9–15.

Christiansen, B., Pfannkuche, O. and Thiel, H. (1990). Vertical distribution and population structure of the necrophagous amphipod *Eurythenes gryllus* in the West European basin. *Marine Ecology Progress Series*, **66**, 35–45.

Christiansen, B., Beckmann, W. and Weikert, H. (2001). The structure and carbon demand of the bathyal benthic boundary layer community: a comparison of two oceanic locations in the NE-Atlantic. *Deep-Sea Research II*, **48**, 2409–2424.

Cohen, D.M., Inada, T., Iwamoto, T. and Scialabba, N. (1990). Gadiform fishes of the world (Order Gadiformes). An annotated and illustrated catalogue of cods, hakes, grenadiers and other gadiform fishes known to date. *FAO Fisheries Synopsis*, **10**(125), 442.

Collins, M.A., Priede, I.G. and Bagley, P.M. (1999). *In situ* comparison of activity in two deep-sea scavenging fishes occupying different depth zones. *Proceedings of the Royal Society London B*, **266**, 2011–2016.

Conan, G., Roux, M. and Sibuet, M. (1980). A photographic survey of a population of the stalked crinoid *Diplocrinus* (*Annacrinus*) *wyvillethomsoni* (Echinodermata) from the bathyal slope of the Bay of Biscay. *Deep-Sea Research*, **28**, 441–453.

Connerney, J.E.P., Acuna, M.H., Wasilewski, P.J. *et al.* (1999). Magnetic lineations in the ancient crust of Mars. *Science*, **284**, 794–798.

Connor, E.F. and McCoy, E.D. (1979). The statistics and biology of the species area relationship. *American Naturalist*, **113**(6) 791–833.

Cook-Anderson, G. and Beasley, D. (2005). NASA details earthquake effects on the Earth. *NASA Press Release*, 10 January 2005.

Cossins, A.R., and MacDonald, A.G. (1984). Homeoviscous theory under pressure: II. The molecular order of membranes from deep-sea fish. *Biochimica et Biophysica Acta (BBA)-Biomembranes*, **776**(1), 144–150.

Cossins, A.R. and MacDonald, A.G. (1989). The adaptation of biological membranes to temperature and pressure: fish from the deep and cold. *Journal of Bioenergetics and Biomembranes*, **21**(1), 15–35.

Cousteau, J.-Y. (1958). *Calypso* explores an undersea canyon. *National Geographic Magazine*, **113**, 373–396.

Craig, J., Jamieson, A.J, Heger, A. and Priede, I.G. (2009). Distribution of bioluminescent organisms in the Mediterranean Sea and predicted effects on a deep-sea neutrino telescope. *Nuclear Instruments and Methods in Physics Research A*, **602**, 224–226.

Craig, J., Jamieson, A.J., Bagley, P.M. and Priede, I.G. (2011a). Seasonal variation of deep-sea bioluminescence in the Ionian Sea. *Nuclear Instruments and Methods in Physics Research A*, **626**, S115–S117.

Craig, J, Jamieson, A.J., Bagley, P.M. and Priede, I.G. (2011b). Naturally occurring bioluminescence on the deep-sea floor. *Journal of Marine Systems*, **88**, 563–567.

Dahl, E. (1959). Amphipoda from depths exceeding 6000 meters. *Galathea Report*, **1**, 211–241.

Dahl, E. (1977). The amphipod functional model and its bearing upon systematics and phylogeny. *Zoologica Scripta*, **6**, 221–228.

Dahl, E. (1979). Deep-sea carrion feeding amphipods: evolutionary patterns in niche adaptation. *Oikos*, **33**, 167–175.

Dahlgren, T., Glover, A.G., Smith, C.R. and Baco, A. (2004). Fauna of whale falls: systematics and ecology of a new polychaete (Annelida: Chrysopetalidae) from the deep Pacific Ocean. *Deep-Sea Research I*, **51**, 1873–1887.

Daito, H., Suzuki, M., Shiihara, J. *et al.* (2012). Impact of the Tohoku earthquake and tsunami on pneumonia hospitalisations and mortality among adults in northern Miyagi, Japan: a multi-centre observational study. *Thorax*, **68**, 544–550.

Dalsgaard, J., St, John, M., Kattner, G., Müller-Navarra, D. and Hagen, W. (2003). Fatty acid trophic markers in the pelagic marine environment. *Advances in Marine Biology*, **46**, 225–340.

Danovaro, R., Gambi, C. and Della Croce, N. (2002). Meiofauna hotspot in the Atacama Trench (Southern Pacific Ocean). *Deep-Sea Research I*, **49**, 843–857.

Danovaro, R., Della Croce, N., Dell'Anno, A. and Pusceddu, A. (2003). A depocenter of organic matter at 7800m depth in SE Pacific Ocean. *Deep-Sea Research I*, **50**, 1411–1420.

Danovaro, R., Dell'Anno, A. and Pusceddu, A. (2004). Biodiversity response to climate change in a warm deep sea. *Ecology Letters*, **7**, 821–828.

DaSilva, E.J. (2004). The colours of biotechnology: science, development and humankind. *Electronic Journal of Biotechnology*, **7**(3), 01–02.

De Broyer, C. and Thurston, M.H. (1987). New Atlantic material and redescription of the type specimens of the giant abyssal amphipod *Alicella gigantea* Chevreux (Crustacea). *Zoologica Scripta*, **16**(4), 335–350.

De Broyer, C., Nyssen, F. and Dauby, P. (2004). The crustacean scavenger guild in Antarctic shelf, bathyal and abyssal communities. *Deep-Sea Research II*, **51**(14–16), 1733–1752.

De La Rocha, C.L. and Passow, U. (2007). Factors influencing the sinking of POC and the efficiency of the biological carbon pump. *Deep-Sea Research I*, **54**, 639–658.

De Leo, F.C., Smith, C.R., Rowden, A.A., Bowden, D.A. and Clarke, M.R. (2010). Submarine canyons: hotspots of benthic biomass and productivity in the deep sea. *Proceedings of the Royal Society London B*, **277**, 2783–2792.

DeLong, E.F. (1986). Adaptations of deep-sea bacteria to the abyssal environment. PhD thesis, University of California, San Diego.

DeLong, E.F. and Yayanos, A.A. (1985). Adaptation of the membrane lipids of a deep-sea bacterium to changes in hydrostatic pressure. *Science*, **228**, 1101–1103.

DeLong, E.F. and Yayanos, A.A. (1986). Biochemical function and ecological significance of novel bacterial lipids in deep-sea procaryotes. *Applied Environmental Microbiology*, **51**, 730–737.

Demhardt, I.J. (2005). Alfred Wegener's hypothesis on continental drift and its discussion in *Petermanns Geographishe Mitteilungen* (1912–1942). *Polarforschung*, **75**, 29–35.

Deming, J.W., Somers, L.K., Straube, W.L., Swartz, D.G. and MacDonell, M.T. (1988). Isolation of an obligately barophilic bacterium and description of a new genus, *Colwellia* gen. nov. *Systematic and Applied Microbiology*, **10**, 152–160.

Denton, E.J. (1990). Light and vision at depths greater than 200 metres. In *Light and Life in the Sea*, ed. P.J. Herring, A.K. Campbell, M. Whitfield and L. Maddock. Cambridge: Cambridge University Press, pp. 127–148.

Deuser, W.G. and Ross, E.H. (1980). Seasonal change in the flux of organic carbon to the deep Sargasso Sea. *Nature*, **283**, 364–365.

Dietz, R.S. (1961). Continent and ocean basin evolution by spreading of the sea floor. *Nature*, **190**, 854–857.

Dobzhansky, T. (1951). *Genetics and the Origin of Species*, 3rd edn. New York: Columbia University Press.

Doebeli, M. and Dieckmann, U. (2003). Speciation along environmental gradients. *Nature*, **421**, 259–264.

Domanski, P. (1986). The near-bottom shrimp faunas (Decapoda: Natantia) at two abyssal sites in the Northeast Atlantic Ocean. *Marine Biology*, **93**, 171–180.

Drazen, J.C. and Seibel, B.A. (2007). Depth-related trends in metabolism of benthic and bentho-pelagic deep-sea fishes. *Limnology and Oceanography*, **52**, 2306–2316.

Drazen, J.C., Yeh, J., Friedman, J. and Condon, N. (2011). Metabolism and enzyme activities of hagfish from shallow and deep water of the Pacific Ocean. *Comparative Biochemistry and Physiology A*, **159**(2), 182–187.

Duarte, C.M., Middelburg, J.J. and Caraco, N. (2005). Major role of marine vegetation on the oceanic carbon cycle, *Biogeosciences*, **2**(1), 1–8.

Duineveld, G, Lavaleye, M., Berghuis, E. and de Wilde, P. (2001). Activity and composition of the benthic fauna in the Whittard Canyon and the adjacent continental slope (NE Atlantic). *Oceanologica Acta*, **24**, 69–83.

Dunn, D.F. (1983). Some Antarctic and Sub-Antarctic sea anemones (Coelenterata: Ptychodac-tiaria and Actiniaria). *Antarctic Research Series*, **39**, 1–67.

Eckman, J.E. and Thistle, D. (1991). Effects of flow about a biologically produced structure on harpacticoid copepods in San Diego Trough. *Deep-Sea Research A*, **38**(11), 1397–1416.

Eleftheriou, A. and McIntyre, A.D. (2005). *Methods for the Study of the Marine Benthos*. 3rd edn. Oxford: Blackwell Scientific Publications.

Eliason, A. (1951). Polychaeta. *Reports of the Swedish Deep-Sea Expedition, 2, Zoology*, **11**, 131–148.

Elliott, A.J. and Thorpe, S.A. (1983). Benthic observations on the Madeira Abyssal Plain. *Oceanologica Acta*, **6**, 463–466.

Ellis-Evans, C. and Walter, N. (2008). Coordination Action for Research activities on life in EXtreme environments: the CAREX project. *Journal of Biological Research-Thessaloniki*, **9**, 11–15.

Eloe, E.A., Shulse, C.N., Fadrosh, D.W. *et al.* (2010). Compositional differences in particle-associated and free-living microbial assemblages from an extreme deep-ocean environment. *Environmental Microbiology Reports*, **3**(4), 449–458.

Eloe, E.A., Malfatti, F., Gutierrez, J. *et al.* (2011). Isolation and characterization of a psychropiezo-philic alphaproteobacterium. *Applied and Environmental Microbiology*, **77**(22), 8145–8153.

Emery, K.O. (1952). Submarine photography with the benthograph. *Science Monthly*, **75**, 3–11.

Emery, K.O., Merrill, A.S. and Trumbull, J.V.A. (1965). Geology and biology of the sea floor as deduced from simultaneous photographs and samples. *Limnology and Oceanography*, **10**(1), 1–21.

Emiliani, C. (1961). The temperature decrease of surface sea-water in high latitudes and of abyssal-hadal water in open oceanic basins during the past 75 million years. *Deep-Sea Research*, **8**(2), 144–147.

Endo, H. and Okamura, O. (1992). New records of the abyssal grenadiers *Coryphaenoides armatus*. *Japanese Journal of Ichthyology*, **38**, 433–437.

England, P. and Jackson, J. (2011). Uncharted seismic risk. *Nature Geoscience*, **4**, 348–349.

Eustace, R.M., Kilgallen, N.M., Lacey, N.C. and Jamieson, A.J. (2013). Population structure of the hadal amphipod *Hirondellea gigas* from the Izu-Bonin Trench (NW Pacific; 8173–9316 m). *Journal of Crustacean Biology*, **33**(6), 793–801.

Ewing, M. and Heezen, B.C. (1955). Puerto-Rico Trench topographic and geophysical data. *Special Paper: Geological Society of America*, **62**, 255–267.

Ewing, M., Vine, A. and Worzel, J.L. (1946). Photography of the ocean bottom. *Journal Optical Society of America*, **36**, 307–321.

Fabiano, M., Pusceddu, A., Dell'Anno, A. *et al.* (2001). Fluxes of phytopigments and labile organic matter to the deep ocean in the NE Atlantic Ocean. *Progress in Oceanography*, **50**, 89–104.

Faccenna, C., Becker, T.W., Pio Lucente, F., Jolivet, L. and Rossetti, F. (2001). History of subduction and back-arc extension in the central Mediterranean. *Geophysical Journal International*, **145**, 809–820.

Fang, J. and Kato, C. (2010). Deep-sea piezophilic bacteria: geomicrobiology and biotechnology. In *Geomicrobiology: Biodiversity and Biotechnology*, ed. S.K. Jain, A.A. Khan and M.K. Rai. Boca Raton, FL: CRC Press, pp. 47–77.

Fang, J., Barcelona, M.J., Nogi, Y. and Kato, C. (2000). Biochemical implications and geochemical significance of novel phospholipids of the extremely barophilic bacteria from the Marianas Trench at 11 000 m. *Deep-Sea Research I*, **47**, 1173–1182.

Feely, R.A., Sabine, C.L., Lee, K. *et al.* (2004). Impact of anthropogenic CO_2 on the $CaCO_3$ system in the oceans. *Science*, **305**, 362–366.

Fisher, C.R. (1990). Chemoautotrophic and methanotrophic symbioses in marine invertebrates. *Review of Aquatic Science*, **2**, 399–436.

Fisher, R.L. (1954). On the sounding of trenches. *Deep-Sea Research*, **2**, 48–58.

Fisher, R.L. (2009). Meanwhile, back on the surface: further notes on the sounding of trenches. *Marine Technology Society Journal*, **43**(5), 16–19.

Fisher, R.L. and Hess, H.H. (1963). Trenches. In *The Sea*, ed. M.N. Hill. New York: Wiley, pp. 411–436.

Fletcher, B., Bowen, A., Yoerger, D.R. and Whitcomb, L.L. (2009). Journey to the Challenger Deep: 50 years later with the Nereus Hybrid remotely operated vehicle. *Marine Technology Society Journal*, **43**(5), 65–76.

Fluery, A.G. and Drazen, J.C. (2013). Abyssal scavenging communities attracted to Sargassum and fish in the Sargasso Sea. *Deep-Sea Research I*, **72**, 141–147.

Fofonoff, N.P. (1977). Computation of potential temperature of seawater for an arbitrary reference pressure. *Deep-Sea Research*, **24**, 489–491.

Fofonoff, N.P. and Millard, R.C. (1983). Algorithms for computation of fundamental properties of seawater. *UNESCO Technical Papers in Marine Science*, **44**, 53.

Forbes, E. (1844). Report on the Mollusca and Radiata of the Aegean Sea, and their distribution, considered as bearing on geology. *Report (1843) to the 13th Meeting of the British Association for the Advancement of Science*, pp. 30–193.

France, S.C. (1993). Geographic variation among three isolated populations of the hadal amphipod *Hirondellea gigas* (Crustacea: Amphipoda: Lysianassoidea). *Marine Ecology Progress Series*, **92**, 277–287.

France, S.C. (1994). Genetic population structure and gene flow among deep-sea amphipods, *Abyssorchomene* spp., from six California continental Borderland basins. *Marine Biology*, **118**, 67–77.

France, S.C. and Kocher, T.D. (1996). Geographic and bathymetric patterns of mitochondrial 16S rRNA sequence divergence among deep-sea amphipods, *Eurythenes gryllus*. *Marine Biology*, **126**, 633–643.

Frankenberg, D. and Menzies, R.J. (1968). Some quantitative analyses of deep-sea benthos off Peru. *Deep-Sea Research*, **15**(5), 623–626.

Fraser, P.J. (2001). Statocysts in crabs: short-term control of locomotion and long-term monitoring of hydrostatic pressure. *Biological Bulletin*, **200**, 155–159.

Fraser, P.J. (2006). Review. Depth, navigation and orientation in crabs: angular acceleration, gravity and hydrostatic pressure sensing during path integration. *Marine and Freshwater Behaviour and Physiology*, **39**(2), 87–97.

Fraser, P.J. and MacDonald, A.G. (1994). Crab hydrostatic pressure sensors. *Nature*, **371**, 383–384.

Fraser, P.J. and Shelmerdine, R.L. (2002). Fish physiology: dogfish hair cells sense hydrostatic pressure. *Nature*, **415**, 495–496.

Fraser, P.J., MacDonald, A.G., Cruickshank, S.F. and Schraner, M.P. (2001). Integration of hydrostatic pressure information by identified interneurones in the crab *Carcinus maenas* (L.); long-term recordings. *Journal of Navigation*, **54**, 71–79.

Fraser, P.J., Cruickshank, S.F. and Shelmerdine, R.L. (2003). Hydrostatic pressure effects on vestibular hair cell afferents in fish and crustacea. *Journal of Vestibular Research*, **13**, 235–242.

Froese, R. and Pauly, D. (2009). *FishBase*. Available at: www.fishbase.org. Accessed 28 August 2009.

Fryer, P., Becker, N., Applegate, B. *et al.* (2002). Why is Challenger Deep so deep? *Earth and Planetary Science Letters*, **211**, 259–269.

Forman, W. (2009). From Beebe and Barton to Piccard and Trieste. *Marine Technology Society Journal*, **43**(5), 27–36.

Fujii, T., Jamieson, A.J., Solan, M., Bagley, P.M. and Priede, I.G. (2010). A large aggregation of liparids at 7703 m depth and a reappraisal of the abundance and diversity of hadal fish. *BioScience*, **60**(7), 506–515.

Fujii, T., Kilgallen, N.M., Rowden, A.A. and Jamieson, A.J. (2013). Amphipod community structure across abyssal to hadal depths in the Peru–Chile and the Kermadec Trenches. *Marine Ecology Progress Series*, **492**, 125–138.

Fujikura, K., Kojima, S., Tamaki, K. *et al.* (1999). The deepest chemosynthesis-based community yet discovered from the hadal zone, 7326 m deep, in the Japan Trench. *Marine Ecology Progess Series*, **190**, 17–26.

Fujimoto, H., Fujiwara, T., Kong, L. and Igarashi, C. (1993). Sea-beam survey over the Challenger Deep, revisited. In *Preliminary Report of the Hakuho-Maru Cruise (KH-92–1)*. Tokyo: Ocean Research Institute, University of Tokyo, pp. 26–27.

Fujio, S. and Yanagimoto, D. (2005). Deep current measurements at 38°N east of Japan. *Journal of Geophysical Research C*, **110**(C2), C02010.

Fujio, S., Yanagimoto, D. and Taira, K. (2000). Deep current structure above the Izu-Ogasawara Trench. *Journal of Geophysical Research*, **105**(C3), 6377–6386.

Fujioka, K., Takeuchi, A., Horiuchi, K. *et al.* (1993). Constrated nature between landward and seaward slopes of the Japan Trench off Miyako, Northern Japan. *Proceedings of JAMSTEC Symposium of Deep-Sea Research*, **9**, 1–26.

Fujioka, K., Okino, K., Kanamatsu, T. and Ohara, Y. (2002). Morphology and origin of the Challenger Deep in the southern Mariana Trench. *Geophysical Research Letters*, **19**, 1–4.

Fujiwara, T., Kodaira, S., No, T. *et al.* (2011). The 2011 Tohoku-Oki earthquake: displacement reaching the trench axis. *Science*, **334**, 1240.

Fujiwara, Y., Dato, C., Masui, N., Fujikura, K. and Kojima, S. (2001). Dual symbiosis in the cold-seep thyasirid clam *Maorithyas hadalis* from the hadal zone in the Japan Trench, western Pacific. *Marine Ecology Progress Series*, **214**, 151–159.

Fulton, J. (1973). Some aspects of the life history of *Calanus plumchrus* in the Strait of Georgia. *Journal of the Fisheries Research Board of Canada*, **30**, 811–815.

Furlong, R.R. and Wahlquist, E.J. (1999). US space missions using radioisotope power systems. *Nuclear News*, April, 26–34.

Gage, J.D. (2003). Food inputs, utilization, carbon flow and energetics. In *Ecosystems of the World 28, Ecosystems of the Deep Sea*, ed. P.A. Tyler. Amsterdam: Elsevier, pp. 315–382.

Gage, J.D. and Bett, B.J. (2005). Deep-sea benthic sampling. In *Methods for the Study of the Marine Benthos*, 3rd edn, ed. A. Eleftheriou and A.D. McIntyre. Oxford: Blackwell Scientific Publications, pp. 273–325.

Gage, J.D. and Tyler, P.A. (1991). *Deep-sea Biology: A Natural History of Organisms at the Deep-sea Floor*. Cambridge: Cambridge University Press.

Galgani, F., Leaute, J.P., Moguedet, P. *et al.* (2000). Litter on the sea floor along European coasts. *Marine Pollution Bulletin*, **40**, 516.

Gambi, C., Vanreusal, A. and Danovaro, R. (2003). Biodiversity of nematode assemblages from deep-sea sediments of the Atacama Slope and Trench (South Pacific Ocean). *Deep-Sea Research I*, **50**, 103–117.

Gamô, S. (1984). A new remarkably giant tanaid, *Gigapseudes maximus* sp.nov. (Crustacea) from abyssal depths far off southeast of Mindanao, the Philippines. *Scientific Reports of Yokahoma Natural University Series*, **11**, 1–12.

Garfield, N., Rago, T.A., Schnebele, K.J. and Collins, C.A. (1994). Evidence of a turbidity current in Monterey submarine canyon associated with the 1989 Loma Prieta earthquake. *Continental Shelf Research*, **14**, 673–686.

Gartner, J.V. Jr (1983). Sexual dimorphism in the bathypelagic gulper eel E*urypharynx pelecanoides* (Lyomeri: Eurypharyngidae), with comments on reproductive strategy. *Copia*, **2**, 560–563.

Gaskell, T.F., Swallow, J.C. and Ritchie, G.S. (1953). Further notes on the greatest oceanic sounding and the topography of the Marianas Trench. *Deep-Sea Research*, **1**, 60–63.

Gebruk, A.V. (1993). New records of elasipodid holothurians in the Atlantic sector of Antarctic and Subantarctic. *Transactions of the P.P. Shirshov Institute of Oceanology*, **127**, 228–244.

Genin, A., Dayton, P.K., Lonsdale, P.F. and Speiss, F.N. (1986). Corals on seamount peaks provide evidence of current acceleration over deep-sea topography. *Nature*, **323**, 59–61.

George, R.Y. (1979). What adaptive strategies promote immigration and speciation in deep-sea environments. *Sarsia*, **64**(1–2), 61–65.

George, R.Y. and Higgins, R.P. (1979). Eutrophic hadal benthic community in the Puerto-Rico Trench. *Ambio Special Report*, **6**, 51–58.

Gerdes, D. (1990). Antarctic trials of the multi-box corer, a new device for benthos sampling. *Polar Record*, **26**(156), 35–38.

Giere, O. (2009). *Meiobenthology*, 2nd edn. Berlin: Springer.

Gislén, T. (1956). Crinoids from depths exceeding 6000 meters. *Galathea Report*, **2**, 61–62.

Gilchrist, I. and MacDonald, A.G. (1980). Techniques for experiments with deep-sea organisms at high pressure. In *Experimental Biology at Sea*, ed. A.G. MacDonald and I.G. Priede. London: Academic Press, pp. 234–276.

Gillett, M.B., Suko, J.R. Santoso, F.O. and Yancey, P.H. (1997). Elevated levels of trimethylamine oxide in muscles of deep-sea gadiform teleosts: a high-pressure adaptation? *Journal of Experimental Zoology*, **279**, 386–391.

Gillibrand, E.J.V., Jamieson, A.J., Bagley, P.M., Zuur, A.F. and Priede, I.G. (2007a). Seasonal development of a deep pelagic bioluminescent layer in the temperate northeast Atlantic Ocean. *Marine Ecology Progress Series*, **341**, 37–44.

Gillibrand, E.J.V., Bagley P.M., Jamieson, A.J. *et al.* (2007b). Deep sea benthic bioluminescence at artificial food falls, 1000 to 4800m depth, in the Porcupine Seabight and Abyssal Plain, North East Atlantic Ocean. *Marine Biology*, **150**, 1053–1060.

Glover, A.G., Wiklund, H., Taboada, S. *et al.* (2013). Bone-eating worms from the Antarctic: the contrasting fate of whale and wood remains on the Southern Ocean seafloor. *Proceedings of the Royal Society London B*, **280**, 20131390.

Glud, R.N., Ståhl, H., Berg, P., Wenzhöfer, F., Oguri, K. and Kitazato, H. (2009). *In situ* microscale variation in distribution and consumption of O_2: a case study from a deep ocean margin sediment (Sagami Bay, Japan). *Limnology and Oceanography*, **54**(1), 1–12.

Glud, R.N., Wenzhöfer, F., Middelboe, M. *et al.* (2013). High rates of microbial carbon turnover in sediments in the deepest oceanic trench on Earth. *Nature Geoscience*, **6**, 284–288.

Godbold, J.A., Rosenberg, R. and Solan, M. (2009). Species-specific traits rather than resource partitioning mediate diversity effects on resource use. *PLoS ONE*, **4**, e7423.

Godbold, J.A., Bulling, M.T. and Solan, M. (2011). Habitat structure mediates biodiversity effects on ecosystem properties. *Proceedings of the Royal Society London B*, **278**, 1717–2510.

Gooday, A.J., Holzmann, M., Cornelius, N. and Pawlowski, J. (2004). A new monothalamous foraminiferan from 1000–6300 m water depth in the Weddell Sea: morphological and molecular characterisation. *Deep-Sea Research II*, **51**, 1603–1616.

Gooday, A.J., Cedhagen, T., Kamenskaya, O.E. and Cornelius, N. (2007). The biodiversity and biogeography of komokiaceans and other enigmatic foraminiferan-like protists in the deep Southern Ocean. *Deep-Sea Research II: Topical Studies in Oceanography*, **54**(16), 1691–1719.

Gooday, A.J., Todo, Y., Uematsu, K. and Kitazato, H. (2008). New organic-walled Foraminifera (Protista) from the ocean's deepest point, the Challenger Deep (western Pacific Ocean). *Zoological Journal of the Linnean Society*, **153**, 399–423.

Gooday, A.J., Uematsu, K., Kitazato, H., Toyofuku, T. and Young, J.R. (2010). Traces of dissolved particles, including coccoliths, in the tests of agglutinated Foraminifera from the Challenger Deep (10 897 m water depth, western equatorial Pacific). *Deep-Sea Research I*, **57**(2), 239–247.

Gonzalez-Leon, J.A., Acar, M.H., Ryu, S.-W., Ruzette, A.-V.J. and Mayes, A.M. (2003). Low-temperature processing of 'baroplastics' by pressure-induced flow. *Nature*, **426**, 424–428.

Gore, R.H. (1985a). Abyssobenthic and abyssopelagic penaeoidean shimp (families Aristaeidae and Penaeidae) from the Venezuela Basin, Carribean Sea. *Crustaceana*, **49**, 119–138.

Gore, R.H. (1985b). Bright colours in the realm of eternal light. *Sea Frontiers*, **31**, 264–271.

Gould, W.J. and McKee, W.D. (1973). Vertical structure of semi-diurnal currents in the Bay of Biscay. *Nature*, **244**, 88–91.

Gracia, A., Ardila, N.E., Rachello, P. and Diaz, J.M. (2005). Additions to the scaphopod fauna (Mollusca: Scaphopoda) of the Colombian Caribbean. *Caribbean Journal of Science*, **41**(2), 328–334.

Graeve, M., Hagen, W. and Kattner, G. (1994). Herbivorous or omnivorous? On the significance of lipid compositions as trophic markers in Antarctic copepods. *Deep-Sea Research*, **41**, 915–924.

Graeve, M., Kattner, G. and Piependurgo, D. (1997). Lipids in Arctic benthos: does the fatty acid and alcohol composition reflect feeding and trophic interactions? *Polar Biology*, **18**, 53–61.

Grimes, D.J., Singleton, F.L., Stemmler, J. *et al*. (1984). Microbiological effects of wastewater effluent discharge into coastal waters of Puerto Rico. *Water Research*, **18**, 613–619.

Guennegan, Y. and Rannou, M. (1979). Semi-diurnal rhythmic activity in deep sea benthic fishes in the Bay of Biscay. *Sarsia*, **64**, 113–116.

Gupta, N., Woldesenbet, E. and Sankaran, S. (2001). Studies on compression failure features in syntactic foam material. *Journal of Materials Science*, **36**, 4485–4491.

Haddock, S.H.D., Moline, M.A. and Case, J.F. (2010). Bioluminescence in the sea. *Annual Review of Marine Science*, **2**, 443–493.

Haedrich, R.L., Rowe, G.T. and Polloni, P.T. (1980). The megabenthic fauna in the deep sea south of New England, USA. *Marine Biology*, **57**, 165–179.

Haefner, B. (2003). Drugs from the deep: marine natural products as drug candidates. *Drug Discovery Today*, **8**(2), 536–544.

Hahn, J. (1950). Some aspects of deep sea underwater photography. *Photographic Society of America Journal, Section B*, **16**(6), 27–29.

Hallock, Z.R. and Teague. W.J. (1996). Evidence for a North Pacific deep western boundary current. *Journal of Geophysical Research*, **101**, 6617–6624.

Hansen, B. (1957). Holothurioidea from depths exceeding 6000 metres. *Galathea Report*, **2**, 33–54.

Hansen, B. (1972). Photographic evidence of a unique type of walking in deep-sea holothurians. *Deep-Sea Research*, **19**, 461–462.

Hanson, P.P., Zenkevich, N.L., Sergeev, U.V. and Udintsev, G.B. (1959). Maximum depths of the Pacific Ocean. *Priroda*, **6**, 84–88. (In Russian.)

Hardy, K., Olsson, M., Yayanos, A.A., Prsha, J. and Hagey, W. (2002). Deep ocean visualisation experimenter (DOVE): low cost 10 km camera and instrument platform. *OCEANS'02 MTS/IEEE*, **4**, 2390–2394.

Hargrave, B.T., Phillips, G.A., Prouse, N.J. and Cranford, P.J. (1995). Rapid digestion and assimilation of bait by the deep-sea amphipod *Eurythenes gryllus*. *Deep-Sea Research I*, **42**(11/12), 1905–1921.

Harper, A.A., MacDonald, A.G., Wardle, C.S. and Pennec, J.-P. (1987). The pressure tolerance of deep-sea fish axons: results of Challenger cruise 6B/85. *Comparative Biochemistry and Physiology Part A*, **88A**, 647–653.

Harrison, C.S., Hida, T.S. and Seki, M.P. (1983). Hawaiian seabird feeding ecology. *Wildlife Monographs*, **85**, 1–71.

Hartmann, A.C. and Levin, L.A. (2012). Conservation concerns in the deep. *Science*, **336**, 667–668.

Hasegawa, M., Kurohiji, Y., Takayanagi, S., Sawadaishi, S. and Yao, M. (1986). Collection of fish and amphipoda from abyssal sea-floor at 30°N-147°E using traps tied to 10000m wire of research vessel. *Bulletin of the Tokai Regional Fishery Research Laboratory/TOKAISUIKENHO*, **119**, 65–75. (In Japanese.)

Hashimoto, J. (1998). *Onboard Report of KR98–05 Cruise in the Challenger Deep*. RV KAIREI/ ROV KAIKO. Yokosuka, Japan: JAMSTEC.

Havermans, C., Nagy, Z.T., Sonet, G., De Broyer, C. and Martin, P. (2010). Incongruence between molecular phylogeny and morphological classificationin amphipod crustaceans: a case study of Antarctic lysianassoids. *Molecular Phylogenetics and Evolution*, **55**, 202–209.

Hay, W.W., Sloan, J.L. and Wold, C.N. (1988). Mass/age distribution and composition of sediments on the ocean floor and the global rate of sediment subduction. *Journal of Geophysical Research: Solid Earth (1978–2012)*, **93**(B12), 14933–14940.

Hazel, J.R. and Williams, E.E. (1990). The role of alterations in membrane lipid composition in enabling physiological adaptation of organisms to their physical environment. *Progress in Lipid Research*, **29**, 167–227.

Heezen, B.C. (1960). The rift in the ocean floor. *Scientific American*, **203**(4), 98–110.

Heezen, B.C. and Ewing, M. (1952). Turbidity currents and submarine slumps and the 1929 Grand Banks earthquake. *American Journal of Science*, **250**, 849–878.

Heezen, B.C. and Hollister, C.D. (1971). *The Face of the Deep*. Oxford: Oxford University Press.

Heezen, B.C. and Johnson, G.L. (1965). The South Sandwich Trench. *Deep-Sea Research*, **12**, 185–197.

Heezen, B.C. and McGregor, I.D. (1973). The evolution f the Pacific. *Scientific American*, **229**, 102–112.

Heezen, B.C., Bunce, E.T., Hersey, J.B. and Tharp, M. (1964). Chain and Romanche fracture zones. *Deep-Sea Research*, **11**, 11–33.

Henriques, C., Priede, I.G. and Bagley, P.M. (2002). Baited camera observations of deep-sea demersal fishes of the northeast Atlantic Ocean at 15–28°N off West Africa. *Marine Biology*, **141**, 307–314.

Herdman, H.F.P., Wiseman, J.D.H. and Ovey, C.D. (1956). Proposed names of features on the deep-sea floor, 3. Southern or Antarctic Ocean. *Deep-Sea Research*, **3**, 258–261.

Herring, P.J. (2002). *The Biology of the Deep Ocean*. Oxford: Oxford University Press.

Hess, H.H. and Buell, H.W. (1950). The greatest depth in the oceans. *Transactions of the American Geophysics Union*, **31**, 401–405.

Hessler, R.R. and Jumars, P.A. (1974). Abyssal community analysis from replicate box cores in the central North Pacific. *Deep-Sea Research*, **21**, 185–209.

Hessler, R.R. and Sanders, H.L. (1967). Faunal diversity in the deep-sea. *Deep-Sea Research*, **14**, 65–78.

Hessler, R.R. and Strömberg, J.-O. (1989). Behavior of Janiroidean isopods (Asellota), with special reference to deep-sea genre. *Sarsia*, **74**, 145–159.

Hessler, R.R., Isaacs, J.D. and Mills, E.L. (1972). Giant amphipod from the abyssal Pacific Ocean. *Science*, **175**(4022), 636–637.

Hessler, R.R., Ingram, C.L., Yayanos, A.A. and Burnett, B.R. (1978). Scavenging amphipods from the floor of the Philippine Trench. *Deep-Sea Research*, **25**, 1029–1047.

HessIer, R.R., Wilson, G.D.F. and Thistle, D. (1979). The deep-sea isopods: a biogeographic and phylogenetic overview. *Sarsia*, **64**, 67–75.

Hinrichs, K.U. and Inagaki, F. (2012). Downsizing the deep biosphere. *Science*, **338**(6104), 204–205.

Hochachka, P.W. and Somero, G.N. (1984). Temperature adaptation. In *Biochemical Adaptation: Mechanism and Process in Physiological Evolution*. ed. P.W. Hochachka and G.N. Somero. Oxford: Oxford University Press, pp. 355–449.

Hochachka, P.W. and Somero, G.N. (2002). *Biochemical Adaptation: Mechanism and Process in Physiological Evolution*. Oxford: Oxford University Press.

Hollister, C.D. and McCave, I.N. (1984). Sedimentation under deep sea storms. *Nature*, **309**, 220–225.

Holt, R.D. (2010). 2020 visions: ecology. *Nature*, **463**, 32.

Holzer, T.L. and Savage, J.C. (2013). Global earthquake fatalities and population. *Earthquake Spectra*, **29**(1), 155–175.

Honda, C.M., Kusakabe, M., Nakabayashi, S., Manganini, S.J. and Honjo, S. (1997). Change in pCO_2 through biological activity in the marginal seas of the western North Pacific: the efficiency of the biological pump estimated by a sediment trap experiment. *Journal of Oceanography*, **53**, 645–662.

Horibe, S. (1982). Technique and studies of marine environmental assessment. Results of tests of automatic floating deep-sea sampling device in deep water (6000 m). Biological and collecting experiments. *Special Report of the Ocean Research Institute*, Tokyo University, March, p. 23.

Horikoshi, K. and Bull, A.T. (2011). Prologue: definition, categories, distribution, origin and evolution, pioneering studies, and emerging fields of extremophiles. In *Extremophiles Handbook*, ed. K. Horikoshi, G. Antranikian, A.T. Bull, F.T. Robb and K.O. Stetter. London: Springer, pp. 4–14.

Horikoshi, K. and Grant, W.D. (1998). *Extremophiles. Microbial Life in Extreme Environments.* New York: Wiley-Liss.

Horikoshi, K., Antranikian, G., Bull, A.T., Robb, F.T. and Stetter, K.O. (2011). *Extremophiles Handbook*. London: Springer.

Horikoshi, M., Fujita, T. and Ohta, S. (1990). Benthic associations in bathyal and hadal depths off the Pacific coast of north eastern Japan: physiognomies and site factors. *Progress in Oceanography*, **24**, 331–339.

Horne, D.J. (1999). Ocean circulation modes of the Phanaerozoic: implications for the antiquity of deep-sea benthonic invertebrates. *Crustaceana*, **72**, 999–1018.

Hoskin, C.J., Higgie, M., McDonald, K.R. and Moritz, C. (2005). Reinforcement drives rapid allopatric speciation. *Nature*, **437**(27), 1353–1356.

Howell, D.G. (1989). *Tectonics of Suspect Terranes: Mountain Building and Continental Growth*. London: Chapman and Hall.

Howell, D.G. and Murray, R.W. (1986). A budget for continental growth and denudation. *Science*, **233**(4762), 446–449.

Hsu, K.J. (1992). *Challenger at Sea: A Ship that Revolutionized Earth Science*. Princeton, NJ: Princeton University Press.

Huber, J.A., Mark Welch, D.B., Morrison, H.G. *et al.* (2007). Microbial population structures in the deep marine biosphere. *Science*, **318**, 97–100.

Hudson, I.R., Pond, D.W., Billett, D.S.M. *et al.* (2004). Temporal variations in fatty acid composition of deep-sea holothurians: evidence of bentho-pelagic coupling. *Marine Ecology Progress Series*, **281**, 109–120.

Humphris, S.E. (2010). Vehicles for deep sea exploration. In *Marine Ecological Processes: A Derivative of the Encyclopedia of Ocean Sciences*, ed. J.H. Steele, S.A. Thorpe and K.K. Turekian. London: Academic Press, pp. 197–209.

Hydrographic Department, Japan Marine Safety Agency (1984). Mariana Trench survey by the 'Takuyo'. *International Hydrographic Bulletin*, 351–352.

Imai, E., Honda, H., Hatori, K., Brack, A. and Matsuno, K. (1999). Elongation of oligopeptides in a simulated submarine hydrothermal system. *Science*, **283**, 831–833.

Ingram, C.L. and Hessler, R.R. (1983). Distribution and behavior of scavenging amphipods from the central North Pacific. *Deep-Sea Research*, **30**(7A), 683–706.

Ingram, C.L. and Hessler, R.R. (1987). Population biology of the deep-sea amphipod *Eurythenes gryllus* inferences from instar analyses. *Deep-Sea Research A*, **34**(12) 1889–1910.

Isaacs, J.D. and Schick, G.B. (1960). Deep-sea free instrument vehicle. *Deep-Sea Research*, **7**(1), 61–67.

Isaacs, J.D. and Schwartzlose, R.A. (1975). Active animals of the deep sea floor. *Scientific American*, **233**(4), 84–91.

Itoh, K., Inoue, T., Tahara, J. *et al.* (2008). Sea trials of the new ROV ABISMO to explore the deepest parts of oceans. *Proceedings of the Eighth ISOPE Pacific/Asia Offshore Mechanics Symposium.*

Itou, M., Matsumura, I. and Noriki, S. (2000). A large flux of particulate matter in the deep Japan Trench observed just after the 1994 Sanriku-Oki earthquake. *Deep-Sea Research I*, **47**, 1987–1998.

Itoh, M., Kawamura, K., Kitahashi, T. *et al.* (2011). Bathymetric patterns of meiofaunal abundance and biomass associated with the Kuril and Ryukyu trenches, western North Pacific Ocean. *Deep-Sea Research*, **58**, 86–97.

Ivanov, A.V. (1963). Vertical and geographical dissemination of Pogonophora. *Proceedings of the XVI International Congress on Zoology,* Washington, DC, **1**, 97.

Iwamoto, T. and Stein, D.L. (1974). A systematic review of the rattail fishes (Macrouridae: Gadiformes) from Oregon and adjacent waters. *Occasional Papers of the California Academy of Sciences*, **111**, 1–79.

Jacobs, D.K. and Lindberg, D.R. (1998). Oxygen and evolutionary patterns in the sea: onshore/ offshore trends and recent recruitment of deep-sea faunas. *Proceedings of the National Academy of Sciences, USA*, **95**, 9396–9401.

Jamieson, A.J. and Fujii, T. (2011). Trench connection. *Biology Letters*, **7**, 641–643.

Jamieson, A.J. and Yancey, P.H. (2012). On the validity of the *Trieste* flatfish; dispelling the myth. *Biological Bulletin*, **222**, 171–175.

Jamieson, A.J., Fujii, T., Solan, M. *et al.* (2009a). Liparid and Macrourid fishes of the hadal zone: *in situ* observations of activity and feeding behaviour. *Proceedings of the Royal Society of London B*, **276**, 1037–1045.

Jamieson, A.J., Fujii, T., Solan, M. *et al.* (2009b). First findings of decapod crustacea in the hadal-zone. *Deep-Sea Research I*, **56**, 641–647.

Jamieson, A.J., Fujii, T., Solan, M. and Priede, I.G. (2009c). HADEEP: free-falling landers to the deepest places on Earth. *Marine Technology Society Journal*, **43**(5), 151–159.

Jamieson, A.J., Solan, M. and Fujii, T. (2009d). Imaging deep-sea life beyond the abyssal zone. *Sea Technology*, **50**(3), 41–46.

Jamieson, A.J., Fujii, T., Mayor, D.J., Solan, M. and Priede, I.G. (2010). Hadal trenches: the ecology of the deepest places on Earth. *Trends in Ecology and Evolution*, **25**(3), 190–197.

Jamieson, A.J., Kilgallen, N.M., Rowden, A.A. *et al.* (2011a). Bait-attending fauna of the Kermadec Trench, SW Pacific Ocean: evidence for an ecotone across the abyssal-hadal transition zone. *Deep-Sea Research I*, **58**, 49–62.

Jamieson, A.J., Gebruk, A., Fujii, T. and Solan, M. (2011b). Functional effects of the hadal sea cucumber *Elpidia atakama* (Holothuroidea, Elasipodida) reflect small-scale patterns of resource availability. *Marine Biology*, **158**(12), 2695–2703.

Jamieson, A.J., Fujii, T., Bagley, P.M. and Priede, I.G. (2011c). The scavenging dependency of the deepwater eel *Synaphobranchus kaupii* on the Portuguese dogfish *Centroscymnus coelolepis*. *Journal of Fish Biology*, **79**, 205–216.

Jamieson, A.J., Lörz, A.-N., Fujii, T. and Priede, I.G. (2012a). *In situ* observations of trophic behaviour and locomotion of *Princaxelia* amphipods (Crustacea, Pardaliscidae) at hadal depths in four West Pacific trenches. *Journal of the Marine Biology Association of the United Kingdom*, **91**(1), 143–150.

Jamieson, A.J., Fujii, T. and Priede, I.G. (2012b). Locomotory activity and feeding strategy of the hadal munnopsid isopod *Rectisura* cf. *herculea* (Crustacea: Asellota) in the Japan Trench. *Journal of Experimental Biology*, **215**, 3010–3017.

Jamieson, A.J., Priede, I.G. and Craig, J. (2012c). Distinguishing between the abyssal macrourids *Coryphaenoides yaquinae* and *C. armatus* from *in situ* photography. *Deep-Sea Research I*, **64**, 78–85.

Jamieson, A.J., Lacey, N.C., Lörz, A.-N., Rowden, A.A. and Piertney, S.B. (2013). The super-giant amphipod *Alicella gigantea* (Crustacea: Alicellidae) from hadal depths in the Kermadec Trench, SW Pacific Ocean. *Deep-Sea Research II*. **92**, 107–113.

Jannasch, H.W. and Taylor, C.D. (1984). Deep-sea microbiology. *Annual Review of Microbiology*, **38**, 487–514.

Janβen, F., Treude, T. and Witte, U. (2000). Scavenger assembleges under differing trophic conditions: a case study in the deep Arabian Sea. *Deep-Sea Research II*, **47**, 2999–3026.

Jenik, J. (1992). Ecotone and ecocline: two questionable concepts in ecology. *Ekologia*, **11**(3), 243–250.

Jones, D.O.B., Bett, B.J., Wynn, R.B. and Masson, D.G. (2009). The use of towed camera platforms in deep-water science. *Underwater Technology*, **28**(2), 41–50.

Johnson, G.C. (1998). Deep water properties, velocities, and dynamics over ocean trenches. *Journal of Marine Research*, **56**(2), 239–347.

Jørgensen, B.B. (2012). Shrinking majority of the deep biosphere. *Proceedings of the National Academy of Sciences, USA*, **109**(40), 15976–15977.

Jørgensen, B.B. and D'Hondt, S. (2006). A starving majority deep beneath the seafloor. *Science*, **314**, 932–934.

Jumars, P.A. (1975). Environmental grain and polychaete species' diversity in a bathyal benthic community. *Marine Biology*, **30**(3), 253–266.

Jumars, P.A. and Hessler, R.R. (1976). Hadal community structure: implications from the Aleutian Trench. *Journal of Marine Research*, **34**, 547–560.

Kaartvedt, S., Van Dover, C.L., Mullineaux, L.S. Wiebe, P.H. and Bollens, S.M. (1994). Amphipods on a deep-sea hydrothermal treadmill. *Deep-Sea Research I*, **41**(1), 179–195.

Kaiser, M.J. and Moore, P.G. (1999). Obligate marine scavengers: do they exist? *Journal of Natural History*, **33**, 475–481.

Kamenskaya, O.E. (1981). The amphipods (Crustacea) from deep-sea trenches in the western part of the Pacific Ocean. *Transactions of the P.P. Shirshov Institute of Oceanology*, **115**, 94–107. (In Russian.)

Kamenskaya, O.E. (1989). Peculiarities of the vertical distribution of komokiaceans in the Pacific Ocean. *Transactions of the P.P. Shirshov Institute of Oceanology*, **123**, 55–58.

Karan, P.P. and Cotton Mather (1985). Tourism and environment in the Mount Everest region. *Geographical Review*, **75**(1), 93–95.

Karig, D.E. and Sharman, G.F. (1975). Subduction and accretion in trenches. *Earth Planetary Science Letters*, **86**, 377–389.

Kato, C. (2011). Distribution of piezophiles. In *Extremophiles Handbook*, ed. K. Horikoshi, G. Antranikian, A.T. Bull, F.T. Robb and K.O. Stetter. London: Springer, pp. 644–653.

Kato, C. and Bartlett, D.H. (1997). The molecular biology of barophilic bacteria. *Extremophiles*, **1**, 111–116.

Kato, C. and Horikoshi, K. (1996). Gene expression under high pressure. In *Progress in Biotechnology 13, High Pressure Bioscience and Biotechnology*, ed. R. Hayashi and C. Balny. Amsterdam: Elsevier Science, pp. 59–66.

Kato, C. and Nogi, Y. (2001). Correlation between phylogenetic structure and function: examples from deep-sea *Shewanella*. *FEMS Microbiology Ecology*, **35**(3), 223–230.

Kato, C. and Qureshi, M.H. (1999). Pressure response in deep-sea piezophilic bacteria. *Journal of Molecular Microbiology and Biotechnology*, **1**(1), 87–92.

Kato, C., Sato, T. and Horikoshi, K. (1995a). Isolation and properties of barophilic and barotolerant bacteria from deep-sea mud samples. *Biodiversity and Conservation*, **4**, 1–9.

Kato, C., Smorawinska, M., Sato, T. and Horikoshi, K. (1995b). Cloning and expression in *Escherichia coli* of a pressure-regulated promoter region from a barophilic bacterium, stain DB6705. *Journal of Marine Biotechnology*, **2**, 125–129.

Kato, C., Suzuki, S., Hata, S., Ito, T. and Horikoshi, K. (1995c). The properties of a protease activated by high pressure from *Sprosarcina* sp. strain DSK25 isolated from deep-sea sediment. *JAMSTEC Research*, **32**, 7–13.

Kato, C., Masui, N. and Horikoshi, K. (1996). Properties of obligately barophilic bacteria isolated from a sample of deep-sea sediment from the Izu-Bonin Trench. *Journal of Marine Biotechnology*, **4**, 96–99.

Kato, C., Li, L., Tamaoka, J. and Horikoshi, K. (1997). Molecular analyses of the sediment of the 11000-m deep Mariana Trench. *Extremophiles*, **1**, 117–123.

Kato, C., Li, L., Nogi, Y. *et al.* (1998). Extremely barophilic bacteria isolated from the Mariana Trench, Challenger Deep, at a depth of 11,000 meters. *Applied Environmental Microbiology*, **64**, 1510–1513.

Kaufmann, R.S. (1994). Structure and function of chemoreceptors in scavenging lysianassoid amphipods. *Journal of Crustacean Biology*, **14**(1), 54–71.

Kaufmann, R.S. and Smith, K.L. (1997). Activity patterns of mobile epibenthic megafauna at an abyssal site in the eastern North Pacific: results from a 17-month time-lapse photographic study. *Deep-Sea Research*, **44**, 559–579.

Kawabe, M. (1993). Deep water properties and circulation in the western North Pacific. In *Deep Ocean Circulation: Physical and Chemical Aspects*, ed. T. Teramoto. Amsterdam: Elsevier, pp. 17–37.

Kawabe, M. and Fujio, S. (2010). Pacific Ocean circulation based on observation. *Journal of Oceanography*, **66**, 389–403.

Kawabe, M., Fujio, S. and Yanagimoto, D. (2003). Deep-water circulation at low latitudes in the western North Pacific. *Deep-Sea Research I*, **50**(5), 631–656.

Kearey, P. and Vine, F.J. (1990). *Global Tectonics*. Oxford: Blackwell Science.

Keller, N., Naumov, D. and Pasternak, F. (1975). Bottom deep-sea Coelenterata from the Gulf and Caribbean. *Trudy Instituta Okeanologii*, **100**, 147–159.

Kelly, R.H. and Yancey, P.H. (1999). High contents of trimethylamine oxide correlating with depth in deep-sea teleost fishes, skates, and decapod crustaceans. *Biological Bulletin*, **196**, 18–25.

Kemp, K.M., Jamieson, A.J., Bagley, P.M. *et al.* (2006). Consumption of a large bathyal food fall, a six month study in the north-east Atlantic. *Marine Ecology Progress Series*, **310**, 65–76.

Kendall, V.J. and Haedrich, R.L. (2006). Species richness in Atlantic deep-sea fishes assessed in terms of themed-domain effect and Rapoport's rule. *Deep-Sea Research I*, **53**(3), 506–515.

Kennedy, H., Beggins, J., Duarte, C.M. *et al.* (2010). Seagrass sediments as a global carbon sink: isotopic constraints. *Global Biogeochemical Cycles*, **24**, GB4026.

Kiilerich, A. (1955). Bathymetric features of the Philippine Trench. *Galathea Report*, **1**, 155–172.

Kikuchi, T. and Nemoto, T. (1991). Deep-sea shrimps of the genus *Benthesicymus* (Decapoda: Dendrobranchiata) from the western North Pacific. *Journal of Crustacean Biology*, **11**(1), 64–89.

Kilgallen, N.M. (in press). Three new species of *Hirondellea* (Crustacea, Amphipoda, Hirondelleidae) from hadal depths of the Peru–Chile Trench. *Marine Biology Research*.

Kiraly, S., Moore, J.A. and Jasinski, P.H. (2003). Deepwater and other sharks of the US Atlantic Ocean Exclusive Economic Zone. *Marine Fisheries Review*, **62**, 1483–1491.

Kirkgaard, J.B. (1956). Benthic polychaeta from depths exceeding 6000 meters. *Galathea Report*, **2**, 63–78.

Kirsch, P.E., Iverson, S.J. and Bowen, W.D. (2000). Effect of a low-fat diet on body composition and blubber fatty acids of captive juvenile harp seals (*Phoca groenlandica*). *Physiological and Biochemical Zoology*, **73**, 45–59.

Kitahashi, T., Kawamura, K., Veit-Köhler, G. *et al.* (2012). Assemblages of Harpacticoida (Crustacea: Copepoda) from the Ryukyu and Kuril Trenches, north-west Pacific Ocean. *Journal of the Marine Biological Association of the United Kingdom*, **92**, 275–286.

Kitahashi, T., Kawamura, K., Kojima, S. and Shimanaga, M. (2013). Assemblages gradually change from bathyal to hadal depth: a case study on harpacticoid copepods around the Kuril Trench (north-west Pacific Ocean). *Deep-Sea Research I*, **74**, 39–47.

Kitazato, H., Uematsui, K., Todo, Y. and Gooday, A.J. (2009). New species of *Leptohalysis* (Rhizaria, Foraminifera) from an extreme hadal site in the western Pacific Ocean. *Zootaxa*, **2059**, 23–32.

Klages, M., Vopel, K., Bluhm, H. *et al.* (2001). Deep-sea food falls: first observation of a natural event in the Arctic Ocean. *Polar Biology*, **24**, 292–295.

Knauss, J.A. (1997). *Introduction to Physical Oceanography*, 2nd edn. Upper Saddle River, NJ: Prentice Hall.

Knudsen, J. (1964). Scaphopoda and Gastropoda from depths exceeding 6000 m. *Galathea Report*, **7**, 1–12.

Knudsen, J. (1970). The systematics and biology of abyssal and hadal Bivalvia. *Galathea Report*, **11**, 1–241.

Kobayashi, H., Hatada, Y., Tsubouchi, T., Nagahama, T. and Takami, H. (2012). The hadal amphipod *Hirondellea gigas* possessing a unique cellulase for digesting wooden debris buried in the deepest seafloor. *PLoS ONE*, **7**(8), e42727.

Koltun, V.M. (1970). Sponges of the Arctic and Antarctic: a faunistic review. *Symposia of the Zoological Society of London*, **25**, 285–297.

Koehler, R. (1909). Echinodermes provenant des campagnes du yacht 'Princesse Alice'. *Résultats des Campagnes Scientifiques du Prince Albert Ier, Monaco*, **34**, 1–317.

Kohnen, W. (2009). Human exploration of the deep seas: fifty years and the inspiration continues. *Marine Technology Society Journal*, **43**(5), 42–62.

Kramp, P.L. (1956). Hydroids from depths exceeding 6000 meters. *Galathea Report*, **2**, 17–20.

Kramp, P.L. (1959). *Stephanoscyphus* (Scyphozoa). *Galathea Report*, **1**, 173–188.

Krylova, E.M. and Sahling, H. (2010). Vesicomyidae (Bivalvia): current taxonomy and distribution. *PLoS ONE*, **5**(4), e9957.

Kudiniva-Pasternak, R.K. (1978). Tanaidacea (Crustacea, Malacostraca) from the deep-sea trenches of the western part of the Pacific. *Trudy Instituta Okeanologii*, **108**, 115–135.

Kullenberg, B. (1956). The technique of trawling. In *The Galathea Deep Sea Expedition*, ed. A.F. Bruun, S. Greve, H. Mielche and R. Spärk. London: George Allen and Unwin, pp. 112–118.

Kyo, M., Miyazaki, E., Tsukioka, S. *et al.* (1995). The sea trial of 'KAIKO', the full ocean depth research ROV. *OCEANS '95*, **3**, 1991–1996.

Lacey, N.C., Jamieson, A.J. and Søreide, F. (2013). Successful capture of ultradeep sea animals from the Puerto Rico Trench. *Sea Technology*, **54**(3), 19–21.

Laist, D.W. (1987). Overview of the biological effects of lost and discarded plastic debris in the marine environment. *Marine Pollution Bulletin*, **18**, 319.

Lallemand, S. (1999). La subduction océanique. *Pour la Science*, **259**, 108.

Lambshead, P.J.D. (2003). Marine nematode deep-sea biodiversity – hyperdiverse or hype? *Journal of Biogeography*, **30**(4), 475–485.

Lampitt, R.S. (1985). Evidence for the seasonal deposition of detritus to the deep-sea floor and its subsequent resuspension. *Deep-Sea Research*, **32**, 885–897.

Larsen, K. and Simomura, M. (2007a). Tanaidacea (Crustacea: Peracarida) from Japan III. The deep trenches; the Kuril–Kamchatka Trench and Japan Trench (Foreword). *Zootaxa*, **1599**, 5–12.

Larsen, K. and Simomura, M. (2007b). Tanaidacea (Crustacea: Peracarida) from Japan II. Tanaidomorpha from the East China Sea, the West Pacific, and the Nansei Islands. *Zootaxa*, **1341**, 29–48.

Laubier, L. (1985). Une contribution française aux recherches écologiques en mer profonde: bilan des plongées en bathyscaphes. *Tethys*, **11**(3–4), 255–263.

Lauro, F.M., Chastain, R.A., Blankenship, L.E., Yayanos, A.A. and Bartlett, D.H. (2007). The unique 16S rRNA genes of piezophiles reflect both phylogeny and adaptation. *Applied Environmental Microbiology*, **73**, 838–845.

Laxson, C.J., Condon, N.E., Drazen, J.C. and Yancey, P.H. (2011). Decreasing urea:trimethylamine n-oxide ratios with depth in Chondrichthyes: a physiological depth limit? *Physiological and Biochemical Zoology*, **84**(5), 494–505.

Lay, T., Kanamori, H., Ammon, C.J. *et al.* (2005). The Great Sumatra-Andaman earthquake of 26 December 2004. *Science*, **308**(5725), 1127–1133.

Le Pichon, X. (1968). Sea-floor spreading and continental drift. *Journal of Geophysical Research*, **73**(12), 3661–3697.

Leal, J.H. and Harasewych, M.G. (1999). Deepest Atlantic molluscs: hadal limpets (Mollusca, Gastropoda, Cocculiniformia) from the northern boundary of the Caribbean Plate. *Invertebrate Biology*, **118**(2), 116–136.

Lebrato, M. and Jones, D.O.B. (2009). Mass deposition event of *Pyrosoma atlanticum* carcasses off Ivory Coast (West Africa). *Limnology and Oceanography*, **54**, 1197–1209.

Lebrato, M., Pitt, K.A., Sweetman, A.K. *et al.* (2012). Jelly-falls historic and recent observations: a review to drive future research directions. *Hydrobiologica*, **690**(1), 227–245.

Lecroq, B., Gooday, A.J., Cedhagen, T., Sabbatini, A. and Pawlowski, J. (2009a). Molecular analyses reveal high levels of eukaryote richness associated with enigmatic deep-sea protists (Komokiacea). *Marine Biodiversity*, **39**, 45–55.

Lecroq, B., Gooday, A.J., Tsuchiya, M. and Pawlowski, J. (2009b). A new genus of xenophyophores (Foraminifera) from Japan Trench: morphological description, molecular phylogeny and elemental analysis. *Zoological Journal of the Linnean Society*, **156**(3), 455–464.

Lee, J.E. (2012). Ocean's deep, dark trenches to get their moment in the spotlight. *Science*, **336**, 141–142.

Lee, R.F., Hagen, W. and Kattner, G. (2006). Lipid storage in marine zooplankton. *Marine Ecology Progress Series*, **307**, 273–306.

Lee, W.Y. and Arnold, C.R. (1983). Chronic toxicity of ocean-dumped pharmaceutical wastes to the marine amphipod *Amphithoe valida*. *Marine Pollution Bulletin*, **14**, 150–153.

Lehtonen, K.K. (1996). Ecophysiology of the benthic amphipod *Monoporeia affinis* in an open-sea of the northern Baltic Sea: seasonal variations in body composition, with bioenergetic considerations. *Marine Ecology Progress Series*, **143**, 87–98.

Lemche, H. (1957). A new living deep-sea mollusc of the Cambro-Devonian class Monoplacophora. *Nature*, **179**, 413–416.

Lemche, H., Hansen, B., Madsen, F.J. Tendal, O.S. and Wolff, T. (1976). Hadal life as analysed from photographs. *Videnskabelige Meddelelser Fra Dansk Naturhistorik Forening*, **139**, 263–336.

Lerche, D. and Nozaki, Y. (1998). Rare earth elements of sinking particulate matter in the Japan Trench. *Earth and Planetary Science Letters*, **159**, 71–86.

Levin. L.A. (1991). Interactions between metazoans and large, agglutinated protozoans: implications for the community structure of deep-sea benthos. *American Zoologist*, **31**, 886–900.

Levin, L.A. (2005). Ecology of cold seep sediments: interactions of fauna with flow, chemistry, and microbes. *Oceanography and Marine Biology*, **43**, 1–46.

Levin, L.A. and Dayton, P.K. (2009). Ecological theory and continental margins: where shallow meets deep. *Trends in Ecology and Evolution*, **24**(11), 606–617.

Li, Z.H., Xu, Z.Q. and Gerya, T.V. (2001). Flat versus steep subduction: contrasting modes for the formation and exhumation of high- to ultrahigh-pressure rocks in continental collision zones. *Earth and Planetary Science Letters*, **301**, 65–77.

Litzov, M.A., Bailey, M.A., Prahl, F.G. and Heintz, R. (2006). Climate regime shifts and reorganization of fish communities: the essential fatty acid limitation hypothesis. *Marine Ecology Progress Series*, **315**, 1–11.

Liu, F., Cui, W.C. and Li, X.Y. (2010). China's first deep manned submersible, JIAOLONG. *Science China Earth Science*, **53**, 1407–1410.

Longhurst, A. (2007). *Ecological Geography of the Sea*, 2nd edn. London: Academic Press.

Longhurst, A., Sathyendranath, S., Platt, T. and Caverhill, C. (1995). An estimate of global primary production in the ocean from satellite radiometer data. *Journal of Plankton Research*, **17**(6), 1245–1271.

Lörz, A.-N. (2010). Trench treasures: the genus *Princaxelia* (Pardaliscidae, Amphipoda). *Zoologica baetica*, **21**, 65–84.

Lörz, A.-N., Berkenbusch, K., Nodder, S. *et al.* (2012). A review of deep-sea benthic biodiversity associated with trench, canyon and abyssal habitats below 1500 m depth in New Zealand waters. *New Zealand Aquatic Environment and Biodiversity Report*, **92**, 133p.

Lutz, M.J., Caldeira, K., Dunbar, R.B. and Behrenfeld, M.J. (2007). Seasonal rhythms of net primary production and particulate organic carbon flux to depth describe the efficiency of biological pump in the global ocean. *Journal of Geophysical Research: Oceans*, **112**, C10.

Lutz, R.A. and Falkowski, P.G. (2012). A dive to Challenger Deep. *Science*, **336**, 301–302.

MacDonald, A.G. (1978). Further studies on the pressure tolerance of deep-sea crustacean, with observations using a new high pressure trap. *Marine Biology*, **45**, 9–21.

MacDonald, A.G. (1984a). The effect of pressure on the molecular structure and physiological functions of cell membranes. *Philosophical Transactions of the Royal Society London B*, **304**, 47–68.

MacDonald, A.G. (1984b). Homeoviscous theory under pressure. I. The fatty acid composition of *Tetrahymena pyriformis* NT-l grown at high pressure. *Biochimica Biophysica Acta*, **775**, 141–149.

MacDonald, A.G. (1997). Hydrostatic pressure as an environmental factor in life processes. *Comparative Biochemistry and Physiology*, **116A**, 291–297.

MacDonald, A.G. and Cossins, A.R. (1985). The theory of homeoviscous adaptation of membranes applied to deep-sea animals. *Society of Experimental Biology Symposium*, **39**, 301–322.

MacDonald, A.G. and Fraser, P.J. (1999). The transduction of very small hydrostatic pressures. *Comparative Biochemistry and Physiology*, **122**, 13–36.

MacDonald, A.G. and Gilchrist, I. (1980). Effects of hydraulic decompression and compression on deep sea amphipods. *Comparative Biochemistry and Physiology*, **67A**, 149–153.

MacDonald, A.G. and Gilchrist, I. (1982). Pressure tolerance of deep-sea amphipods collected at their ambient pressure. *Comparative Biochemistry and Physiology*, **71A**, 349–352.

Macdonald, K.S.I.I.I., Yampolsky, L. and Duffy, J.E. (2005). Molecular and morphological evolution of the amphipod radiation in Lake Baikal. *Molecular and Phylogenetic Evolution*, **35**, 323–343.

MacDonell, M.T. and Colwell, R.R. (1985). Phylogeny of the Vibrionaceae, and recommendation for two new genera, *Listonella* and *Shewanella*. *Systematic and Applied Microbiology*, **6**(2), 171–182.

MacElroy, R.D. (1974). Some comments on the evolution of extremophiles. *Biosystems*, **6**, 74–75.

Machida, Y. (1989). Record of *Abyssobrotula galathea* (Ophidiidae: Ophidiiformes) from the Izu-Bonin Trench, Japan. *Bulletin of Marine Science and Fisheries, Kochi University, Japan*, **11**, 23–25.

Machida, Y. and Tachibana, Y. (1986). A new record of *Bassozetus zenkevitchi* (Ophidiidae, Ophidiiformes) from Japan. *Japanese Journal of Ichthyology*, **32**, 437–439.

Madigan, M.T. and Marrs, B.L. (1997). Extremophiles. *Scientific American*, **276**, 82–87.

Madsen, F.J. (1955). Holothurioidea. *Reports on the Swedish Deep-Sea Expedition, 2, Zoology*, **12**, 151–173.

Madsen, F.J. (1956). The Echinoidea, Asteroidea, and Ophiuroidea at depths exceeding 6000 metres. *Galathea Report*, **2**, 23–32.

Madsen, F.J. (1961). On the zoogeography and origin of the abyssal fauna in view of the knowledge of the Porcellanasteridae. *Galathea Report*, **4**, 177–218.

Magaard, L. and McKee, W.D. (1973). Semi-diurnal tidal currents at 'site D'. *Deep-Sea Research*, **30**, 805–833.

Mahaut, M.L., Sibuet, M. and Shirayama, Y. (1995). Weight-dependent respiration rates in deep-sea organisms. *Deep-Sea Research I*, **42**, 1575–1582.

Malyutina, M.V. (2003). Revision of *Storthyngura* Vanhöffen, 1914 (Crustacea: Isopods: Munnopsidae) with descriptions of three new genera and four new species from the deep South Atlantic. *Organisms, Diversity and Evolution*, **13**, 1–101.

Mantovani, R. (1909). L'Antarctide, *Je m'instruis. La science pour tous*, **38**, 595–597.

Mantyla, A.W. and Reid, J.L. (1978). Measurements of water characteristics at depths greater than 10 km in the Marianas Trench. *Deep-Sea Research*, **25**, 169–173.

Mantyla, A.W. and Reid, J.L. (1983). Abyssal characteristics of the world ocean waters. *Deep-Sea Research*, **30**, 805–833.

Markle, D.F. and Olney, J.E. (1990). Systematics of the pearlfishes (Pisces: Carapidae). *Bulletin of Marine Science*, **47**, 269–410.

Marshall, N.B. (1954). *Aspects of Deep Sea Biology*. New York: Philosophical Library.

Marteinsson, V.T., Reysenbach, A.-L., Birrien, J.-L. and Prieur, D. (1999). A stress protein is induced in the deep-sea barophilic hyperthermophile *Thermococcus barophilus* when grown under atmospheric pressure. *Extremophiles*, **3**, 277–282.

Martin, D.D., Bartlett, D.H. and Roberts, M.F. (2002). Solute accumulation in the deep-sea bacterium *Photobacterium profundum*. *Extremophiles*, **6**, 507–514.

Martin, J. and Miquel, J.-C. (2010). High downward flux of mucilaginous aggregates in the Ligurian Sea during summer 2002. Similarities with the mucilage phenomenon in the Adriatic Sea. *Marine Ecology*, **31**, 393–406.

Maruyama, A., Honda, D., Yamamoto, H., Kitamura, K. and Higashihara, T. (2000). Phylogenetic analysis of psychrophilic bacteria isolated from the Japan Trench, including a description of the deep-sea species *Psychrobacter pacificensis* sp. nov. *International Journal of Systematic and Evolutionary Microbiology*, **50**, 835–846.

Masson, D.G. (2001). Sedimentary processes shaping the eastern slope of the Faeroe-Shetland Channel. *Continental Shelf Research*, **21**, 825–857.

Masuda, H., Amaoka, K., Araga, C., Uyeno, T. and Yoshino, T. (1984). *The Fishes of the Japanese Archipelago*, Vol. 1. Tokyo: Tokai University Press.

Mayer, A.M., Glaser, K.B., Cuevas, C. *et al.* (2010). The odyssey of marine pharmaceuticals: a current pipeline perspective. *Trends in Pharmacological Sciences*, **31**(6), 255–265.

McCain, C.M. (2009). Global analysis of bird elevational diversity. *Global Ecology and Biogeography*, **18**, 346–360.

McCain, C.M. (2010). Global analysis of reptile elevational diversity. *Global Ecology and Biogeography*, **19**, 541–553.

McClain, C.R. and Etter, R.J. (2005). Mid-domain models as predictors of species diversity patterns: bathymetric diversity gradients in the deep sea. *Oikos*, **109**, 555–566.

McClain, C.R. and Hardy, S.M. (2010). The dynamics of biogeographic ranges in the deep sea. *Proceedings of the Royal Society of London B*, **277**, 3533–3546.

McClain, C.R., Johnson, N.A. and Rex, M.A. (2004). Morphological disparity as a biodiversity metric in lower bathyal and abyssal gastropod assemblages. *Evolution*, **58**, 338–348.

Meek, R.P. and Childress, J.J. (1973). Respiration and the effect of pressure in the mesopelagic fish *Anoplogaster cornuta* (Beryciformes). *Deep-Sea Research*, **20**, 1111–1118.

Menard, H.W. (1958). Development of median elevations in ocean basins. *Bulletin of the Geological Society of America*, **69**, 1179–1186.

Menard, H.W. (1966). Fracture zones and offsets of the East Pacific Rise. *Journal of Geophysics Research*, **71**, 682–685.

Menard, H.W. and Smith, S.M. (1966). Hypsometry of ocean basins. *Journal of Geophysical Research*, **71**, 4305–4325.

Menzies, R.J. (1965). Conditions for the existence of life at the abyssal sea floor. *Oceanography and Marine Biology Annual Review*, **3**, 195–210.

Menzies, R.J. and George, R.Y. (1967). A re-evaluation of the concept of hadal or ultra-abyssal fauna. *Deep-Sea Research*, **14**, 703–723.

Menzies, R.J., Ewing, M., Worzel, J.L. and Clarke, A.H. (1959). Ecology of the Recent Monoplacophora. *Internationale Revue der gesamten Hydrobiologie und Hydrographie*, **48**(4), 529–545.

Menzies, R.J., Smith, L. and Emery, K.O. (1963). A combined underwater camera and bottom grab: a new tool for investigation of deep-sea benthos. *Oikos*, **10**(2), 168–182.

Menzies, R.J., George, R.Y. and Rowe, G.T. (1973). *Abyssal Environment and Ecology of the World's Oceans*. New York: John Wiley and Sons.

Merrett, N.R. (1987). A zone of faunal change in assemblages of abyssal demersal fish in the eastern North Atlantic: a response to seasonality in production? *Biological Oceanography*, **5**, 137–151.

Merrett, N.R. and Haedrich, R.L. (1997). *Deep-sea Demersal Fish and Fisheries*. London: Chapman and Hall.

Mironov, A.N. (2000). New taxa of stalked crinoids from the suborder Bourgueticrinina (Echinodermata, Crinoidea). *Zoologichesky Zhurnal*, **79**, 712–728. (In Russian.)

Messing, C.G., Neumann, A.C. and Lang, J.C. (1990). Biozonation of deepwater lithoherms and associated hardgrounds in the northeastern Straits of Florida. *Palaios*, **5**, 15–33.

Mikagawa, T. and Aoki, M. (2001). An outline of R/V Kairei and recent activity of the multichannel seismic reflection survey system (MCS) and ROV *Kaikō*. *Journal of Marine Science and Technology*, **6**, 42–49.

Momma, H., Watanbe, M., Hashimoto, K. and Tashiro, S. (2004). Loss of the full ocean depth ROV *Kaikō*, Part 1: ROV *Kaikō*, a review. *Proceedings of the 14th International Offshore and Polar Engineering Conference Volume II*, 191–193.

Monaco, A., Biscaye, P., Soyer, J., Pocklington, R. and Heussner, S. (1990). Particle fluxes and ecosystem response on a continental margin: the 1985}1988 Mediterranean ECOMARGE experiment. *Continental Shelf Research*, **10**, 809–839.

Monastersky, R. (2012). Dive master. *Nature*, **486**, 194–196.

Moore, D.R. (1963). Turtle grass in the deep sea. *Science*, **139**, 1234–1235.

Morgan, J.W. (1968). Rises, trenches, great faults, and crustal blocks. *Journal of Geophysical Research*, **73**(6), 1959–1982.

Morton, J.E. (1959). The habits and feeding organs of *Dentalium entalis*. *Journal of the Marine Biological Association of the United Kingdom*, **38**, 225–238.

Mountfort, D.O., Rainey, F.A., Burghardt, J., Kaspar, H.F. and Stackebrandt, E. (1998). *Psychromonas antarctica* gen. nov., sp. nov., a new aerotolerant anaerobic, halophilic psychrophile isolated from pond sediment of the McMurdo ice shelf, Antarctica. *Archives of Microbiology*, **169**, 231–238.

Murashima, T., Nakajoh, H., Yoshida, H., Yamauchi, N. and Sezoko, H. (2004). 7000 m class ROV KAIKŌ7000. *Proceedings of the OCEANS '04MTS/IEEE*, **2**, 812–817.

Murashima, T., Kakajoh, H. and Takami, H. (2009). 11,000m class free fall mooring system. *Oceans 2009: Europe*, 1–5.

Murray, J. (1888). On the height of the land and the depth of the ocean. *Scottish Geographic Magazine*, **4**, S.1.

Murray, J. and Hjort, J. (1912). *The Depths of the Oceans*. London: Macmillan and Company.

Murray, J.W. (2007). Biodiversity of living benthic Foraminifera: how many species are there? *Marine Micropaleontology*, **64**(3–4), 163–176.

Nakajoh, H., Murashima, T. and Yoshida, H. (2005). 7000 m operable deep-sea ROV system KAIKO7000. *Proceedings of the OMAE 2005*, Halkidiki, Greece.

Nakanishi, M. and Hashimoto, J. (2011). A precise bathymetric map of the world's deepest seafloor, Challenger Deep in the Mariana Trench. *Marine Geophysics Research*, **32**, 455–463.

Nakasone, K., Ikegami, A., Kato, C., Usami, R. and Horikoshi, K. (1998). Mechanisms of gene expression controlled by pressure in deep-sea microorganisms. *Extremophiles*, **2**, 149–154.

Nakasone, K., Ikegami, A., Kawano, H. *et al.* (2002). Transcriptional regulation under pressure conditions by the RNA polymerase s54 factor with a two component regulatory system in *Shewanella violacea*. *Extremophiles*, **6**, 89–95.

Nanba, N., Morihana, H., Nakamura, E. and Watanabe, N. (1990). Development of deep submergence research vehicle 'SHINKAI 6500'. *Technology Review Mitsubishi Heavy Industry Ltd*, **27**, 157–168.

Naoi, M., Seko, M. and Sumita, K. (2009). Earthquake risk and housing prices in Japan: evidence before and after massive earthquakes. *Regional Science and Urban Economics*, **39**, 658–669.

Naylor, E. (1985). Tidally rhythmic behaviour of marine animals. *Symposium of the Society of Experimental Biology*, **39**, 63–93.

Newman, K.R., Cormier, M.-H., Weissel, J.K. *et al.* (2008). Active methane venting observed at giant pockmarks along the US mid-Atlantic shelf break. *Earth and Planetary Science Letters*, **267**(1–2), 341–352.

Nichols, D.S., Nichols, P.D. and McMeekin, T.A. (1993). Polyunsaturated fatty acids in Antarctic bacteria. *Antarctic Science*, **5**, 149–160.

Nicholson, W.L., Munakata, N., Horneck, G., Melosh, H.J. and Setlow, P. (2000). Resistance of *Bacillus* endospores to extreme terrestrial and extraterrestrial environments. *Microbiolology and Molecular Biology Reviews*, **64**, 548–572.

Nicol, J.A.C., Lee, W.Y. and Hannebaum, N. (1978). Toxicity of Puerto Rican organic waste materials on marine invertebrates. Final Report to National Oceanic and Atmospheric Administration/Ocean Dumping and Monitoring Program, Marine Science Institute, University of Texas, Port Aransas. 37 pp.

Nielsen, J.G. (1964). Fishes from depths exceeding 6000 meters. *Galathea Report*, **7**, 113–124.

Nielsen, J.G. (1975). A review of the oviparous ophidioid fishes of the genus *Leucicorus*, with description of a new Atlantic species. *Trudy Instituta Oceanologii*, **100**, 106–123.

Nielsen, J.G. (1977). The deepest living fish *Abyssobrotula galathea*: a new genus and species of oviparous ophidiids (Pisces, Brotulidae). *Galathea Report*, **14**, 41–48.

Nielsen, J.G. (1986). Ophidiidae. In *Fishes of the North-eastern Mediterranean*, ed. P.J.P. Whitehead, M.L. Bauchot, J.C. Hureau, J. Nielsen and E. Tortonese. Paris: UNESCO, Chaucer, pp. 1158–1166.

Nielsen, J.G. and Merrett, N.R. (2000). Revision of the cosmopolitan deep-sea genus *Bassozetus* (Pisces: Ophidiidae) with two new species. *Galathea Report*, **18**, 7–56.

Nielson, J.G. and Munk, C. (1964). A hadal fish (*Bassogigas profundissimus*) with a functional swimbladder. *Nature*, **204**, 594–595.

Nielsen, J.G., Cohen, D.M., Markle, D.F. and Robins, C.R. (1999). Ophidiiform fishes of the world (Order Ophidiiformes): an annotated and illustrated catalogue of pearlfishes, cusk-eels, brotulas and other ophidiiform fishes known to date. *FAO Fisheries Synopsis*, **125**, 18.

Newell, I.M. (1967). Abyssal Halacaridae (Acari) from the southeast Pacific. *Pacific Insects*, **9**(4), 693–708.

Nogi, Y. and Kato, C. (1999). Taxonomic studies of extremely barophilic bacteria isolated from the Mariana Trench and description of *Moritella yayanosii* sp. nov., a new barophilic bacterial isolate. *Extremophiles*, **3**, 71–77.

Nogi, Y., Kato, C. and Horikoshi, K. (1998a). *Moritella japonica* sp. nov., a novel barophilic bacterium isolated from a Japan Trench sediment. *Journal of General and Applied Microbiology*, **44**, 289–295.

Nogi, Y., Kato, C. and Horikoshi, K. (1998b). Taxonomic studies of deep-sea barophilic *Shewanella* species, and *Shewanella violacea* sp. nov., a new barophilic bacterial species. *Archives of Microbiology*, **170**, 331–338.

Nogi, Y., Kato, C. and Horikoshi, K. (2002). *Psychromonas kaikoae* sp. nov., a novel piezophilic bacterium from the deepest cold-seep sediments in the Japan Trench. *International Journal of Systematic and Evolutionary Microbiology*, **52**, 1527–1532.

Nogi, Y., Hosoya, S., Kato, C. and Horikoshi, K. (2004). *Colwellia piezophila* sp. nov., a novel piezophilic species from deep-sea sediments of the Japan Trench. *International Journal of Systematic and Evolutionary Microbiology*, **54**, 1627–1631.

Nogi, Y., Hosoya, S., Kato, C. and Horikoshi, K. (2007). *Psychromonas hadalis* sp. nov., a novel piezophilic bacterium isolated from the bottom of the Japan Trench. *International Journal of Systematic and Evolutionary Microbiology*, **57**, 1360–1364.

Nozaki, Y. and Ohta, Y. (1993). Rapid and frequent trubidite accumulation in the bottom of Izu-Ogasawara Trench: chemical and radiochemical evidence. *Earth and Planetary Science Letters*, **120**, 345–360.

Nozaki, Y., Yamada, M., Nakanishi, T. *et al.* (1998). The distribution of radionuclides and some trace metals in the water columns of the Japan and Bonin trenches. *Oceanologica acta*, **21**(3), 469–484.

Nybelin, O. (1951). Introduction and station list. *Reports of the Swedish Deep-Sea Expedition, 2, Zoology*, **1**, 1–28.

Nyssen, F., Brey, T., Lepoint, G. *et al.* (2002). A stable isotope approach to the eastern Weddell Sea trophic web: focus on benthic amphipods. *Polar Biology*, **25**, 280–287.

Oguri, K., Kawamura, K., Sakaguchi, A. *et al.* (2013). Hadal disturbance in the Japan Trench induced by the 2011 Tohoku-Oki Earthquake. *Scientific Reports*, **3**, 1915.

Oji, T., Ogawa, Y., Hunter, A.W. and Kitazawa, K. (2009). Discovery of dense aggregations of stalked crinoids in Izu-Ogasawara Trench, Japan. *Zoological Science*, **26**, 406–408.

Oliphant, A., Thatje, S., Brown, A. *et al.* (2011). Pressure tolerance of the shallow-water caridean shrimp *Palaemonetes varians* across its thermal tolerance window. *Journal of Experimental Biology*, **214**, 1109–1117.

Orr, J.W., Sinclair, E.H. and Walker, W.W. (2005). *Bassozetus zenkevitchi* (Ophidiidae: teleostei) and *Paraliparis paucidens* (Liparidae: teleostei): new records for Alaska from the Bering Sea. *Northwestern Naturalist*, **86**, 65–71.

Ortelius, A. (1596). *Thesaurus Geographicus*. Antwerp: Plantin.

Osborn, K.J., Kuhnz, L.A., Priede, I.G. *et al.* (2012). Diversification of acorn worms (Hemichordata, Enteropneusta) revealed in the deep sea. *Proceedings of the Royal Society London B*, **279**(1733), 1646–1654.

Osterberg, C., Carey, A.G. and Curl, H. (1963). Acceleration of sinking rates of radionucleides in the ocean. *Nature*, **200**, 1276–1277.

Otosaka, S. and Noriki, S. (2000). REEs and Mn/Al ratio of settling particles: horizontal transport or particulate material in the northern Japan Trench. *Marine Chemistry*, **72**, 329–342.

Owens, W.B. and Warren, B.A. (2001). Deep circulation in the Pacific Ocean. *Deep-Sea Research I*, **48**(4), 959–993.

Owre, H.B. and Bayer, F.M. (1970). The deep-sea gulper *Eurypharynx pelecanoides* Vaillant 1882 (order Lyomeri) from the Hispaniola basin. *Bulletin of Marine Science*, **20**, 186–192.

Palmer, J.D. and Williams, B.G. (1986). Comparative studies of tidal rhythms: II. The duel clock control of the locomotor rhythms of two decapod crustaceans. *Marine Behaviour and Physiology*, **12**, 269–278.

Palumbi, S.R. (1994). Genetic divergence, reproductive isolation, and marine speciation. *Annual Review of Ecology and Systematics*, **25**, 547–572.

Panzeri, D., Caroli, P. and Haack, B. (2013). Sagarmatha Park (Mt Everest) porter survey and analysis. *Tourism Management*, **36**, 26–34.

Paterson, G.L.J., Doner, S., Budaeva, N. *et al.* (2009). A census of abyssal polychaetes. *Deep-Sea Research II*, **56**, 1739–1746.

Pathom-aree, W., Nogi, Y., Sutcliffe, I.C. *et al.* (2006). *Dermacoccus abyssi* sp. nov., a piezo-tolerant actinomycete isolated from the Mariana Trench. *International Journal of Systematic and Evolutionary Microbiology*, **56**, 1233–1237.

Paul, A.Z. (1973). Trapping and recovery of living deep-sea amphipods from the Arctic Ocean floor. *Deep-Sea Research*, **20**, 289–290.

Pausch, S., Below, D. and Hardy, K. (2009). Under high pressure: spherical glass flotation and instrument housings in deep ocean research. *Marine Technology Society Journal*, **43**(5), 105–109.

Pavlov, D.S., Sadkovskii, R.V., Kostin, V.V. and Lupandin, A.I. (2000). Experimental study of young fish distribution and behaviour under combined influence of baro-, photo- and thermo-gradients. *Journal of Fish Biology*, **57**, 69–81.

Peck, L. and Chapelle, G. (1999). Amphipod gigantism dictated by oxygen availability? A reply to John I. Spicer and Kevin J. Gaston. *Ecology Letters*, **2**, 401–403.

Peck, L.S., Webb, K.E. and Bailey, D.M. (2004). Extreme sensitivity of biological function to temperature in Antarctic marine species. *Functional Ecology*, **18**, 625–630.

Peele, E.R., Singleton, F.L., Deming, J.W., Cavari, B. and Colwell, R.R. (1981). Effects of pharmaceutical wastes on microbial populations in surface waters at the Puerto Rico dump site in the Atlantic Ocean. *Applied Environmental Microbiology*, **41**, 873–879.

Pennec, J.-P., Wardle, C.S., Harper, A.A. and MacDonald, A.G. (1988). Effects of high hydro-static pressure on the isolated hearts of shallow water and deep-sea fish: results of Challenger cruise 6BI 85. *Comparative Biochemistry and Physiology*, **89A**, 215–218.

Pérès, J.M. (1965). Apercu sur les resultats de deux plongees effectuees dans le ravin de Puerto-Rico par le bathyscaphe *Archimède*. *Deep-Sea Research*, **12**, 883–891.

Perrone, F.M., Dell'Anno, A., Danovaro, R., Della Croce, N. and Thurston, M.H. (2002). Population biology of *Hirondellea* sp nov (Amphipoda: Gammaridea: Lysianassoidea) from the Atacama Trench (south-east Pacific Ocean). *Journal of the Marine Biology Association of the United Kingdom*, **82**(3), 419–425.

Perrone, F.M., Della Croce, N. and Dell'Anno, A. (2003). Biochemical composition and trophic strategies of the amphipod *Eurythenes gryllus* at hadal depths (Atacama Trench, South Pacific). *Chemistry and Ecology*, **19**(6), 441–449.

Pettersson, H. (1948). The Swedish Deep-Sea expedition. *Nature*, **162**, 324–325.

Phillips, R.J. and Hansen, V.L. (1998). Geological evolution of Venus: rises, plains, plumes, and plateaus, *Science*, **279**, 1492–1497.

Phleger, C.F. and Soutar, A. (1971). Free vehicles and deep-sea biology. *American Zoologist*, **11**, 409–418.

Piccard, J. and Dietz, R.S. (1961). *Seven Miles Down*. London: Longman.

Pineda, J. (1993). Boundary effects on the vertical ranges of deepsea benthic species. *Deep-Sea Research I*, **40**, 2179–2192.

Porebski, S.J., Meischner, D. and Görlich, K. (1991). Quaternary mud turbidites from the South Shetland Trench (West Antarctica): recognition and implications for turbidite facies modelling. *Sedimentology*, **38**, 691–715.

Pörtner, H.O. (2002). Climate variations and the physiological basis of temperature dependent biogeography: systemic to molecular hierarchy of thermal tolerance in animals. *Comparative Biochemistry and Physiology A*, **132**, 739–761.

Pradillon, F. and Gaill, F. (2007). Pressure and life: some biological strategies. *Review of Environmental Science and Biotechnology*, **6**, 181–195.

Pratt, R.M. (1962). The ocean bottom. *Science*, **138**, 492–495.

Priede, I.G. and Merrett, N.R. (1998). The relationship between numbers of fish attracted to baited cameras and population density: studies on demersal grenadiers *Coryphaenoides* (*Nematonurus*) *armatus* in the abyssal NE Atlantic Ocean. *Fisheries Research*, **36**(2–3), 133–137.

Priede, I.G. and Smith, K.L. (1986). Behaviour of the abyssal grenadier, *Coryphaenoides yaquinae*, monitored using ingestible acoustic transmitters in the Pacific Ocean. *Journal of Fish Biology*, **29**, 199–206.

Priede, I.G., Bagley, P.M., Armstrong, J.D., Smith, K.L. and Merrett, N.R. (1991). Direct measurement of active dispersal of food-falls by abyssal demersal fishes. *Nature*, **351**, 647–649.

Priede, I.G., Bagley, P.M. and Smith, K.L. (1994). Seasonal change in activity of abyssal demersal scavenging grenadiers *Coryphaenoides* (*Nematonurus*) *armatus* in the eastern North Pacific Ocean. *Limnology and Oceanography*, **39**, 279–285.

Priede, I.G., Deary, A.R., Bailey, D.M. and Smith, K.L. (2003). Low activity and seasonal change in population size structure of grenadiers in oligotrophic abyssal North Pacific Ocean. *Journal of Fish Biology*, **63**, 187–196.

Priede, I.G., Bagley, P.M., Way, S., Herring, P.J. and Partridge, J.C. (2006a). Bioluminescence in the deep sea: free-fall lander observations in the Atlantic Ocean off Cape Verde. *Deep-Sea Research I*, **53**, 1272–1283.

Priede, I.G., Froese, R., Bailey, D.M. *et al.* (2006b). The absence of sharks from abyssal regions of the world's oceans. *Proceedings of the Royal Society London B*, **273**, 1435–1441.

Priede, I.G., Gobold, J.A., King, N.J. *et al.* (2010). Deep-sea demersal fish species richness in the Porcupine Seabight, NE Atlantic Ocean: global and regional patterns. *Marine Ecology*, **31**(1), 247–260.

Prior, D.B., Bornhold, B.D., Wiseman, W.J. and Lowe, D.R. (1987). Turbidity current activity in a British Columbia fjord. *Science*, **237**, 1330–1333.

Pytkowicz, R.M. (1970). On the carbonate compensation depth in the Pacific Ocean. *Geochimica et Cosmochimica Acta*, **34**, 836–839.

Querellou, J., Borresen, T., Boyen, C. *et al.* (2010). Marine biotechnology: a new vision and strategy for Europe. Marine Board-ESF Position Paper **15**, 1–96.

Radchenkco, V.I. (2007). Mesopelagic fish community supplies 'biological pump'. *Raffles Bulletin of Zoology*, **14**, 265–271.

Rahmstorf, S. (2006). Thermohaline ocean circulation. In *Encyclopaedia of Quaternary Sciences*, ed. S.A. Elias. Amsterdam: Elsevier, pp. 739–750.

Ramaswany, V., Kumar, B.V., Parthiban, G., Ittekkot, V. and Nair, R.R. (1997). Lithogenic fluxes in the Bay of Bengal measured by sediment traps. *Deep-Sea Research*, **44**, 793–810.

Ramirez-Llodra, E., Tyler, P.A., Baker, M.A. *et al.* (2011). Man and the last great wilderness: human impact on the deep sea. *PLoS ONE*, **6**(7), e22588.

Rass, T.S., Grigorash, V.A., Spanovskaya, V.D. and Shcherbachev, Y.N. (1955). Deep-sea bottom fishes caught on the 14th cruise of NLS Akademik Kurchatov. *Trudy Instituta Okeanologii*, **100**, 337–347.

Rathburn, A.E., Levin, L.A., Tryon, M. *et al.* (2009). Geological and biological heterogeneity of the Aleutian margin (1965–4822m). *Progress in Oceanography*, **80**, 22–50.

Raupach, M. J., Mayer, C., Malyutina, M. and Wägele, J.-W. (2009). Multiple origins of deep-sea Asellota (Crustacea: Isopoda) from shallow waters revealed by molecular data. *Proceedings of the Royal Society of London B*, **276**, 799–808.

Reid, D.G. and Naylor, E. (1990). Entrainment of biomodal circatidal rhythms in the shore crab *Carcinus maenus*. *Journal of Biological Rhythms*, **5**, 333–347.

Reinhardt, S.B. and Van-Vleet, E.S. (1985). Lipid composition of Antarctic midwater invertebrates. *Antarctic Journal of the United States*, **19**(5), 139–141.

Rex, M.A., McClain, C.R., Johnston, N.A. *et al.* (2005). A source–sink hypothesis for abyssal biodiversity. *American Naturalist*, **165**, 163–178.

Reymer, A. and Schubert, G. (1984). Phanerozoic addition rates to the continental crust and crustal growth, *Tectonics*, **3**, 63–77.

Rice, A.L., Aldred, R.G., Billett, D.S.M. and Thurston, M.H. (1979). The combined use of an epibenthic sledge and a deep-sea camera to give quantitative relevance to macro-benthos samples. *Ambio Special Report*, **6**, 59–72.

Rice, A.L., Billett, D.S.M., Fry, J. *et al.* (1986). Seasonal deposition of phytodetritus to the deep-sea floor. *Proceedings of the Royal Society of London B*, **88**, 265–279.

Richardson, M.D., Briggs, K.B., Bowles, F.A. and Tietjen, J.H. (1995). A depauperate benthic assemblage from the nutrient poor sediments of the Puerto-Rico Trench. *Deep-Sea Research I*, **42**(3), 351–364.

Ridgwell, A. and Zeebe, R. (2005). The role of the global carbonate cycle in the regulation and evolution of the Earth system. *Earth and Planetary Science Letters*, **234**, 299–315.

Rittmann, B.E. and McCarty, P.L. (2001). *Environmental Biotechnology*. New York: McGraw-Hill.

Robison, B.H., Reisenbichler, K.R. and Sherlock, R.E. (2005). Giant larvacean houses, rapid carbon transport to the deep seafloor. *Science*, **308**, 1609–1611.

Rocha-Olivares, A., Fleeger, J.W. and Foltz, D.W. (2001). Decoupling of molecular and morphological evolution in deep lineages of a meiobenthic harpacticoid copepod. *Molecular Biology and Evolution*, **18**, 1088–1102.

Rogers, A.D. (2000). The role of the oceanic oxygen minima in generating biodiversity in the deep sea. *Deep-Sea Research II*, **47**, 119–148.

Romankevich, E.A., Vetrov, A.A. and Peresypkin, V.I. (2009). Organic matter of the world ocean. *Russian Geology and Geophysics*, **50**, 299–307.

Romm, J. (1994). A new forerunner for continental drift. *Nature*, **367**, 407–408.

Rosen, B.R. (1988). Biogeographical patterns: a perceptual overview. In *Analytical Biogeography; An Integrated Approach to the Study of Animal and Plant Distributions*, ed. A.A. Myers and P.S. Giller. London: Chapman and Hall, pp. 23–55.

Rothschild, L.I. and Mancinelli, R.L. (2001). Life in extreme environments. *Nature*, **409**, 1092–1101.

Roule, L. (1913). N otice préliminaire sur *Grimaldichthys profundissimus* nov. gen., nov. sp. Poisson abyssal recueilli `a 6.035 m`etres de profondeur dans l'Océan Atlantique par S.A.S. le Prince de Monaco. *Bulletin de l'Institut Oceanographique (Monaco)*, **261**, 1–8.

Rouse, G.W. (2001). A cladistic analysis of Siboglinidae Caullery, 1914 (Polychaeta, Annelida): formerly the phyla Pogonophora and Vestimentifera. *Zoological Journal of the Linnean Society*, **132**(1), 55–80.

Roussel, E.G., Bonavita, M.A.C., Querellou, J. *et al.* (2008). Extending the sub-sea-floor biosphere. *Science*, **320**(5879), 1046.

Rowe, G.T. and Clifford, C.H. (1973). Modifications of the Birge–Ekman box corer for use with SCUBA or deep submergence research vessels. *Limnology and Oceanography*, **18**, 172–175.

Ruhl, H.A. (2007). Abundance and size distribution dynamics of abyssal epibenthic megafauna in the northeast Pacific. *Ecology*, **88**(5), 1250–1262.

Ruhl, H.A. and Smith Jr, K.L. (2004). Shifts in deep-sea community structure linked to climate and food supply. *Science*, **305**, 513–515.

Ruhl, H.A., Ellena, J.A. and Smith, K.L. (2008). Connections between climate, food limitation, and carbon cycling in abyssal sediment communities. *Proceedings of the National Academy of Sciences, USA*, **105**(44), 17006–17011.

Ruhl, H.A., André, M., Beranzoli, L. *et al.* (2011). Societal need for improved understanding of climate change, anthropogenic impacts, and geo-hazard warning drive development of ocean observatories in European seas. *Progress in Oceanography*, **91**(1), 1–33.

Sabbatini, A., Morigi, C., Negri, A. and Gooday, A.J. (2002). Soft-shelled benthic Foraminifera from a hadal site (7800m water depth) in the Atacama Trench (SE Pacific): preliminary observations. *Journal of Micropalaeontology*, **21**, 131–135.

Saidova, Kh.M. (1970). Benthic foraminifers of the Kuril–Kamchatka Trench area. In *Fauna of the Kuril–Kamchatka Trench and its Environment. Academy of Sciences of the USSR. Proceedings of the Shirshov Institute of Oceanology*, Bogorov, V.G. ed. **86**, 144–173.

Saidova, Kh.M. (1975). *Benthic Foraminifera of the Pacific Ocean*, Vol. 3. Moscow: Institut Okeanologii P.P. Shirshova.

Sainte-Marie, B. (1992). Foraging of scavenging deep-sea lysianassoid amphipods. In *Deep-sea Food Chains in the Global Carbon Cycle*, ed. G.T. Rowe and V. Pariente. Dordrecht, the Netherlands: Kluwer Academic Publishers, pp. 105–124.

Sainte-Marie, B. and Hargrave, B.T. (1987). Estimation of scavenger abundance and distance of attraction to bait. *Marine Biology*, **94**, 431–443.

Samerotte, A.L., Drazen, J.C., Brand, G.L., Seibel, B.A. and Yancey, P.H. (2007). Correlation of trimethylamine oxide and habitat depth within and among species of teleost fish: an analysis of causation. *Physiological and Biochemical Zoology*, **80**, 197–208.

Sanders, H.L. (1968). Marine benthic diversity: a comparative study. *American Naturalist*, **102**, 243–282.

Scarratt, D.J. (1965). Oredation on lobsters (*Homarus americanus*) by *Anonyx* sp. (Crustacea, Amphipoda). *Journal of the Fisheries Research Board of Canada*, **22**, 1103–1104.

Scheidegger, A.E. (1953). Examination of the physics of theories of orogenesis. *GSA Bulletin*, **64**, 127–150.

Schizas, N.V., Street, G.T., Coull, B.C., Chandler, G.T. and Quattro, J.M. (1999). Molecular population structure of the marine benthic copepod *Microarthridion littorale* along the southeastern and Gulf coasts of the USA. *Marine Biology*, **135**, 399–405.

Schlacher, T.A., Schlacher-Hoenlinger, M.A, Williams, A. *et al.* (2007). Richness and distribution of sponge megabenthos in continental margin canyons off southeastern Australia. *Marine Ecology Progress Series*, **340**, 73–88.

Schmidt, W.E. and Siegel, E. (2011). Free descent and on bottom ADCM measurements in the Puerto-Rico Trench, 19.77°N, 67.40°W. *Deep-Sea Research I*, **58**(9), 970–977.

Schmitz, W.J. (1995). On the interbasin-scale thermohaline circulation. *Reviews of Geophysics*, **33**(2), 151–173.

Scholl, D.W., Christensen, M.N., yon Huene, R. and Marlow, M.S. (1970). Peru–Chile trench sediments and sea-floor spreading. *Geology Society of America Bulletin*, **81**, 1339–1360.

Schotte, M., Kensley, B.F. and Shilling, S. (1995 onwards). *World List of Marine, Freshwater and Terrestrial Crustacea Isopoda*. Washington DC: National Museum of Natural History Smithsonian Institution. Available at: http://invertebrates.si.edu/isopod.

Schwabe, E. (2008). A summary of abyssal and hadal Monoplacophora and Polyplacophora (Mollusca). *Zootaxa*, **1866**, 205–222.

Seibel, B.A. and Drazen, J.C. (2007). The rate of metabolism in marine animals: environmental constraints, ecological demands and energetic opportunities. *Philosophical Transactions of the Royal Society London B*, **362**, 2061–2078.

Sexton, E.W. (1924). The moulting and growth-stages of *Gammarus*, with descriptions of the normals and intersexes of G. *cheureuxi*. *Journal of the Marine Biological Association of the United Kingdom*, **13**, 340–401.

Shcherbachev, Y.N. and Tsinovsky, V.D. (1980). New finds of deep-sea brotulids *Abyssobrotula galathea* Nielsen, *Acanthonus armatus* Günther, and *Typhlonus nasus* Günther (Pisces, Ophidiiformes) in the Pacific and Indian Oceans. *Bulletin of the Moscow Society Natural Experiments, Biology Department*, **85**, 53–57.

Shirayama, Y. (1984). The abundance of deep-sea meiobenthos in the western Pacific in relation to environmental factors. *Oceanologica Acta*, **7**(1), 113–121.

Shirayama, Y. and Fukushima, T. (1995). Comparisons of deep-sea sediments and overlying water collected using multiple corer and box corer. *Journal of Oceanography*, **51**, 75–82.

Shulenberger, E. and Hessler, R.R. (1974). Scavenging abyssal benthic amphipods trapped under oligotrophic Central North Pacific Gyre waters. *Marine Biology*, **28**, 185–187.

Siebenaller, J.F., Somero, G.N. and Haedrich, R.L. (1982). Biochemical characteristics of macrourid fishes differing in their depths of distribution. *Biological Bulletin*, **163**, 240–249.

Siedler, G., Holfort, J., Zenk, W., Muller, T.J. and Csernok, T. (2004). Deep-water flow in the Mariana and Caroline Basins. *Journal of Physical Oceanography*, **34**(3), 566–581.

Simonato, F., Campanaro, S., Lauro, F.M. *et al.* (2006). Piezophilic adaptation: a genomic point of view. *Journal of Biotechnology*, **126**, 11–25.

Simpson, D.C., O'Connor, T.P. and Park, P.K. (1981). Deep-ocean dumping of industrial wastes. In *Marine Environmental Pollution, Vol. 2, Dumping and Mining*, ed. R.A. Geyer. New York: Elsevier Scientific, pp. 379–400.

Sirenko, B.I. (1977). Vertical distribution of chitons of the genus *Lepidopleurus* (Lepidopleuridae) and its new ultraabyssal species. *Zoologiceskij Zurnal*, **56**(7), 1107–1110.

Sirenko, B.I. (1988). A new genus of deep sea chitons Ferreiraella gen. n. (Lepidopleurida, Leptochitonidae) with a description of a new ultra-abyssal species. *Zoologiceskij Zurnal*, **67**(12), 1776–1786.

Sluiter, C.-P. (1912). Gephyriens (Sipunculides et Echiurides) provenant des campagnes de la Princess-Alice (1989–1910). *Résultats des campagnes scientifiques accomplis par le Prince Albert I*, **36**, 1–27.

Smith, C.R. and Baco, A.M. (2003). Ecology of whale falls at the deep-sea floor. *Oceanography and Marine Biology Annual Review*, **41**, 311–354.

Smith, C.R. and Demopoulos, A.W.J. (2003). The deep Pacific Ocean floor. In *Ecosystems of the World 28, Ecosystems of the Deep Sea*, ed. P.A. Tyler. Amsterdam: Elsevier, pp. 179–218.

Smith, C.R., Kukert, H., Wheatcroft, R.A., Jumars, P.A. and Deming, J.W. (1989). Vent fauna on whale remains. *Nature*, **341**, 27–28.

Smith, C.R., De Leo, F.C., Bernardino, A.F., Sweetman, A.K. and Arbizu, P.M. (2008). Abyssal food limitation, ecosystem structure and climate change. *Trends in Ecology and Evolution*, **23**, 518–528.

Smith, K.L. (1992). Benthic boundary layer communities and carbon cycling at abyssal depths in the central North Pacific. *Limnology and Oceanography*, **37**, 1034–1056.

Smith, K.L. and Hessler, R.R. (1974). Respiration of benthopelagic fishes: *in-situ* measurements at 1230 meters. *Science*, **184**, 72–73.

Smith, K.L. and Howard, J.D. (1972). Comparison of a grab sampler and large volume corer. *Limnology and Oceanography*, **28**, 882–898.

Smith, K.L. and Baldwin, R.J. (1984). Vertical distribution of the necrophagous amphipods, *Eurythenes gryllus*, in the North Pacific: spatial and temporal variation. *Deep-Sea Research*, **31**(10), 1179–1196.

Smith, K.L., White, G.A., Laver, M.B., McConnaughey, R.R. and Meador, J.P. (1979). Free vehicle capture of abyssopelagic animals. *Deep-Sea Research*, **26A**, 57–64.

Smith, K.L., Kaufmann, R.S. and Wakefield, W.W. (1993). Mobile megafaunal activity monitored with a time-lapse camera in the abyssal North Pacific. *Deep-Sea Research*, **40**, 2307–2324.

Smith, K.L., Kaufmann, R.S., Baldwin, R.J. and Carlucci, A.F. (2001). Pelagic-benthic coupling in the abyssal eastern North Pacific: an 8-year time-series study of food supply and demand. *Limnology and Oceanography*, **46**, 543–556.

Smith, K.L., Holland, N.D. and Ruhl, H.A. (2005). Enteropneust production of spiral fecal trails on the deep-sea floor observed with time-lapse photography. *Deep-Sea Research I*, **52**(7), 1228–1240.

Smith, K.L., Baldwin, R.J., Ruhl, H.A. *et al.* (2006). Climate effect on food supply to depths greater than 4000 meters in the northeast Pacific. *Limnology and Oceanography*, **51**(1), 166–176.

Smith, K.L., Ruhl, H.A., Bett, B.J. *et al.* (2009). Climate, carbon cycling and deep-ocean ecosystems. *Proceedings of the National Academy of Sciences, USA*, **106**, 19211–19218.

Snelgrove, P.V.R. (2010). *Discoveries of the Census of Marine Life, Making Ocean Life Count*. Cambridge: Cambridge University Press.

Soltwedel, T., von Juterzenka, K., Premke, K. and Klages, M. (2003). What a lucky shot! Photographic evidence for a medium-sized natural food-fall at the deep-seafloor. *Oceanologica Acta*, **26**, 623–628.

Somero, G.N. (1992). Adaptations to high hydrostatic pressure. *Annual Review of Physiology*, **54**, 557–577.

Somero, G.N. and Siebenaller, J.F. (1979). Inefficient lactate dehydrogenases of deep-sea fishes. *Nature*, **282**, 100–102.

Soong, K. and Mok, H.K. (1994). Size and maturity stage observations of the deep-sea isopod *Bathynomus doederleini* Ortmann, 1894 (Flabellifera: Cirolanidae), in Eastern Taiwan. *Journal of Crustacean Biology*, **14**, 72–79.

Søreide, F. (2012). Ultradeep-sea exploration in the Puerto-Rico Trench. *Sea Technology*, **53**(12), 54–57.

Søreide, F. and Jamieson, A.J. (2013). Ultradeep-sea exploration in the Puerto Rico Trench. *Proceedings of the Oceans '13*, MTS/IEEE, San Diego.

Spengler, A. and Costa, M.F. (2008). Methods applied in studies of benthic marine debris. *Marine Pollution Bulletin*, **56**(2), 226–230.

Spicer, J.I. and Gaston. K.J. (1999). Amphipod gigantism dictated by oxygen availability? *Ecology Letters*, **2**, 397–403.

Staiger, J.C. (1972). *Bassogigas profundissimus* (Pisces; Brotulidae) from the Puerto Rico Trench. *Bulletin of Marine Science*, **22**, 26–33.

Starr, M., Therriault, J.-C., Conan, G.Y., Comeau, M. and Robichaud, G. (1994). Larval release in the sub-euphotic zone invertebrate triggered by sinking phytoplankton particles. *Journal of Plankton Research*, **16**, 1137–1147.

Steele, D.H. and Steele, V.J. (1991). The structure and organization of the gills of gammaridean Amphipoda. *Journal of Natural History*, **25**(4), 1247–1258.

Steele, V.J. and Steele, D.H. (1970). The biology of *Gammarus* (Crustacea, Amphipoda) in the northwestern Atlantic II. *Gammarus setosus* Dementieva. *Canadian Journal of Zoology*, **38**, 659–671.

Stein, D.L. (1985). Towing large nets by single warp at abyssal depths: methods and biological results. *Deep-Sea Research*, **32**, 183–200.

Stein, D.L. (2005). Descriptions of four new species, redescription of *Paraliparis membranaceus*, and additional data on species of the fish family Liparidae (Pisces, Scorpaeniformes) from the west coast of South America and the Indian Ocean. *Zootaxa*, **1019**, 1–25.

Stern, R.J. (2002). Subduction zones, *Reviews of Geophysics*, **40**(4), 1012.

Stockton, W.L. (1982). Scavenging amphipods from under the Ross Ice Shelf, Antarctica. *Deep-Sea Research*, **29**, 819–835.

Stockton, W.L. and DeLaca, T.E. (1982). Food falls in the deep sea: occurrence, quality, and significance. *Deep-Sea Research*, **29**, 157–169.

Stoddart, H.E. and Lowry, J.K. (2004). The deep-sea lysianassoid genus *Eurythenes* (Crustacea, amphipoda, Eurytheneidae n. fam.). *Zoosystema*, **26**(3), 425–468.

Stommel, H. (1958). The abyssal circulation. *Deep-Sea Research*, **5**, 80–82.

Stowasser, G., McAllen, R., Pierce, G.J. *et al.* (2009). Trophic position of deep-sea fish – assessment through fatty acid and stable isotope analysis. *Deep-Sea Research I*, **56**, 812–826.

Strong, E.E. and Harasewych, M.G. (1999). Anatomy of the hadal limpet *Macleaniella moskalevi* (Gastropoda, Cocculinoidea). *Invertebrate Biology*, **118**(2), 137–148.

Suess, E., Bohrmann, G., von Huene, R. *et al.* (1998). Fluid venting in the eastern Aleutian subduction zone. *Journal of Geophysical Research*, **103**, 2597–2614.

Sullivan, K.M. and Smith, K.L. (1982). Energetics of sablefish, *Anoplopoma fimbria*, under laboratory conditions. *Canadaian Journal of Fisheries and Aquatics Sciences*, **39**, 1012–1020.

Sullivan, K.M. and Somero, G.N. (1980). Enzyme activities of fish skeletal muscle and brain as influenced by depth of occurrence and habits of feeding and locomotion. *Marine Biology*, **60**, 91–99.

Svavarsson, J., Strömberg, J.-O. and Brattegard, T. (1993). The deep-sea asellote (Isopoda, Crustacea) fauna of the Northern Seas: species composition, distributional patterns and origin. *Journal of Biogeography*, **20**, 537–555.

Sweetman, A.K. and Chapman, A. (2011). First observations of jelly-falls at the seafloor in a deep-sea fjord. *Deep-Sea Research I*, **58**, 1206–1211.

Taft, B.A., Hayes, S.P., Friedrich, G.E. and Codispoti, L.A. (1991). Flow of abyssal water into the Samoa Passage. *Deep-Sea Research*, **38**, 128–130.

Taira, K. (2006). Super-deep CTD measurements in the Izu-Ogasawara Trench and a comparison of geostrophic shears with direct measurements. *Journal of Oceanography*, **62**, 753–758.

Taira, K., Kitagawa, S., Yamashiro, T. and Yanagimoto, D. (2004). Deep and bottom currents in the Challenger Deep, Mariana Trench, measured with super-deep current meters. *Journal of Oceanography*, **60**, 919–926.

Taira, K., Yanagimoto, D. and Kitagawa, S. (2005). Deep CTD casts in the Challenger Deep, Mariana Trench. *Journal of Oceanography*, **61**(3), 446–454.

Takagawa. S. (1995). Advanced technology used in Shinkai 6500 and full ocean depth ROV Kaikō. *Marine Technology Society Journal*, **29**(3), 15–25.

Takagawa, S., Aoki, T. and Kawana, I. (1997). Diving to Mariana Trench by *Kaikō*. *Recent Advances in Marine Science and Technology*, **96**, 89–96.

Takahashi, T. and Broecker, W.S. (1977). Mechanisms for calcite dissolution on the sea floor. In *The Fate of Fossil Fuel CO₂ in the Oceans. Marine Science*, Vol. 6, ed. N.R. Anderson and A. Malahoff. New York: Plenum, pp. 455–477.

Takami, H., Inoue, A., Fuji, F. and Horikoshi, K. (1997). Microbial flora in the deepest sea mud of the Mariana Trench. *FEMS Microbiology Letters*, **152**(2), 279–285.

Takashima, R., Nishi, H., Huber, B.T. and Leckie, R.M. (2006). Greenhouse world and the mesozoic ocean. *Oceanography*, **19**(4) 82–92.

Tamburri, M.N. and Barry, J.P. (1999). Adaptations for scavenging by three diverse bathyal species, *Eptatretus stouti*, *Neptunea amianta* and *Orchomebe obtusus*. *Deep-Sea Research I*, **46**, 2079–2093.

Tashiro, S., Watanbe, M. and Momma, H. (2004). Loss of the full ocean depth ROV *Kaikō*, Part 2: search for the ROV *Kaikō* vehicle. *Proceedings of the 14th International Offshore and Polar Engineering Conference*, **2**, 194–198.

Teitjen, J.H., Deming, J.W., Rowe, G.T., Macko, S. and Wilke, R.J. (1989). Meiobenthos of the Hatteras Abyssal Plain and Puerto-Rico Trench: abundance, biomass and associations with bacteria and particulate fluxes. *Deep-Sea Research*, **36**(10) 1567–1577.

Tendal, O.S. (1972). A monograph of the Xenophyoporia. *Galathea Report*, **12**, 7–100.

Tendal, O.S. and Gooday, A.J. (1981). Xenophyophoria (Rhizopoda, Protozoa) in bottom photographs from the bathyal and abyssal NE Atlantic. *Oceanologica Acta*, **4**, 415–422.

Tendal, O.S. and Hessler, R.R. (1977). An introduction to the biology and systematics of Komokiacea (Textulariina, Foraminiferida). *Galathea Report*, **14**, 165–194.

Tengberg, A., De Bovee, F., Hall, P. *et al.* (1995) Benthic chamber and profiling landers in oceanography: a review of design, technical solutions and functioning, *Progress in Oceanography*, **35**, 253–294.

Tengberg, A., Andersson, U., Hall, P. *et al.* (2005). Intercalibration of benthic flux chambers II: hydrodynamic characterization and flux comparisons of 14 different designs. *Marine Chemistry*, **94**, 147–173.

Thiel, H. (1966). Quantitative Untersuchungen über die Meiofauna des Tiefseebodens. *Veröffentlichungen des Instituts für Meeresforschung Bremerhaven, Sonderband*, **2**, 131–148.

Thiel, H. (1972). Meiofauna und struktur der benthischen Lebens gemeinschaft des Iberischen Tiefseebeckens. *'Meteor' Forschungsergebnisse*, **12**, 36–51.

Thistle, D. (2003). The deep-sea floor: an overview. In *Ecosystems of the World 28, Ecosystems of the Deep Sea*, ed. P.A. Tyler. Amsterdam: Elsevier, pp. 5–37.

Thompson, R.C. (2006). Plastic debris in the marine environment: consequences and solutions. In *Marine Nature Conservation in Europe*, ed. J.C. Krause, H. Nordheim and S. Brager. Stralsund, Germany: Bundesamt fur Naturschutz, pp.107–115.

Thomson, C.W. (1873). *The Depths of the Sea*. London: MacMillan.

Thomson, C.W. and Murray, J. (1895). *Report on the Results of the Voyage of H.M.S. Challenger during the Years 1873–76, Narrative*, Vol. A(1). London: HM Stationery Office.

Thornburg, T.M. and Kulm, L.D. (1987). Sedimentation in the Chile Trench: depositional morphologies, lithofacies, and stratigraphy. *Geological Society of America Bulletin*, **98**, 33–52.

Thorson, G. (1957). Sampling the benthos. In *Treatise on Marine Ecology and Paleoecology*, ed. J. Hedgepeth. New York: Geological Society of America, pp. 61–86.

Thunell, R., Tappa, E., Varela, R. *et al.* (1999). Increased marine sediment suspension and fluxes following an earthquake. *Nature*, **398**, 233–236.

Thurston, M.H. (1979). Scavenging abyssal amphipods from the north-east Atlantic Ocean. *Marine Biology*, **51**, 55–68.

Thurston, M.H. (1990). Abyssal necrophagous amphipods (Crustacea: Amphipoda) in the northeast and tropical Atlantic Ocean. *Progress in Oceanography*, **24**, 257–274.

Thurston, M.H., Bett, B.J. and Rice, A.L. (1995). Abyssal megafaunal necrofages: latitudinal differences in the eastern North Atlantic Ocean. *Internationale Revue der gesamten Hydrobiologie*, **80**(2), 267–286.

Thurston, M.H., Petrillo, M. and Della Croce, N. (2002). Population structure of the necrophagous amphipod *Eurytythenes gryllus* (Amphipods: Gammaridea) from the Atacama Trench (south-east Pacific Ocean). *Journal of the Marine Biological Association of the United Kingdom*, **82**, 205–211.

Tiefenbacher, L. (2001). Recent samples of mainly rare decapod crustacean taken from the deep-sea floor of the southern West Europe Basin. *Hydrobiologia*, **449**, 59–70.

Tietjen, J.H. (1989). Ecology of deep-sea nematodes from the Puerto Rico Trench area and Hatteras Abyssal Plain. *Deep-Sea Research A*, **36**(10), 1579–1594.

Tietjen, J.H., Deming, J.W., Rowe, G.T., Mackie, S. and Wilke, R.J. (1989). Meiobenthos of the Hatteras Abyssal Plain and Puerto Rico Trench: abundance, biomass and associations with bacteria and particulate fluxes. *Deep-Sea Research I*, **36**, 1567–1577.

Tilston, H. (2011). Biogeography of deep-sea trenches. MSc thesis, University of Southampton, UK.

Tobriner, S. (2006). *Bracing for Disaster: Earthquake-resistant Architecture and Engineering in San Francisco, 1838–1933*. Berkeley, CA: Heyday Books.

Todo, Y., Kitazato, H., Hashimoto, J. and Gooday, A.J. (2005). Simple Foraminifera flourish at the ocean's deepest point. *Science*, **307**, 689.

Toggweiler, J.R., Russell, J.L. and Carson, S.R. (2006). Midlatitude westerlies, atmospheric CO_2, and climate change. *Paleoceanography*, **21**(2), PA2005.

Tomczak, M. and Godfrey, J.S. (1994). *Regional Oceanography: An Introduction*. London: Pergamon.

Tosatto, M. (2009). Charting a course from the Marianas Trench Marine National Monument. *Marine Technology Society Journal*, **43**(5), 161–163.

Truede, T., Janssen, F., Queisser, W. and Witte, U. (2002). Metabolism and decompression tolerance of scavenging lysianassoid deep-sea amphipods. *Deep-Sea Research I*, **49**, 1281–1289.

Tselepides, A. and Lampadariou, N. (2004). Deep-sea meiofaunal community structure in the Eastern Mediterranean: are trenches benthic hotspots? *Deep-Sea Research I*, **51**, 833–847.

Turner, J.T. (2002). Zooplankton faecal pellets, marine snow and sinking phytoplankton blooms. *Aquatic Microbiology and Ecology*, **27**, 57–102.

Turner, R.D. (1973). Wood-boring bivalves, opportunistic species in the deep sea. *Science*, **180**, 1377–1379.

Turnewitsch, R., Falahat, S., Stehlikova, J. *et al.* (in prep). Recent sediment dynamics in hadal trenches: evidence for the influence of higher-frequency (tidal, near-inertial) fluid dynamics. *Deep-Sea Research I*.

Tyler, P.A. (1995). Conditions for the existence of life at the deep-sea floor: an update. *Oceanography and Marine Biology: Annual Review*, **33**, 221–244.

Tyler, P.A. (2003). Epilogue: exploration, observation and experimentation. In *Ecosystems of the World 28, Ecosystems of the Deep Sea*, ed. P.A. Tyler. Amsterdam: Elsevier, pp. 473–476.

Tyler, P.A. and Young, C.M. (1998). Temperature and pressure tolerances in dispersal stages of the genus *Echinus* (Echinodermata: Echonoidea): prerequisties for deep-sea invasion and speciation. *Deep-Seu Research II*, **45**, 253–277.

Tyler, P., Amaro, T., Arzola, R. *et al.* (2009). Europe's Grand Canyon: Nazaré Submarine Canyon. *Oceanography*, **22**, 46–57.

UNESCO (2009). *Global Open Oceans and Deep Seabed (GOODS) – Biogeographic Classification*. IOC Technical Series, 84. Paris: UNESCO-IOC.

Ushakov, P.V. (1952). Study of deep-sea fauna. *Priroda*, **6**, 100–102.

Van der Maarel, E. (1990). Ecotones and ecoclines are different. *Journal of Vegetation Science*, **1**(1), 135–138.

Van Dover, C.L. and Fry, B. (1994). Microorganisms as food resources at deep-sea hydrothermal vents. *Limnology and Oceanography*, **39**(1), 51–57.

Vardaro, M.F., Ruhl, H.A. and Smith, K.L. (2009). Climate variation, carbon flux, and bioturbation in the abyssal North Pacific. *Limnology and Oceanography*, **54**(6), 2081–2088.

Vetter, E.W. and Dayton, P.K. (1998). Macrofaunal communities within and adjacent to a detritus-rich submarine canyon system. *Deep-Sea Research II*, **45**, 25–54.

Villalobos, F.B., Tyler, P.A. and Young, C.M. (2006). Temperature and pressure tolerance of embryos and larvae of the Atlantic seastars *Asterias rubens* and *Marthasterias glacialis* (Echinodermata: Asteroidea): potential for deep-sea invasion. *Marine Ecology Progress Series*, **314**, 109–117.

Vine, F.J. and Matthews, D.H. (1963). Magnetic anomalies over oceanic ridges. *Nature* **199**(4897), 947–949.

Vinogradov, M.E. (1962). Quantitative distribution of deep-sea plankton in the western Pacific and its relation to deep-water circulation. *Deep-Sea Research*, **8**, 251–258.

Vinogradova, N.G. (1979). The geographical distribution of the abyssal and hadal (ultra-abyssal) fauna in relation to the vertical zonation of the ocean. *Sarsia*, **64**(1–2), 41–49.

Vinogradova, N.G. (1997). Zoogeography of the abyssal and hadal zones. *Advances in Marine Biology*, **32**, 325–387.

Vinogradova, N.G., Gebruk, A.V. and Romanov, V.N. (1993a). Some new data on the Orkney Trench ultra abyssal fauna. *The Second Polish Soviet Antarctic Symposium*, 213–221.

Vinogradova, N.G., Belyaev, G.M., Gebruk, A.V. *et al.* (1993b). Investigations of Orkney Trench in the 43rd cruise of R/V *Dmitriy Mendeleev*. Geomorphology and bottom sediments, benthos. In *The Deep-sea Bottom Fauna in the Southern Part of the Atlantic Ocean*, ed. N.G. Vinogradova. Moscow: Nauka, pp. 127–253.

Vogel, S. (1981). *Life in Moving Fluids*. Boston, MA: Willard Grant Press.

Von Huene, R. and Scholl, D.W. (1991). Observations at convergent margins concerning sediment subduction, subduction erosion, and the growth of continental crust. *Reviews of Geophysics*, **29**, 279–316.

Von Huene, R. and Shor, G.G. (1969). The structure and tectonic history of the eastern Aleutian Trench. *Geology Society of America Bulletin*, **80**, 1889–1902.

Waelbroeck, C., Labeyrie, L., Michel, E. *et al.* (2001). Sea-level and deep water temperature changes derived from benthic Foraminifera isotopic records. *Quarterly Scientific Review*, **21**, 295–305.

Wagner, H.-J., Kemp, K., Mattheus, U. and Priede, I.G. (2007). Rhythms at the bottom of the deep sea: cyclic current flow changes and melatonin patterns in two species of demersal fish. *Deep-Sea Research I*, **54**, 1944–1956.

Wakeham, S.G., Lee, C., Farrington, J.W. and Gagosian, R.B. (1984). Biogeochemistry of particulate organic matter in the oceans: results from sediment trap experiments. *Deep-Sea Research A*, **31**, 509–528.

Wakeham, S.G., Hedges, J.I., Lee, C., Peterson, M.L. and Hernes, P.J. (1997). Compositions and transport of lipid biomarkers through the water column and surficial sediments of the equatorial Pacific Ocean. *Deep-Sea Research II*, **44**, 2131–2162.

Walsh, D. (2009). In the beginning... A personal view. *Marine Technology Society Journal*, **43**, 9–14.

Wann, K.T. and MacDonald, A.G. (1980). The effects of pressure on excitable cells. *Comparative Biochemistry and Physiology*, **66**, 1–12.

Warrant, E.J. and Locket, N.A. (2004). Vision in the deep-sea. *Biological Reviews*, **79**, 671–712.

Warren, B.A. (1981). Deep circulation of the world ocean. In *Evolution of Physical Oceanography*, ed. B. Warren and C. Wunsch. Boston, MA: Massachusetts Institute of Technology, pp. 6–40.

Warren, B.A., and Owens, W.B. (1985). Some preliminary results concerning deep northern-boundary currents in the North Pacific. *Progress in Oceanography*, **14**, 537–551.

Warren, B.A. and Owens, W.B. (1988). Deep currents in the central subarctic Pacific Ocean. *Journal of Physical Oceanography*, **18**(4), 529–551.

Watanbe, M., Tashiro, S. and Momma, H. (2004). Loss of the full ocean depth ROV *Kaikō*. Part 3: the cause of secondary cable fracture. *Proceedings of the 14th International Offshore and Polar Engineering Conference*, **2**, 199–202.

Watling, L., Guinotte, J., Clarke, M.R. and Smith, C.R. (2013). A proposed biogeography of the deep ocean floor. *Progress in Oceanography*, **111**, 91–112.

Webb, T.J., Berghe, E.V. and O'Dor, R. (2010). Biodiversity's big wet secret: the global distribution of marine biological records reveals chronic under-exploration of the deep pelagic ocean. *PLoS ONE*, **5**(8), e10223.

Weber, G. and Drickamer, H.G. (1999). The effect of high pressure upon proteins and other biomolecules. *Quarterly Review of Biophysics*, **16**, 89–112.

Wegener, A. (1912). Die Entstehung der Kontinente: Dr. A. Petermanns Mitteilungen aus Justus Perthes. *Geographischer Anstalt*, **63**, 185–195, 253–256, 305–309.

Weiser, W. (1956). Free-living marine nematodes III. Axonolaimoidea and Monhysteroidea. *Acta Universitatis Lund*, **52**(13), 1–115.

Welch, T.J., Farewell, A., Neidhardt, F.C. and Bartlett, D.H. (1993). Stress response in *Escherichia coli* induced by elevated hydrostatic pressure, *Journal of Bacteriology*, **175**, 7170–7177.

White, B.N. (1987). Oceanic anoxic events and allopatric speciation in the deep sea. *Biological Oceanography*, **5**, 243–259.

White, D.A., Roeder, D.H., Nelson, T.H. and Crowell, J.C. (1970). Subduction. *Geological Society of America Bulletin*, **81**, 3431–3432.

Whitman, W.B., Coleman, D.C. and Wiebe, W.J. (1998). Prokaryotes: the unseen majority. *Proceedings of the National Academy of Sciences, USA*, **95**, 6578–6583.

Whitworth, T., Warren, B.A., Nowlin, W.D., Rutz, S.B., Pillsbury, R.D. and Moore, M.I. (1999). On the deep western-boundary current in the Southwest Pacific Ocean. *Progress in Oceanography*, **43**(1), 1–54.

Wickramasinghe, N., Wallis, J. and Wallis, D. (2013). Panspermia: evidence from astronomy to meteorites. *Modern Physics Letters A*, **28**(14), 1330009.

Wigham, B.D., Hudson, I.R., Billett, D.S.M. and Wolff, G.A. (2003). Is long-term change in the abyssal Northeast Atlantic driven by qualitative changes in export flux? Evidence from selective feeding in deep-sea holothurians. *Progress in Oceanography*, **59**, 409–441.

Wigley, R.L. (1967). Comparative efficiencies of Van Veen and Smith–McIntyre grab samplers as revealed by motion pictures. *Ecology*, **48**, 168–169.

Williams, J.T. and Machida, Y. (1992). *Echiodon anchipterus*: a valid western Pacific species of the pearlfish family Carapidae with comments on *Eurypleuron*. *Japanese Journal of Ichthyology*, **38**, 367–373.

Wilson, G.D.F. (1999). Some of the deep-sea fauna is ancient. *Crustaceana*, **72**, 1019–1030.

Wilson, G.D.F. and Hessler, R.R. (1987). Speciation in the deep sea. *Annual Review of Ecology and Systematics*, **18**, 185–207.

Wilson, G.D.F. and Thistle, D. (1985). *Amuletta*, a new genus for *llyarachna abyssorum* Richardson, 1911 (Isopoda: Asellota: Eurycopidae). *Journal of Crustacean Biology*, **5**, 350–360.

Wilson, R.R. and Smith, K.L. (1984). Effect of near-bottom currents on detection of bait by the abyssal grenadier fishes, *Coryphaenoides* spp. recorded *in situ* with a video camera on a free vehicle. *Marine Biology*, **84**, 83–91.

Wilson. R.R. and Waples, R.S. (1983). Distribution, morphology, and biochemical genetics of *Coryphaenoides armatus* and *C. yaquinae* (Pisces: Macrouridae) in the central and eastern North Pacific. *Deep-Sea Research*, **30**, 1127–1145.

Wilson, T.J. (1965). A new class of faults and their bearing on continental drift. *Nature*, **207**(4995), 343–347.

Wingstrand, K.G. (1985). On the anatomy and relationships of recent Monoplacophora. *Galathea Report*, **16**, 7–94.

Wiseman, J.D.H. and Ovey, C.D. (1953). Definitions of features on the deep-sea floor. *Deep-Sea Research*, **1**, 11–16.

Wiseman, J.D.H. and Ovey, C.D. (1954). Proposed names of features on the deep-sea floor, 1. The Pacific Ocean. *Deep-Sea Research*, **2**, 93–106.

Wishner, K., Levin, L., Gowing, M. and Mullineaux, L. (1990). Involvement of the oxygen minimum in benthic zonation on a deep seamount. *Nature*, **346**, 57–59.

Wolff, T. (1956). Crustacea Tanaidacea from depths exceeding 6000 meters. *Galathea Report*, **2**, 187–241.

Wolff, T. (1960). The hadal community, an introduction. *Deep-Sea Research*, **6**, 95–124.

Wolff, T. (1961). The deepest recorded fishes. *Nature*, **190**, 283–284.

Wolff, T. (1962). The systematics and biology of bathyal and abyssal Isopoda Asellota. *Galathea Report*, **6**, 1–320.

Wolff, T. (1970). The concept of the hadal or ultra-abyssal fauna. *Deep Sea Research*, **17**, 983–1003.

Wolff, T. (1976). Utilization of seagrass in the deep sea. *Aquatic Botany*, **2**, 161–174.

Wong, Y.M. and Moore, P.G. (1995). Biology of feeding in the scavenging isopod *Natatolana borealis* (Isopoda: Cirolanidae). *Ophelia*, **43**(3), 181–196.

Worthington, L.V. (1976). *On the North Atlantic Circulation. Johns Hopkins Oceanographic Studies Vol. VI*. Baltimore, MD and London: The Johns Hopkins University Press.

Worzel, J.L. and Ewing, M. (1954). Gravity anomalies and structure of the West Indies – 2. *Bulletin of the Geological Society of America*, **65**, 195–200.

Yamamoto, J., Nobetsu, T., Iwamori, T. and Sakurai, Y. (2009). Observations of food falls off the Shiretoko Peninsula, Japan, using a remotely operated vehicle. *Fisheries Science*, **75**, 513–515.

Yancey, P.H. (2005). Organic osmolytes as compatible, metabolic, and counteracting cytoprotectants in high osmolarity and other stresses. *Journal of Experimental Biology*, **208**, 2819–2830.

Yancey, P.H. and Siebenaller, J.F. (1999). Trimethylamine oxide stabilizes teleost and mammalian lactate dehydrogenases against inactivation by hydrostatic pressure and trypsinolysis. *Journal of Experimental Biology*, **202**, 3597–3360.

Yancey, P.H., Fyfe-Johnson, A.L., Kelly, R.H., Walker, V.P. and Aunon, M.T. (2001). Trimethylamine oxide counteracts effects of hydrostatic pressure on proteins of deep-sea teleosts. *Journal of Experimental Zoology*, **289**, 172–176.

Yancey, P.H., Rhea, M.D., Kemp, K.M. and Bailey, D.M. (2004). Trimethylamine oxide, betaine and other osmolytes in deep-sea animals: depth trends and effects on enzymes under hydrostatic pressure. *Cellular and Molecular Biology*, **50**, 371–376.

Yancey, P.H., Gerringer, M.E., Drazen, J.C., Rowden, A.A. and Jamieson, A.J. (in press). Marine fish are biochemically constrained from inhabiting deepest ocean depths. *Proceedings of the National Academy of Sciences, USA*.

Yang, T.-H. and Somero, G.N. (1993). The effects of feeding and food deprivation on oxygen consumption, muscle protein concentration and activities of energy metabolism enzymes in muscle and brain of shallow-living (*Scorpaena guttata*) and deep-living (*Sebastolobus alascanus*) scorpaenid fishes. *Journal of Experimental Biology*, **181**, 213–232.

Yano, Y., Nakayama, A., Ishihara, K. and Saito, H. (1998). Adapative changes in membrane lipids of barophilic bacteria in repsonse to changes in growth pressure. *Applied and Environmental Microbiology*, **64**(2), 479–485.

Yayanos, A.A. (1976). Determination of the pressure-volume-temperature (PVT) surface of Isopar-M: a quantitative evaluation of its use to float deep-sea instruments. *Deep-Sea Research*, **23**, 989–993.

Yayanos, A.A. (1977). Simply actuated closure for a pressure vessel: design for use to trap deep-sea animals. *Review of Scientific Instruments*, **48**, 786–789.

Yayanos, A.A. (1978). Recovery and maintence of live amphipods at a pressure of 508 bars from an ocean depth of 5700 metres. *Science*, **200**, 1056–1059.

Yayanos, A.A. (1981). Reversible inactivation of deep-sea amphipods (*Paralicella caperesa*) by a decompression from 601 bars to atmospheric pressure. *Comparative Biochemistry and Physiology*, **69A**, 563–565.

Yayanos, A.A. (1986). Evolutional and ecological implications of the properties of deep-sea barophilic bacteria. *Proceedings of the National Academy of Sciences, USA*, **83**, 9542–9546.

Yayanos, A.A. (1995). Microbiology to 10 500 meters in the deep sea. *Annual Review of Microbiology*, **49**, 777–805.

Yayanos, A.A. (2009). Recovery of live amphipods at over 102 MPa from the Challenger Deep. *Marine Technology Society Journal*, **43**(5), 132–136.

Yayanos, A.A. and Dietz, A.S. (1983). Death of a hadal deep-sea bacterium after decompression. *Science*, **220**, 497–498.

Yayanos, A.A. and Nevenzel, J.C. (1978). Rising-particle hypothesis: rapid ascent of matter from the deep ocean. *Naturwissenschaften*, **65**, 255–256.

Yayanos, A.A., Dietz, A.S. and Van Boxtel, R. (1979). Isolation of a deep-sea barophilic bacterium and some of its growth characteristics. *Science*, **205**(4408), 808–810.

Yayanos, A.A., Dietz, A.S. and Van Boxtel, R. (1981). Depemdamce of reproduction rate on pressure as a hallmark of deep-sea bacteria. *Applied Environmental Microbiology*, **78**, 5212–5215.

Yayanos, A.A., Dietz, A.S. and Van Boxtel, R. (1982). Obligately barophilic bacterium from the Mariana trench. *Proceedings of the National Academy of Sciences, USA*, **44**(6), 1356.

Yeh, J. and Drazen, J.C. (2009). Depth zonation and bathymetric trends of deep-sea megafaunal scavengers of the Hawaiian Islands. *Deep-Sea Research I*, **56**, 251–266.

Yoshida, H., Ishibashi, S., Watanabe, Y. *et al.* (2009). The ABISMO mud and water sampling ROV for surveys at 10,000 m depth. *Marine Technology Society Journal*, **43**(5), 87–96.

Young, C.M., Tyler, P.A. and Fenaux, L. (1997). Potential for deep-sea invasion by Mediterranean shallow water echinoids: pressure and temperature as stage-specific dispersal barriers. *Marine Ecology Progress Series*, **154**, 197–209.

Zeigler, J.M., Athearn, W.D. and Small, H. (1957). Profiles across the Peru–Chile Trench. *Deep-Sea Research*, **4**, 238–249.

Zenkevich, L.A. (1954). Erforschungen der Tiefseefauna im nordwestlichen Teil des Stillen Ozeans. *Union of Antarctic Science and Biology, Series B*, **16**, 72–85.

Zenkevich, L.A. (1967). *Study of the Fauna of the Seas and Oceans. Development of Biology in the USSR*. Moscow: Nauka.

Zenkevitch, L.A. and Birstein, J.A. (1953). On the problem of the antiquity of the deep-sea fauna. *Deep-Sea Research*, **7**, 10–23.

Zenkevitch, L.A. and Birstein, J.A. (1956). Studies of the deep water fauna and related problems. *Deep-Sea Research*, **4**(1), 54–65.

Zenkevitch, L.A., Birstein, Y.A. and Beliaev, G.M. (1955). Studies of Kuril–Kamchatka Basin benthic fauna. *Trudy Instituta Okeanologii*, **12**, 345–381.

Zezina, O.N. (1997). Biogeography of the bathyal zone. *Advances in Marine Biology*, **32**, 389–426.

ZoBell, C.E. (1952). Bacterial life at the bottom of the Philippine Trench. *Science*, **115**(2993), 507–508.

ZoBell, C.E. and Johnson, F.H. (1949). The influence of hydrostatic pressure on the growth and viability of terrestrial and marine bacteria. *Journal of Bacteriology*, **57**, 179–189.

Index

Printed in the United States
by Baker & Taylor Publisher Services